# Stochastic Finance

*Stochastic Finance* provides an introduction to mathematical finance that is unparalleled in its accessibility. Through classroom testing, the authors have identified common pain points for students, and their approach takes great care to help the reader to overcome these difficulties and to foster understanding where comparable texts often do not.

Written for advanced undergraduate students, and making use of numerous detailed examples to illustrate key concepts, this text provides all the mathematical foundations necessary to model transactions in the world of finance. A first course in probability is the only necessary background.

The book begins with the discrete binomial model and the finite market model, followed by the continuous Black–Scholes model.

It studies the pricing of European options by combining financial concepts such as arbitrage and self-financing trading strategies with probabilistic tools such as sigma algebras, martingales and stochastic integration.

All these concepts are introduced in a relaxed and user-friendly fashion.

**Amanda Turner** is Professor of Statistics at the University of Leeds. She received her Ph.D. from the University of Cambridge in Scaling Limits of Stochastic Processes in 2007. Before moving to Leeds, she taught probability and stochastic processes for finance at Lancaster University and the University of Geneva for over 15 years. She is a founding member of the Royal Statistical Society's Applied Probability Section and is heavily involved in the London Mathematical Society, including as a member of council since 2021. When not doing mathematics, she enjoys mountaineering and skiing.

**Dirk Zeindler** is Senior Lecturer in Pure Mathematics at Lancaster University. He holds a Ph.D. in random matrix theory from the University of Zurich. He has taught probability courses at Lancaster University and at the University of Bielefeld for over 10 years. His teaching includes introductory first-year probability to advanced financial mathematics, for mathematics, accounting and finance students. His research interests are in probability and number theory. In particular, he and his co-authors have proven that at least 41.7% of the zeros of the Riemann zeta lie on the critical line, which is the current world record.

# Stochastic Finance

## An Introduction with Examples

**Amanda Turner**
University of Leeds

**Dirk Zeindler**
Lancaster University

Shaftesbury Road, Cambridge CB2 8EA, United Kingdom

One Liberty Plaza, 20th Floor, New York, NY 10006, USA

477 Williamstown Road, Port Melbourne, VIC 3207, Australia

314–321, 3rd Floor, Plot 3, Splendor Forum, Jasola District Centre, New Delhi – 110025, India

103 Penang Road, #05–06/07, Visioncrest Commercial, Singapore 238467

Cambridge University Press is part of Cambridge University Press & Assessment,
a department of the University of Cambridge.

We share the University's mission to contribute to society through the pursuit of
education, learning, and research at the highest international levels of excellence.

www.cambridge.org
Information on this title: www.cambridge.org/highereducation/isbn/9781316511251
DOI: 10.1017/9781009049672



First published 2023

Printed in the United Kingdom by TJ Books Limited, Padstow, Cornwall, 2023

*A Cataloging-in-Publication data record for this book is available from the Library of Congress.*

ISBN 978-1-316-51125-1 Hardback
ISBN 978-1-009-04894-1 Paperback

Additional resources for this publication at www.cambridge.org/turner-zeindler

# Contents

# Preface

Traders in the financial industry make decisions every day about whether to buy or sell securities such as stocks, gold and oil. While they know the prices of these securities today, their values tomorrow could be higher or lower. Another difficulty that arises is that some financial products depend on the future prices of other financial products. Examples are put and call options. The question here is how do you determine the current price of such a financial product. Financial mathematics, which is a branch of probability theory, allows us to answer questions like this and the answer is usually not the price one would expect at first glance. In this book, we introduce the subject of financial mathematics at a level that will be accessible to undergraduate students. One of our main aims is to highlight the underlying ideas and concepts and to motivate each step. We will also present these ideas and concepts as clearly and simply as possible, illustrating them with examples and pointing out typical misunderstandings. A solid acquaintance with probability is a prerequisite, so this textbook will most likely be of interest to third- or fourth-year students. Basic knowledge of stochastic processes and measure theory would be a helpful foundation, but is not essential as we will introduce the necessary definition and results.

In this book we examine several stochastic models for a stock market. We start with the simple binomial model and develop more complexity as the chapters progress, finally concluding with the Black–Scholes model, which is a reasonably realistic model for a real stock market. Each chapter is clearly explained so as to be accessible to a reader with no previous experience of the topic, and illustrated with numerous examples that explain both the motivation, and how it is used to answer concrete problems.

This textbook consists of two parts and, aside from Chapter 1, which is a prerequisite for both parts, each one can be read largely independently of the other. In Part I, we discuss discrete time models: the binomial model, the finite market model and the discrete Black–Scholes model. We introduce the risk neutral measure and prove the two fundamental theorems of asset pricing. The probability theory required for Part I is covered in Chapter 2. This chapter may be skipped by those who already have a good grounding in probability. Part II describes continuous-time models. We introduce Brownian motion, explain stochastic integration and discuss the Black–Scholes model. In particular, we derive the Nobel Prize-winning Black–Scholes formula for the pricing of call and put options. The probability theory required for Part II is covered in Chapter 6. Again, this chapter is optional for those familiar with the prerequisites.

# Acknowledgements

We would like to thank everyone who helped us with this book. Special thanks to Reda Chhaibi, Shane Turnbull and David Lucy.

# Part I

# Discrete-Time Models for Finance

# 1 Introduction to Finance

In this chapter we introduce some basic notions from finance. We explain the assumptions on financial markets which will be used in the rest of the book, and define key concepts such as arbitrage.

## 1.1 Motivation

One of the main aims of this book is to answer questions such as:

*What is the correct price for a financial product like a* call option*?*

Surprisingly, the answer turns out not to be the average payoff for the option. We illustrate this with a simple example. Suppose you would like to buy a *stock*. Let $S_0$ denote the price of the stock today and $S_1$ the price tomorrow. Suppose that $S_0$ is £100 and that tomorrow $S_1$ will be either £99 or £101 with

$$\mathbb{P}[S_1 = £99] = 0.3 \text{ and } \mathbb{P}[S_1 = £101] = 0.7. \tag{1.1.1}$$

If you would like to buy the stock today, what price should you pay? Suppose a trader offered you the stock for the price $P = \mathbb{E}[S_1] = 100.4$. Would you accept this? As the market price for the stock is £100, you should not pay £100.4 as you can get the stock cheaper directly from the market. The average price is clearly the wrong answer and the correct price is, of course, £100 as this is the market price today. In this situation, we know the price today so can easily answer the question. However, suppose you would like to buy a financial product that tomorrow gives the payoff

$$\Phi_1 = \begin{cases} £2, & \text{if } S_1 = £101, \\ £0, & \text{if } S_1 = £99. \end{cases} \tag{1.1.2}$$

What price should you pay for this product? In this case the product cannot be bought directly from the market, so we do not immediately know the price. We can compute $\mathbb{E}[\Phi_1] = 1.4$. Is this the correct price? Consider a different situation. Suppose you buy one stock today by borrowing £99, which you need to return tomorrow, and investing £1 of your own money. Assume that there is no interest rate for borrowing money. Tomorrow your portfolio will have the value

$$S_1 - 99 = \begin{cases} £2, & \text{if } S_1 = £101, \\ £0, & \text{if } S_1 = £99. \end{cases} \tag{1.1.3}$$

This trading strategy gives the same payoff as the product (1.1.2), for an initial investment of £1. We should therefore definitely not pay $\mathbb{E}[\Phi_1] = 1.4$ for the financial product in (1.1.2) as we can achieve the same payoff for less money.

At this point we see that the approach with average prices does not work. Instead, one prices products in financial markets so that neither the investor nor the trader can make a risk-free profit. An example of a risk-free profit is buying a security in one market and simultaneously selling it in another market at a higher price. A chance to make a risk-free profit is called an arbitrage opportunity. Intuition tells us that the price for the financial product in (1.1.2) has to be £1, otherwise this would create an arbitrage opportunity in the market. This is indeed correct and will be justified in Example 3.3.9.

## 1.2 Financial Markets

The term 'financial market' is a general term for all marketplaces where trading in financial products takes place. This includes, in particular, all markets in which stocks, bonds, foreign exchange and derivatives are traded. One of the reasons for the existence of financial markets is to facilitate the flow of capital. If, for instance, a company needs money to build a new factory then it can sell shares of stocks to investors and these investors will receive as a future reward dividends or a rise in the stock price. The financial market can be divided into the primary and secondary markets.

### 1.2.1 Primary Markets

On a primary market one trades securities such as

- **Stocks**: risky assets whose future value is unpredictable. We will generally denote the value of a single stock at time $t$ by $S_t$. This value is commonly called the **spot** price. If we have more than one stock, we will use the vector notation $\left(S_t^1, \ldots, S_t^d\right)$.
- **Bonds**: risk-free products with a given or predictable value in the future. In simple terms, a bond is a loan from an investor to a borrower. The borrower pays the investor a fixed rate of return over a specific timeframe. Bond prices will take one of the following two forms in this book.
  - The **risk-free bond** with price $B_t = B_0(1+r)^t$ at time $t$.
  - The bond with **continuously compounded interest rate** with price $B_t = B_0 e^{rt}$ at time $t$.
- **Foreign eXchange (FX)**: pounds, euros, dollars, ....
- **Commodities**: oil, metal, grain, ....
- **Fixed income**: credit products, such as mortgages, and interest rates.

Transactions occur directly between the investors and the companies that issue these securities. We will mainly focus on the assets and financial contracts related to stocks. Thus let us have a brief look at stocks. Stocks are also called shares and are units of ownership in a corporation.

The owner of a stock is called a shareholder. A shareholder participates in profits generated by the company. These profits are distributed in the form of dividends. Stocks can be in the form of a physical paper certificate or these days usually in digital form. The two main types of stocks are common stocks and preferred stocks. We will not discuss the different types of stocks in this book. Also, we will not discuss how to buy or sell stocks. It is sufficient to know that

- We can buy, sell and borrow stocks.
- Stocks have a non-negative value which changes over time.

Also, it is useful to keep in mind that stocks are units of ownership in a corporation. For simplicity, we will assume that

- Stocks do not generate any dividends.
- We can buy, sell and borrow as many stocks (and bonds and any other securities) as we would like.
- We do not have to pay transaction fees and the transactions happen instantly.
- Our interaction with the market has no influence on the pricing in the market.

These assumptions are not entirely realistic, but they make it easier to understand the basic concepts. In more complicated models, these assumptions are replaced by more realistic ones but many of the general principles still hold.

One further term which we would like to introduce at this point is the concept of a **portfolio**. A portfolio is a collection of financial assets like stocks, bonds, commodities and cash. A portfolio can of course also contain contingent claims such as the call and put options, which we define in Section 1.2.2 below. When we speak of our portfolio, we are referring to the collection of all assets we own.

### 1.2.2 Secondary Markets

On a secondary market, investors trade securities which they already hold with each other, as opposed to a primary market on which securities are sold directly by the entities who created them. When a company issues new stocks, it does this on a primary market, and the proceeds of such a sale go to the company. Investors then sell on these stocks to other investors. These sales take place on a secondary market and any proceeds go to the investor, not to the original company. One can think of a secondary market as one on which the securities being traded are 'second-hand'. The **stock market** is one such example.

On a secondary market, securities which originated on a primary market can be repackaged in elaborate ways, for example in the form of contracts, to create new securities. The values of such securities depend on the values of the underlying securities from the primary market and are called **derivatives** or **contingent claims**. Some examples are

- The **forward contract** (or short *forward*) is an agreement to buy or sell an asset at a certain price at a certain future time.

- A **future contract** is similar to the *forward* and both contracts allow people to buy or sell a specific type of asset at a specific time at a given price. However, *future contracts* (or simply *futures*) differ from *forwards* in a number of aspects. The main difference is that *futures* are traded on the stock exchange while *forwards* are not. Thus *futures* are more regulated than *forwards*.
- An **option** is a financial product that gives you the right to buy or sell an asset at a certain price called the **strike price**. The difference from a *forward* is that one has the right, but not the obligation, to buy or sell. A **call option** gives the option holder the right to buy at a given **strike price** $K$ and a **put option** gives the option holder the right to sell at a given **strike price** $K$. The cash flow at the time of exercise is called the **payoff** and determines the option. There are two main types of options.
  - **European options:** Exercising happens at a fixed time $T$ (called **maturity**).
  - **American options:** Exercising can happen at any time until $T$.

We will mainly focus on the assets and financial contracts related to stocks. Such products are called **equities**. This is the standard entry point to mathematical finance.

Sometimes people have trouble seeing the difference between the price of an option and the **strike price** of an option. To better understand the difference, imagine a European call option as a *piece of paper*, on which person A writes the inscription

> The holder of this piece of paper can purchase ⟨**number**⟩ stocks from the company ⟨**company name**⟩ from person A on date ⟨**maturity**⟩ at the price of ⟨**strike price**⟩.

Suppose that person A then sells the call option (piece of paper) to person B. The price of the option is the amount of money person B pays to person A in order to obtain the option in the first place, and the strike price is the amount which, at maturity, person B will pay to person A should they wish to buy the underlying security. The strike price is therefore a fixed value written into the contract. The option price, on the other hand, must be a value that B is willing to pay and A is willing to accept, and may depend on the current price of the stock, the time remaining until maturity, the riskiness of the stock, etc.

The question at this point is therefore: At what price should person A sell this call option? One of the main aims of this book is to answer this question. As a precursor, we should at least convince ourselves that this piece of paper does have some value. Would person A give this *piece of paper* to person B for free? Person B will only use this *piece of paper* if the strike price is below the market price at maturity, otherwise it would be cheaper to just buy the stocks directly on the market on that date. However, should person B buy from person A in this case, person A will lose money as they have to sell their stock at below market value. Therefore, person A can lose money from this contract, but not make money. The answer is therefore no, person A would not give the *piece of paper* to person B for free. Person B has to pay person A some money to get this contract. The European call option therefore has some (positive) value. The following example shows that in general the strike price is also not the correct price.

**Example 1.2.1.** *Assume that person A would like to sell a call option with maturity tomorrow and strike price £5 to person B.*

*Is it a good idea to sell this call option to person B for £5?*

*Suppose that the price of the underlying asset is £100 today. It is very likely that tomorrow the price of this asset will still be around £100. Thus, person B is very likely to exercise this option tomorrow, as he can receive an asset worth £100 for only £5. Person A would lose around £95 in this case. Therefore, person A should not sell this call option for £5.*

It is usual in finance to have in an option the right to buy/sell several units of the underlying security. However, we assume always that *call* and *put options* are to buy/sell only one unit of the security. This assumption only simplifies some formulas and has no influence on the pricing. Also note that in an option only the writer is specified. In the above example this is person A. On the other hand, the holder of the option is not specified. Thus person B can sell this option to somebody else.

### 1.2.3 Payoffs

In this book we mainly study *European contingent claims*, that is **contingent claims** with a fixed **maturity**. When we have a *European contingent claim* like a *call option* then it is natural to ask:

What is the payoff $\Phi_T$ of the *contingent claim* if the *maturity* is $T$ and
what is the profit $\widetilde{\Phi}_T$ of the *contingent claim* if the *maturity* is $T$?

For clarity, the payoff $\Phi_T$ denotes the money we earn from the *contingent claim* at maturity, not taking into account the price we had to pay for the contingent claim. The profit $\widetilde{\Phi}_T$, on the other hand, denotes the amount we earn from the *contingent claim* at maturity minus the price we had to pay for the *contingent claim*.

**Example 1.2.2.** *Suppose you play the lottery and buy five tickets at £10 each today. A week later, one of the tickets wins £200 while the other four win nothing. Then the payoff of the five tickets is £200 while the profit is £150 = £200 – £50.*

Most people are primarily interested in the profit of a *contingent claim*. However, one of the main aims in this book is to determine the initial price of a *contingent claim* in a given model. In this situation, it is much more natural to work with the payoff than with the profit. We thus will work only with the payoff $\Phi_T$ of a *contingent claim*.

Let us consider some examples of payoffs. We use the notation for bonds and stocks we introduced in Section 1.2.1.

**Example 1.2.3.**

- Stocks*: The payoff of one stock at time $T$ is $\Phi_T = S_T$.*
- Bonds*: The payoff of one bond at time $T$ is $\Phi_T = B_T$.*
- Forward contract*: Suppose we have agreed on a forward contract to sell one stock at time $T$ for the price $P_0$. Then the payoff is $\Phi_T = P_0 - S_T$. If we have, on the other hand, agreed on a forward contract to buy one stock at time $T$ for the price $P_0$ then the payoff is $\Phi_T = S_T - P_0$.*
- European call option*: Suppose that the strike price of this call option is $K \in (0, \infty)$ and the maturity is $T$. Then the payoff is*

$$\Phi_T = \max\{S_T - K, 0\} = \begin{cases} S_T - K, & \text{if } S_T \geq K, \\ 0, & \text{if } S_T < K, \end{cases} \tag{1.2.1}$$

*where $S_T$ is the price of the underlying asset at time $T$. Let us briefly explain this. If $S_T > K$ then the holder of the option will exercise it since he can buy the stock for the price $K$ and can then sell it for the price $S_T$ on the market. The payoff per option is then $S_T - K$. However, if $S_T < K$ then the holder of the option will not exercise it as he can get the stock cheaper on the market. Thus the payoff is 0 in this case.*

- *European put option: When the strike price is $K \in (0, \infty)$ then the payoff is*

$$\Phi_T = \max\{K - S_T, 0\} = \begin{cases} 0, & \text{if } S_T \geq K, \\ K - S_T, & \text{if } S_T < K. \end{cases} \tag{1.2.2}$$

Note that one could be tempted to define the payoff $\Phi_T$ of a *contingent claim* as the money one earns when it is exercised. However, if we consider a call or put option then we see it is not always reasonable to exercise it. Thus the payoff has to be viewed as the money we earn from the *contingent claim* by making the 'best' decision, that is the one for which we get the most money out of the *contingent claim*. For instance, in Example 1.2.2 we have the choice between cashing and not cashing the winning ticket. The best choice in this case is of course to cash the winning ticket.

We have to point out here another typical misconception. Suppose you own a European call option with maturity $T$ and would like to sell it at a time $t$ with $t < T$. At what price should you sell it? In view of (1.2.1), one could be tempted to think that one should sell this option for the price $P = \max\{S_t - K, 0\}$. Let us consider an example.

**Example 1.2.4.** *Suppose you own a European call option with maturity in one year and strike price £60, and you would like to sell this call option today. Suppose that the price of the underlying asset is £50 today. In this case*

$$\max\{50 - 60, 0\} = 0.$$

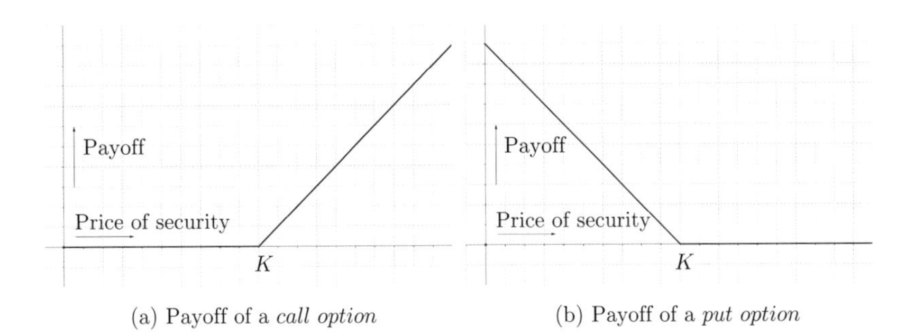


(a) Payoff of a *call option* (b) Payoff of a *put option*

**Figure 1.1** Payoff diagrams for *call* and *put options*.

*This price would make sense if the payoff of the call option would always be* 0*. However, suppose that we have* $90 \leq S_T \leq 110$*, no matter how the market evolves. In this case the call option gives a payoff of at least £30 at maturity. Obviously, you should not sell this call option for £0 (i.e. give it away for free).*

Here is another point to consider. If we own a call option with strike price $K$ and would like to exercise it at maturity then we require money. In view of the assumptions in Section 1.2.1, we can do the following: Borrow £$K$ at maturity, exercise the call option, sell the obtained security and return the borrowed money. Since we only exercise the call option if the price of the underlying security is larger than the strike price, we are always able to return the borrowed money. Thus, we are always able to exercise a call option under the assumptions in Section 1.2.1. For the sake of simplicity, from now on when we say that we are exercising a call option, we will implicitly assume that the holder would follow the procedure above.

### 1.2.4 Payoff Diagrams

Many *contingent claims* depend only on one security at a given time. Examples are *call* and *put options*. The payoff $\Phi_T$ of such *contingent claims* can be illustrated with a graph in $\mathbb{R}^2$. One uses the $x$-axis for the price of the underlying security and the $y$-axis for the payoff of the *contingent claim*. Such a graph is called a *payoff diagram*. The payoff diagrams for *call options* and *put options* are given in Figure 1.1.

There are of course also *contingent claims* depending on more than one security or on the value of a security at different times. An example is the so-called *lookback options*. For instance, the payoff of a *lookback call option* with fixed strike $K$ is

$$\Phi_T = \max\{S_{\max} - K, 0\} \quad \text{with} \quad S_{\max} = \max_{t \leq T} S_t. \tag{1.2.3}$$

## 1.3 Arbitrage Opportunities and Liquid Markets

An important term in finance is **arbitrage opportunity** (AO). An **arbitrage** is an opportunity to make money out of nothing or to make profit without risk. This is of course not a mathematically rigorous definition. We will give precise definitions later. As an illustration, let us consider an example of an arbitrage opportunity.

**Example 1.3.1.** *Suppose the price of an iPhone in the EU is* € 600 *and in the US it is* \$600. *The exchange rate between euros and dollars is* € 1= \$1.15. *An* arbitrage opportunity *can then be used as follows:*

- *Borrow* \$600 *and buy the iPhone on the US market for* \$600.
- *Sell the iPhone on the European market for* € 600.
- *Exchange* € 600→ \$690.
- *Pay back your debt and keep the profit of* \$90.

*Such an* arbitrage opportunity *is called a* **cash-and-carry arbitrage**.

A market is called **liquid** if one can buy and sell large quantities of an asset at any time, the transaction costs are low and transactions happen quickly. In this book, we are only interested in **liquid** markets. To simplify the arguments, we will always assume that we can sell and borrow as much of each security as we want, that we do not have to pay transaction fees and that transactions are immediate.

If a market is **liquid**, prices normally move very fast and eliminate *arbitrage opportunities*. The basic line of reasoning in mathematical finance is that the absence of *arbitrage opportunities* forces relations between prices of *forwards*, *futures*, *calls* and *puts* on a stock. One of the goals of mathematical finance is to establish these relations. However, unlike in physics, very few laws are available. One rule in mathematical finance is that financial products with larger payoffs must have larger prices.

**Proposition 1.3.2.** Suppose we have an arbitrage-free and liquid market, and consider two assets in this market. Suppose that at a future time $T$, the price of the second asset is greater than or equal to the price of the first asset, regardless of how the market evolves. Then this is also true at any prior time. Expressed in formulas,

$$P_T^{(1)} \leq P_T^{(2)} \implies P_t^{(1)} \leq P_t^{(2)} \ \forall 0 \leq t \leq T, \tag{1.3.1}$$

where $P_t^{(1)}$ denotes the price of the first asset and $P_t^{(2)}$ the price of the second asset at time $t$.

Note that in this proposition we only consider liquid markets. From now on we will no longer explicitly state this assumption, but assume that the reader is aware that we consider only liquid markets.

*Proof.* We argue by contradiction. Suppose that the market is arbitrage-free and liquid and (1.3.1) does not hold. Thus there exists a $t \leq T$ with $P_t^{(1)} > P_t^{(2)}$. We must have that $t < T$ since by assumption $P_T^{(1)} \leq P_T^{(2)}$. We now use the following strategy: *Buy cheap and sell expensive.* Explicitly:

- Do nothing until time $t$.
- At time $t$, borrow one unit of the first asset and sell it to buy one unit of the second asset. Since $P_t^{(1)} > P_t^{(2)}$, we have earned $P_t^{(1)} - P_t^{(2)}$. We then do nothing until time $T$.
- At time $T$, sell the unit of the second asset, buy a unit of the first asset and return it to the lender, and keep the remaining money. Since $P_T^{(1)} \leq P_T^{(2)}$, we have earned $P_T^{(2)} - P_T^{(1)}$ and have no debt left.

Our total profit is therefore $(P_t^{(1)} - P_t^{(2)}) + (P_T^{(2)} - P_T^{(1)}) > 0$. Thus this strategy is an arbitrage opportunity, which is a contradiction.

□

We can use Proposition 1.3.2 to give upper and lower bounds for the prices of options.

**Example 1.3.3.** *Consider an arbitrage-free and liquid market. Suppose in this market there is a bond $B_t$ with interest rate $r = 0$ and a stock $S_t$. Now consider a European call option with maturity $T$ and strike price $K$, written on $S_t$. Let $C_t$ denote the price of this call option at time $t$ for $t \leq T$. We then have*

$$S_t - K \leq C_t \leq S_t \text{ for all } t \leq T. \tag{1.3.2}$$

*Let us justify this. The payoff of the call option at time $T$ is $\max\{S_T - K, 0\}$. Thus $C_T = \max\{S_T - K, 0\}$. Furthermore, at time $T$*

$$S_T - K \leq \max\{S_T - K, 0\} \leq S_T. \tag{1.3.3}$$

*To apply Proposition 1.3.2 to this inequality, we have to determine the arbitrage-free price at time $t$ of a contingent claim with payoffs $S_T - K$ and $S_T$.*

- *The arbitrage-free price of the stock at time $t$ is $S_t$.*
- *Since $r = 0$, money does not lose its value over time. Thus the value of £$K$ at time $t$ is the same as its value at time $T$.*

*Combining these two observations, we get that the arbitrage-free price of a contingent claim with a payoff $S_T - K$ is $S_t - K$ at time $t$. Thus Proposition 1.3.2 implies* (1.3.2).

*Similarly, if we consider a European put option with maturity $T$ and strike price $K$ and denote by $P_t$ the price of this put option at time $t$ for $t \leq T$, then*

$$K - S_t \leq P_t \leq K \text{ for all } t \leq T. \tag{1.3.4}$$

The assertion that the arbitrage-free price of £$K$ is always £$K$ only holds because the interest rate $r = 0$. To illustrate this point, let us consider an example.

**Example 1.3.4.** *Consider an arbitrage-free market with a bond $B_t$ of the form*

$$B_t = B_0(1 + r)^t \text{ with } B_0 > 0,\ r \geq 0. \tag{1.3.5}$$

*The time $t = 0$ corresponds to today. Consider £$K$ at time $T$ in the future. This can be viewed as the contingent claim giving a payoff of £$K$ at time $T$, no matter how the market evolves. What is the arbitrage-free price of these £$K$ at time $t$? The answer is*

$$\frac{K}{(1+r)^{T-t}}. \tag{1.3.6}$$

*At first glance, this looks a little bit strange. One might believe that the price of £$K$ is always £$K$. Thus let us illustrate why* (1.3.6) *is the correct answer.*

- *Let us first show that £$K$ is the wrong price. Suppose that we can borrow £$K$ from somebody at time $t$ and have to return these £$K$ at time $T$. Then we invest these £$K$ at time $t$ into bonds and sell these bonds at time $T$. Thus we obtain £$K(1+r)^{T-t}$ at time $T$, which is strictly larger than $K$. We then return the borrowed £$K$ and keep the rest. This is clearly an arbitrage opportunity, which is a contradiction.*
- *Similarly, suppose that we can borrow £$K_t$ with $K_t > K(1+r)^{T-t}$ from somebody at time $t$ and have to return £$K$ at time $T$. In this case we can use the same argument as above.*
- *Finally, suppose that we can lend somebody £$K_t$ with $K_t < K(1+r)^{T-t}$ at time $t$ and will get back £$K$ at time $T$. In this case, at time $t$ we borrow bonds worth $K(1+r)^{t-T}$, sell them and lend £$K_t$. Thus we have earned £$K(1+r)^{T-t} - K_t$. At time $T$, we get £$K$ back. We then use this money to buy bonds and return the borrowed bonds. Thus we have earned in total £$K(1+r)^{T-t} - K_t$ and have no debt left. In other words, we have found an arbitrage opportunity, which is a contradiction.*

*This means that* (1.3.6) *gives the only possible value for which there is no arbitrage opportunity, and hence it must be the time $t$ price in an arbitrage-free market.*

In view of this example, we see that we can interpret the interest of the bond in a financial model as the inflation rate. Proposition 1.3.2 has some further consequences. In particular, it implies that the prices of call and put options are always non-negative since the payoff of call and put options is always non-negative. However, this does not hold for all contingent claims. Suppose that a contingent claim gives a payoff

$$\Phi_T = \min\{K - S_T, 0\}. \tag{1.3.7}$$

This payoff is always non-positive. Thus nobody will pay money to get this contingent claim. Instead, the writer of this contingent claim has to pay somebody money for taking this contingent claim. A further consequence of Proposition 1.3.2 is the *law of one price*.

**Proposition 1.3.5** (Law of one price). Suppose that a market is liquid and arbitrage-free. If two securities have the same value at a (future) time $T$, then they must have the same value at any prior time. Expressed in formulas,

$$P_T^{(1)} = P_T^{(2)} \implies P_t^{(1)} = P_t^{(2)} \ \forall t \leq T, \tag{1.3.8}$$

with $P_t^{(1)}$ and $P_t^{(2)}$ as in Proposition 1.3.2.

This proposition follows immediately from Proposition 1.3.2. A natural question at this point is if the law of one price can also be applied to the past. In other words, if two securities agree at some point in the past, do they also agree prior to that time? The answer is no, they can agree, but they do not have to agree. The reason is that we can apply the law of one price if and only if $P_T^{(1)} = P_T^{(2)}$ holds no matter how the market evolves. This condition is not necessarily fulfilled if two securities have the same price at some point in the past.

The law of one price can also be applied to combinations of securities. An example is the *put–call parity*.

**Example 1.3.6.** *Consider an arbitrage-free market containing a bond $B_t$ with interest rate $r = 0$ and a stock $S_t$. Consider a European call option and a European put option, both with maturity $T$ and strike price $K$, written on $S_t$. Let $C_t$ denote the price of the call option and $P_t$ the price of the put option at time $t$ for $t \leq T$. Then*

$$C_t - P_t = S_t - K \text{ for all } t \leq T. \tag{1.3.9}$$

*Let us justify* (1.3.9). *It is straightforward to see that at time $T$*

$$S_T - K = \max\{S_T - K, 0\} - \max\{K - S_T, 0\} = C_T - P_T. \tag{1.3.10}$$

*The law of one price and the observation in Example 1.3.3 then imply* (1.3.9).

A typical mistake people make when they are asked to show equation (1.3.9) is that they believe $C_t = \max\{S_t - K, 0\}$ and $P_t = \max\{K - S_t, 0\}$. However, in almost all situations

$$C_t \neq \max\{S_t - K, 0\} \text{ and } P_t \neq \max\{K - S_t, 0\} \text{ for } t < T, \tag{1.3.11}$$

see for instance Example 1.2.4. Explicitly, the prices of a call and a put option are determined by the expected behaviour for the price of the underlying security. In the special case of the Black–Scholes model, the prices of call and put options are given by the Black–Scholes formula, see Theorem 5.5.2.

The law of one price can be used in some cases to determine arbitrage-free prices of a security in a given market. Let us consider an example.

**Example 1.3.7.** *Consider an arbitrage-free market containing a stock $S_t$ and a contingent claim $\Phi_T$. This contingent claim $\Phi_T$ has maturity $T$ and is written on the stock $S_T$. Further, the payoff of $\Phi_T$ is*

- £30 *if $S_T$ is greater than or equal to* £20.
- £15 *if $S_T$ is less than or equal to* £5.
- *Interpolated linearly in the interval* [5, 20].

*As illustration, the payoff diagram of $\Phi_T$ can be found in Figure 1.2.*

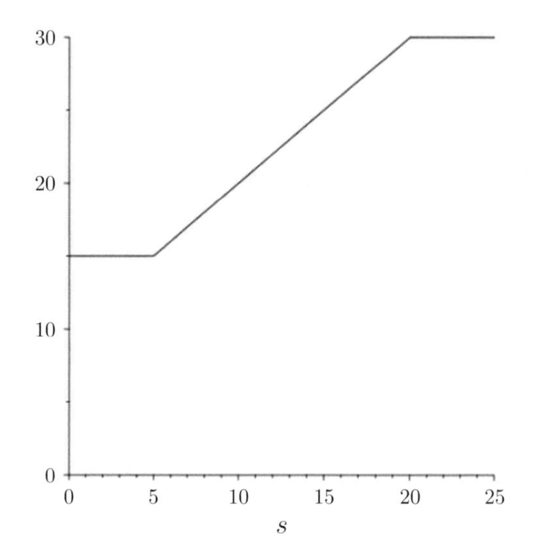


**Figure 1.2** Payoff diagram of the contingent claim in Example 1.3.7.

*Suppose that we can buy and sell put and call options according to Table 1.1.*

**Table 1.1** Option prices in Example 1.3.7

| *Strike price* | *£5* | *£10* | *£15* | *£20* |
|---|---|---|---|---|
| *Call option* | *£12* | *£8* | *£5* | *£4* |
| *Put option* | *£1* | *£2* | *£4* | *£8* |

*All put and call options in Table 1.1 are written on $S_t$ and have maturity $T$. We now determine the arbitrage-free price of the contingent claim $\Phi_T$. For this, we first show that we can reproduce the payoff of $\Phi_T$ by buying and selling put and call options. Then we apply the law of one price.*

*The payoffs of call and put options are* $\max\{S_T - K, 0\}$ *and* $\max\{K - S_T, 0\}$. *Thus their payoffs are continuous functions in $S_T$ and are linear in the intervals* $[0, K]$ *and* $[K, \infty[$. *Since* $\Phi_T = f(S_T)$ *with $f$ continuous and linear in the intervals* $[0, 5]$, $[5, 20]$ *and* $[20, \infty[$, *we use the approach*

$$\Phi_T = a \max\{S_T - 5, 0\} + b \max\{5 - S_T, 0\} + c \max\{S_T - 20, 0\} + d \max\{20 - S_T, 0\},$$

*where a, b, c, d* $\in \mathbb{R}$. *This expression is continuous and linear in the intervals* $[0, 5]$, $[5, 20]$ *and* $[20, \infty[$. *To determine the values of a, b, c, d* $\in \mathbb{R}$, *we enter the values* $S_T = 0$, $S_T = 5$, $S_T = 20$ *and* $S_T = 25$. *This then leads to the system of equations*

$$5b + 20d = 15, \quad 15d = 15, \quad 15a = 30, \quad 20a + 5c = 30. \tag{1.3.12}$$

*A direct computation shows that this system has the solution*

$$a = 2, \quad b = -1, \quad c = -2, \quad d = 1. \tag{1.3.13}$$

*In words, we can achieve the payoff of $\Phi_T$ by buying two call options with strike price 5, selling one put option with strike price 5, selling two call options with strike price 20 and buying one put option with strike price 20. If we would like to produce the payoff of $\Phi_T$ with call and put options then we use Table 1.1 and see that we have to invest*

$$2 \cdot 12 - 1 \cdot 1 - 2 \cdot 4 + 1 \cdot 8 = 23. \tag{1.3.14}$$

*By assumption, the market is liquid and arbitrage-free. Thus the law of one price implies that the arbitrage-free price of $\Phi_T$ is 23.*

The assumption that a market is arbitrage-free and liquid can only be used to derive laws similar to the *law of one price*. These are not, in general, enough to determine the actual prices of financial products. We need to have some additional information about prices, such as in Example 1.3.7, in order to use these laws to determine the prices. Financial products such as *forwards* or *options* are based on the future prices of underlying securities. These prices have a strong influence on whether a financial product makes a profit or a loss. In particular, prices such as in Table 1.1 are determined by the expected behaviour of the underlying security. Thus it is important to have a good stochastic model for the behaviour of the prices in a market and to know how arbitrage-free prices of a financial product are determined in a given model. In this book, we will focus on various stochastic models for a market. The main aim of this book is to show how the prices of financial products are determined in these models, and to develop the necessary tools to do that.

## 1.4 Exercises

**Exercise 1.1.** *Two financial products based on options are the strangle and the butterfly spread.*

- *A* strangle *is a combined option consisting of a long position (we are the* buyer *of the option) in a European call option with strike $K_1$ and a long position in a European put option with strike $K_2 < K_1$.*
- *A butterfly spread is a combined option consisting of a long position in a European call option with strike $K_1$, a long position in a European call option with strike $K_2 \neq K_1$ and short position (we are the* seller *of the option) in two European call options with strike $\frac{K_1+K_2}{2}$.*

*It is assumed that all options in these combined options are written on the same stock and have the same* maturity.

*(a) Determine the payoff of the strangle and the butterfly spread.*
*(b) Draw payoff diagrams for the strangle and the butterfly spread with $K_1 = 15, K_2 = 5$.*

**Exercise 1.2.** *The price of a security is £10 today. Suppose that you are the only person in the world who knows that the price of the security will be £25 in 3 months, while everybody else expects that the price will stay around £10.*

(a) *Suppose you would like to buy a call option today with strike price £15 and maturity 3 months. Explain why this call option will be cheap.*

(b) *Suppose you have £75,000 available. In this situation would you invest the money into call options or buy the security directly? Justify your answer.*
Hint: You can assume the price of the call option is less than £1.

**Exercise 1.3.** *Consider a financial model that runs over the time period* $[0,T]$ *and that contains two bonds* $B_t^1$ *and* $B_t^2$. *Suppose further that for all times* $t \in [0,T]$

$$B_t^1 = (1+r_1)^t \text{ and } B_t^2 = (1+r_2)^t$$

*with* $-1 < r_1 < r_2$. *Show carefully that this model is not arbitrage-free.*

**Exercise 1.4.** *Consider two put options. Both options are written on the same stock and have the same* maturity. *The strike price of the first put option is £10 and the strike price of the second put option is £15. Furthermore, today we can buy/sell the first put option for the price* $P_1$ *and the second put option for the price* $P_2$. *We assume that the stock price at maturity can take any value in* $\mathbb{R}_+$.

(a) *Explain why we must have* $0 \leq P_1 \leq P_2$ *in an arbitrage-free market.*

(b) *Suppose that* $P_1 = 2$ *and* $P_2 = 3$. *Construct an arbitrage opportunity by only buying/selling the given put options.*

(c) *Show that in an arbitrage-free market we must have* $15P_1 - 10P_2 > 0$.

**Exercise 1.5.** *Consider an arbitrage-free market with a bond* $B_t$ *with interest rate* $r = 0$ *and a stock* $S_t$. *The time* $t=0$ *corresponds to today. Suppose that a contingent claim gives a payoff of*

$$\Phi_T = \min\{K - S_T, 0\}$$

*at time* $T$ *in the future, where* $K > 0$. *Show that the arbitrage-free price of this contingent claim is non-positive.*

# 2 Discrete Probability

In this chapter we review basic knowledge of discrete probability theory. We also take a closer look at conditional expectations and filtrations in the discrete case, as these are important tools in financial mathematics. In particular, we will try to illustrate the ideas behind these two concepts and why they lead to the corresponding definitions. A reader having a good understanding of those concepts can skip this chapter.

## 2.1 Basics

In this section we recall some definitions and theorems about discrete probability spaces. We assume the reader is familiar with basic probability theory so we mainly state, without justification, the material we need in the remainder of this book.

**Definition 2.1.1** (Discrete probability space). A *discrete probability space* consists of a pair $(\Omega, \mathbb{P})$, where $\Omega$ is a countable set and $\mathbb{P}$ is a function which assigns to each subset $A \subset \Omega$ a value in $[0, 1]$, such that

- $\mathbb{P}[A] \geq 0$ for all $A \subset \Omega$.
- $\mathbb{P}[\Omega] = 1$.
- If $A_i \subset \Omega$ with $i \in \mathbb{N}$ are *disjoint* (i.e. $A_i \cap A_j = \emptyset$ for $i \neq j$), then

$$\mathbb{P}\left[\bigcup_{i=1}^{\infty} A_i\right] = \sum_{i=1}^{\infty} \mathbb{P}[A_i]. \tag{2.1.1}$$

The property in (2.1.1) is called $\sigma$*-additivity* (spoken 'sigma') or *countable additivity*. We call any $A \subset \Omega$ an event.

We also require the notion of conditional probability.

**Definition 2.1.2.** Let $(\Omega, \mathbb{P})$ be a probability space and $A, B \subset \Omega$ be given with $\mathbb{P}[B] \neq 0$. The *conditional probability* of $A$ given $B$ is

$$\mathbb{P}[A \mid B] := \frac{\mathbb{P}[A \cap B]}{\mathbb{P}[B]}.$$

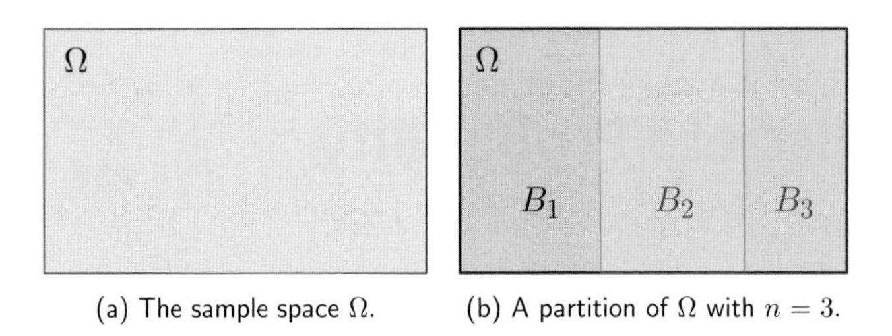


(a) The sample space Ω. (b) A partition of Ω with $n = 3$.

**Figure 2.1** Illustration of partitions and Theorem 2.1.4.

The conditional probability $\mathbb{P}[A|B]$ has interpretation as the probability that '*A occurs if we know that B will occur*'. Note that the conditional probability of a sample point $\omega \in \Omega$ is given by

$$\mathbb{P}[\omega|B] = \frac{\mathbb{P}[\{\omega\} \cap B]}{\mathbb{P}[B]} = \begin{cases} \frac{\mathbb{P}[\omega]}{\mathbb{P}[B]}, & \text{if } \omega \in B, \\ 0, & \text{otherwise.} \end{cases} \quad (2.1.2)$$

Note that we interchangeably use the notation $\mathbb{P}[\{\omega\}]$ and $\mathbb{P}[\omega]$.

**Definition 2.1.3** (Finite partition). Let $(\Omega, \mathbb{P})$ be a (discrete) probability space and $B_1, \ldots, B_n$ be events. We call $B_1, \ldots, B_n$ a *finite partition* of Ω if

- $\Omega = \bigcup_{i=1}^{n} B_i$.
- $B_i \neq \emptyset$ for all $i = 1, \ldots, n$.
- $B_i$ are disjoint (i.e. $B_i \cap B_j = \emptyset$ for all $i \neq j$).

An illustration of a partition of Ω with $n = 3$ can be found in Figure 2.1(b). A finite partition can be interpreted heuristically as follows. We have a sample space Ω and we can split the sample points $\omega$ into $n$ cases. These $n$ cases have to be distinct and have to cover all possibilities. A $\sigma$-*partition* (or *countable partition*) of Ω is defined similarly as a *finite partition*. The only difference is that we take an infinite sequence $B_1, B_2, B_3, \ldots$ instead of a finite one. It will be clear from the context if we consider a finite or countable partition. For simplicity, from now on, we will refer to both as a *partition*.

Partitions can be useful for splitting the computation of the probability of an event into several events.

**Theorem 2.1.4** (Law of total probability). *Let $(\Omega, \mathbb{P})$ be a (discrete) probability space, A be an arbitrary event and $(B_j)_{j=1}^{n}$, with $n \in \mathbb{N} \cup \{\infty\}$, be a partition of Ω. Then*

$$\mathbb{P}[A] = \sum_{\substack{i=1,\ldots,n \\ \mathbb{P}[B_i] \neq 0}} \mathbb{P}[A|B_i] \cdot \mathbb{P}[B_i]. \quad (2.1.3)$$

Let us briefly recall the definition and some properties of discrete random variables.

**Definition 2.1.5.** Let $(\Omega, \mathbb{P})$ be a discrete probability space. A *discrete random variable* is a function

$$\begin{aligned} X : \Omega &\to \mathbb{R}, \\ \omega &\mapsto X(\omega). \end{aligned}$$

We are often interested in events where a random variable $X$ has a certain value. For instance, what is the probability that a stock has a price of £20? For such questions, we introduce the following notation.

**Definition 2.1.6.** Let $(\Omega, \mathbb{P})$ be a (discrete) probability space and $X$ a random variable on this space. For $x \in \mathbb{R}$, define

$$\{X = x\} := \{\omega \in \Omega;\ X(\omega) = x\}.$$

Furthermore, define

$$\mathbb{P}[X = x] := \mathbb{P}[\{X = x\}] = \mathbb{P}[\{\omega \in \Omega;\ X(\omega) = x\}].$$

The event $\{X = x\}$ consists of all $\omega \in \Omega$ with $X(\omega) = x$ and thus can be interpreted heuristically as '*$X$ has the value $x$*' or '*$X$ is equal to $x$*'.

**Definition 2.1.7.** Let $(\Omega, \mathbb{P})$ be a probability space. The *probability mass function* of a discrete random variable $X$ is defined by

$$\begin{aligned} p_X(x) :\ \mathbb{R} &\to [0, 1], \\ x &\mapsto p_X(x) := \mathbb{P}[X = x]. \end{aligned}$$

Note that $\{X = x\} = \emptyset$ for most $x \in \mathbb{R}$ since $\Omega$ is at most countable. We thus have $p_X(x) = 0$ for most $x \in \mathbb{R}$.

The *probability mass function* determines the *distribution* of $X$.

**Definition 2.1.8.** Let $X$ be a random variable. The *cumulative distribution function* $F_X : \mathbb{R} \to [0, 1]$ of $X$ is defined as

$$F_X(x) := \mathbb{P}[X \leq x].$$

Much of the time we will work with more than one random variable. We thus also require the *joint distribution* of random variables.

**Definition 2.1.9.** Let $X$ and $Y$ be discrete random variables on the same probability space $(\Omega, \mathbb{P})$. Set

$$\{X = x, Y = y\} := \{X = x\} \cap \{Y = y\} = \{\omega;\ X(\omega) = x \text{ and } Y(\omega) = y\}.$$

The *joint distribution* of $X, Y$ is defined as

$$\mathbb{P}[X = x, Y = y] := \mathbb{P}[\{X = x, Y = y\}].$$

The *joint distribution* of more than two random variables is similarly defined.

The idea behind the definition of $\{X = x, Y = y\}$ is similar to the idea behind the definition of $\{X = x\}$. More precisely, one puts all $\omega \in \Omega$ together, where $X$ and $Y$ have the same value. Thus $\{X = x, Y = y\}$ can be heuristically interpreted as '*X has the value x and Y has the value y*'.

An important property of random variables is when they are independent.

**Definition 2.1.10.** Let $(\Omega, \mathbb{P})$ be a discrete probability space and $X_1$ and $X_2$ be discrete random variables on this space. The random variables $X_1$ and $X_2$ are called (stochastically) *independent*, if

$$\mathbb{P}[X_1 = x_1, X_2 = x_2] = \mathbb{P}[X_1 = x_1] \cdot \mathbb{P}[X_2 = x_2] \qquad (2.1.4)$$

for all $x_1, x_2 \in \mathbb{R}$. The independence of more than two random variables is defined similarly.

*Remark* 2.1.11. The independence of two random variables is in general defined by

$$\mathbb{P}[X_1 \leq x_1, X_2 \leq x_2] = \mathbb{P}[X_1 \leq x_1] \cdot \mathbb{P}[X_2 \leq x_2] \qquad (2.1.5)$$

for all $x_1, x_2 \in \mathbb{R}$. However, (2.1.4) and (2.1.5) are equivalent on a discrete probability space $(\Omega, \mathbb{P})$, but (2.1.4) is often easier to deal with.

We use the abbreviation i.i.d. for a sequence of *independently and identically distributed* random variables. This is a sequence of random variables $X_1, X_2, \ldots$ for which $F_{X_i} = F_{X_j}$ for all $i,j$ and, for every $n \in \mathbb{N}$ and $x_1, \ldots, x_n \in \mathbb{N}$,

$$\mathbb{P}[X_1 \leq x_1, \ldots, X_n \leq x_n] = \mathbb{P}[X_1 \leq x_1] \cdots \mathbb{P}[X_n \leq x_n].$$

Aside from the *distribution* of a random variable, we are also interested in the typical behaviour of a random variable. One way to quantify this is via the expectation.

**Definition 2.1.12.** Let $(\Omega, \mathbb{P})$ be a discrete probability space and $X$ a random variable on this space. The *expectation* of $X$ is defined as

$$\mathbb{E}[X] = \sum_{\omega \in \Omega} X(\omega) \cdot \mathbb{P}[\omega], \qquad (2.1.6)$$

provided that

$$\mathbb{E}[|X|] := \sum_{\omega \in \Omega} |X(\omega)| \cdot \mathbb{P}[\omega] < \infty. \tag{2.1.7}$$

If (2.1.7) holds, we say that the expectation of $X$ exists.

The expectation can be heuristically interpreted as the mean or average behaviour of a random variable. More precisely, suppose we have an experiment which gives a random outcome, distributed according to a random variable $X$. We now repeat this experiment independently $n$ times with $n$ very large. The $n$ outcomes can be modelled with an i.i.d. sequence of random variables $X_1, \ldots, X_n$, where each $X_j$ has the same distribution as $X$. We expect

$$\frac{1}{n}\sum_{j=1}^{n} X_j \approx \mathbb{E}[X]. \tag{2.1.8}$$

The *law of large numbers* states that our intuition is indeed correct and that (2.1.8) is true with high probability. It is important to keep in mind that the expectation $\mathbb{E}[X]$ is only a reasonable prediction for the average of the $X_i$, but not on the individual $X_i$s! In general

$$X \not\approx \mathbb{E}[X]. \tag{2.1.9}$$

To illustrate this, roll a fair die and let $X$ denote the rolled number. Then

$$X \in \{1,2,3,4,5,6\} \text{ and } \mathbb{E}[X] = 3.5.$$

If we roll the die only once then the expectation $\mathbb{E}[X] = 3.5$ is a poor prediction for $X$. However, if we roll the die 5000 times, then *Chebyshev's inequality* tells us that

$$\mathbb{P}\left[\frac{1}{n}\sum_{j=1}^{n} X_j \in [3.4, 3.6]\right] \geq 0.94.$$

Thus 3.5 is a quite good prediction in this situation. In other words

**$\mathbb{E}[X]$ is a good prediction for the average of $X$, but not for $X$ itself!**

This observation is also important to understand the idea behind the conditional expectation $\mathbb{E}[X|Y]$ in Section 2.2.2. We give another expression for $\mathbb{E}[X]$ in terms of the *probability mass function*.

**Theorem 2.1.13.** *Let $X$ be a discrete random variable with distribution $p_X(x)$ and with $\mathbb{E}[|X|] < \infty$. Then*

$$\mathbb{E}[X] = \sum_{\substack{x \in \mathbb{R} \\ p_X(x) \neq 0}} x \cdot p_X(x) = \sum_{\substack{x \in \mathbb{R} \\ p_X(x) \neq 0}} x \cdot \mathbb{P}[X = x] \tag{2.1.10}$$

*and $\mathbb{E}[X]$ depends only on the distribution of $X$.*

*Further, if $f : \mathbb{R} \to \mathbb{R}$ is an arbitrary function with $\mathbb{E}[|f(X)|] < \infty$, then*

$$\mathbb{E}[f(X)] = \sum_{\substack{x \in \mathbb{R} \\ p_X(x) \neq 0}} f(x) \cdot p_X(x) = \sum_{\substack{x \in \mathbb{R} \\ p_X(x) \neq 0}} f(x) \cdot \mathbb{P}[X = x]. \tag{2.1.11}$$

## 2.2 Conditional Expectation

Consider an event $A$ in some $(\Omega, \mathbb{P})$ and perform $n$ independent trials. The *law of large numbers* states that, when $n$ is large, the proportion of times that the event $A$ occurs is approximately $\mathbb{P}[A]$ (with high probability). However, if we already know that the event $B$ will occur then the picture looks different. In this case the proportion will be around $\mathbb{P}[A \mid B]$ instead. The aim of this section is to extend this idea of conditioning to random variables.

### 2.2.1 Conditioning on an Event $B$

Let $X$ be a random variable with $\mathbb{E}[|X|] < \infty$ on a discrete probability space $(\Omega, \mathbb{P})$. Suppose we have an experiment which gives a random outcome, distributed according to the random variable $X$. When we repeat this experiment many times independently, it is very likely that the average of all the obtained values of $X$ is close to $\mathbb{E}[X]$. The question we would like to address here is: How does the average of $X$ behave if we already know that an event $B$ will occur? For instance, what is the average price of a stock tomorrow if the price today is £190? Let us now consider an example which will motivate the definition of conditional expectation.

**Example 2.2.1.** *Suppose we have an experiment with three possible equally likely outcomes a, b and c, so $\Omega = \{a, b, c\}$ and $\mathbb{P}[a] = \mathbb{P}[b] = \mathbb{P}[c] = \frac{1}{3}$. Consider the random variable $X$ with $X(a) = 1$, $X(b) = 3$ and $X(c) = -5$. We now repeat this experiment 3000 times and work out the average value of $X$ over these experiments. What would we expect this average to be? Since $\mathbb{P}[a] = \mathbb{P}[b] = \mathbb{P}[c] = \frac{1}{3}$, we would expect around 1000 a's, 1000 b's and 1000 c's. Thus the average should be around*

$$\begin{aligned}
\frac{1000 \cdot X(a) + 1000 \cdot X(b) + 1000 \cdot X(c)}{3000} &= X(a) \cdot \frac{1}{3} + X(b) \cdot \frac{1}{3} + X(c) \cdot \frac{1}{3} \\
&= X(a) \cdot \mathbb{P}[a] + X(b) \cdot \mathbb{P}[b] + X(c) \cdot \mathbb{P}[c] \\
&= \mathbb{E}[X] = -\frac{1}{3}.
\end{aligned}$$

*Suppose we know that the event $B = \{a, b\}$ will occur. What do we now expect for the average of $X$? We have $\mathbb{P}[a|B] = \mathbb{P}[b|B] = \frac{1}{2}$ and $\mathbb{P}[c|B] = 0$. We thus expect around 1500 a's, 1500 b's and 0 c's. Thus the average of $X$ should be around*

$$\begin{aligned}
\frac{1500 \cdot X(a) + 1500 \cdot X(b) + 0 \cdot X(c)}{3000} &= X(a) \cdot \frac{1}{2} + X(b) \cdot \frac{1}{2} + X(c) \cdot 0 \\
&= X(a) \cdot \mathbb{P}[a|B] + X(b) \cdot \mathbb{P}[b|B] + X(c) \cdot \mathbb{P}[c|B] \\
&= 2.
\end{aligned}$$

Example 2.2.1 suggests the following definition.

**Definition 2.2.2.** Let $(\Omega, \mathbb{P})$ be a discrete probability space. Further, let $B \subset \Omega$ be an event with $\mathbb{P}[B] > 0$ and $X$ a random variable on this space with $\mathbb{E}[|X|] < \infty$. The *conditional expectation of $X$ given $B$* is defined as

$$\mathbb{E}[X|B] = \sum_{\omega \in \Omega} X(\omega) \cdot \mathbb{P}[\omega|B] = \sum_{\omega \in B} X(\omega) \cdot \mathbb{P}[\omega|B]. \tag{2.2.1}$$

Since $\mathbb{P}[\omega|B] = 0$ for $\omega \notin B$, we see that both expressions in (2.2.1) are equal. Let us consider some examples.

**Example 2.2.3.** *Suppose we roll a fair die. Let $X$ denote the rolled number. What would you expect the average of $X$ to be if you know that we will roll an even number? In this example*

$$\Omega := \{1, 2, 3, 4, 5, 6\} \text{ and } \mathbb{P}[\omega] = \frac{1}{6} \qquad \forall \omega \in \Omega.$$

*We have*

$$\mathbb{P}[2|\text{'}X \text{ is even'}] = \frac{\mathbb{P}[2]}{\mathbb{P}[\{2,4,6\}]} = \frac{\frac{1}{6}}{\frac{3}{6}} = \frac{1}{3}.$$

*Similarly, we get*

$$\mathbb{P}[4|\text{'}X \text{ is even'}] = \mathbb{P}[6|\text{'}X \text{ is even'}] = \frac{1}{3} \text{ and}$$
$$\mathbb{P}[1|\text{'}X \text{ is even'}] = \mathbb{P}[3|\text{'}X \text{ is even'}] = \mathbb{P}[5|\text{'}X \text{ is even'}] = 0.$$

*We thus obtain*

$$\begin{aligned}\mathbb{E}[X|\text{'}X \text{ is even'}] &= \sum_{\omega=1}^{6} X(\omega)\mathbb{P}[\omega|\text{'}X \text{ is even'}] \\ &= 1 \cdot 0 + 2 \cdot \frac{1}{3} + 3 \cdot 0 + 4 \cdot \frac{1}{3} + 5 \cdot 0 + 6 \cdot \frac{1}{3} = 4.\end{aligned}$$

*We get with a similar computation*

$$\mathbb{E}[X|\text{'}X \text{ is odd'}] = 1 \cdot \frac{1}{3} + 2 \cdot 0 + 3 \cdot \frac{1}{3} + 4 \cdot 0 + 5 \cdot \frac{1}{3} + 6 \cdot 0 = 3.$$

*Note that we have*

$$\begin{aligned}\mathbb{E}[X] &= 3.5 = 3 \cdot \frac{1}{2} + 4 \cdot \frac{1}{2} \\ &= \mathbb{E}[X|\text{'}X \text{ is odd'}] \cdot \mathbb{P}[\text{'}X \text{ is odd'}] + \mathbb{E}[X|\text{'}X \text{ is even'}] \cdot \mathbb{P}[\text{'}X \text{ is even'}].\end{aligned}$$

**Example 2.2.4.** *Let us consider a stock and denote the price of this stock tomorrow by $S_1$ and the price in 2 days by $S_2$. Suppose that the price of this stock follows the probability tree in Figure 2.2.*

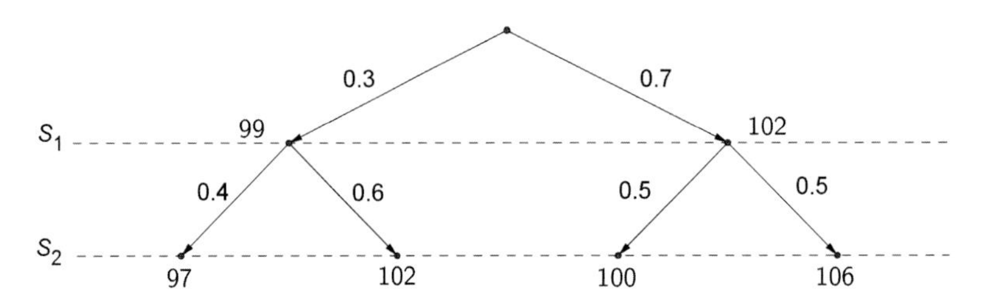


**Figure 2.2** Tree diagram for prices of the stock $S_t$.

(*a*) *What would we expect the average to be for $S_2$? We immediately get*

$$\mathbb{E}[S_2] = 97 \cdot 0.3 \cdot 0.4 + 102 \cdot 0.3 \cdot 0.6 + 100 \cdot 0.7 \cdot 0.5 + 106 \cdot 0.7 \cdot 0.5 = 102.1.$$

(*b*) *Suppose that $S_1 = 99$. What would we expect the average to be for $S_2$? We have by Definition 2.2.2 that*

$$\mathbb{E}[S_2|S_1 = 99] = 97 \cdot 0.4 + 102 \cdot 0.6 + 100 \cdot 0 + 106 \cdot 0 = 100.$$

(*c*) *Suppose that $S_1 = 102$. What would we expect the average to be for $S_2$? As above, we obtain*

$$\mathbb{E}[S_2|S_1 = 102] = 97 \cdot 0 + 102 \cdot 0 + 100 \cdot 0.5 + 106 \cdot 0.5 = 103.$$

*Note that we can make a similar observation here to that in Example 2.2.3, namely that*

$$\begin{aligned}\mathbb{E}[S_2] &= 102.1 = 100 \cdot 0.3 + 103 \cdot 0.7 \\ &= \mathbb{E}[S_2|S_1 = 99] \cdot \mathbb{P}[S_1 = 99] + \mathbb{E}[S_2|S_1 = 102] \cdot \mathbb{P}[S_1 = 102].\end{aligned}$$

Examples 2.2.3 and 2.2.4 suggest that conditional expectations satisfy a similar property to conditional probabilities. More precisely, it suggests that we can compute the expectation $\mathbb{E}[X]$ of a random variable $X$ by splitting it up into different cases as in the *law of total probability*, Theorem 2.1.4. This is indeed true.

**Theorem 2.2.5** (Law of total expectations). *Let $(\Omega, \mathbb{P})$ be a (discrete) probability space, $X$ be a random variable with $\mathbb{E}[|X|] < \infty$ and $(B_j)_{j=1}^n$ with $n \in \mathbb{N} \cup \{\infty\}$ be a partition of $\Omega$. Then*

$$\mathbb{E}[X] = \sum_{\substack{i=1,\ldots,n \\ \mathbb{P}[B_i] \neq 0}} \mathbb{E}[X|B_i] \cdot \mathbb{P}[B_i]. \tag{2.2.2}$$

*Proof.* The idea of the proof is similar to the proof of the *law of total probability*. We have

$$\begin{aligned}
\mathbb{E}[X] &= \sum_{\omega\in\Omega} X(\omega)\cdot\mathbb{P}[\omega] = \sum_{i=1}^{n}\sum_{\omega\in B_i} X(\omega)\cdot\mathbb{P}[\omega] = \sum_{\substack{i=1,\dots,n\\ \mathbb{P}[B_i]\neq 0}}\sum_{\omega\in B_i} X(\omega)\cdot\mathbb{P}[\omega] \\
&= \sum_{\substack{i=1,\dots,n\\ \mathbb{P}[B_i]\neq 0}} \mathbb{P}[B_i]\sum_{\omega\in B_i} X(\omega)\cdot\frac{\mathbb{P}[\omega]}{\mathbb{P}[B_i]} = \sum_{\substack{i=1,\dots,n\\ \mathbb{P}[B_i]\neq 0}} \mathbb{P}[B_i]\sum_{\omega\in B_i} X(\omega)\cdot\mathbb{P}[\omega|B_i] \\
&= \sum_{\substack{i=1,\dots,n\\ \mathbb{P}[B_i]\neq 0}} \mathbb{P}[B_i]\sum_{\omega\in \Omega} X(\omega)\cdot\mathbb{P}[\omega|B_i] = \sum_{\substack{i=1,\dots,n\\ \mathbb{P}[B_i]\neq 0}} \mathbb{E}[X|B_i]\cdot\mathbb{P}[B_i].
\end{aligned}$$

□

**Example 2.2.6.** *Suppose we have two dice of which one is fair and one is manipulated. The manipulated die always gives the number 6 and cannot be visually distinguished from the fair one. We choose a die purely at random and roll it. Denote by $X$ the number obtained. What should we expect the average to be for $X$ if we repeat this experiment very often? Define the events*

$$F = \text{'We have selected the fair die'}.$$
$$M = \text{'We have selected the manipulated die'}.$$

*We have $F \cap M = \emptyset$ and $F \cup M = \Omega$. Thus the sets $F$ and $M$ form a partition of $\Omega$. Furthermore, we clearly have*

$$\mathbb{E}[X|F] = 3.5 \ \text{ and } \ \mathbb{E}[X|M] = 6.$$

*Theorem 2.2.5 implies*

$$\mathbb{E}[X] = \mathbb{E}[X|F]\cdot\mathbb{P}[F] + \mathbb{E}[X|M]\cdot\mathbb{P}[M] = 3.5\cdot\frac{1}{2} + 6\cdot\frac{1}{2} = \frac{19}{4}. \tag{2.2.3}$$

*We could also compute $\mathbb{E}[X]$ directly without Theorem 2.2.5. We use the sample space*

$$\Omega := \big\{(s,x);\ s\in\{m,f\},\ x\in\{1,2,3,4,5,6\}\big\},$$

*where m stands for 'manipulated' and f for 'fair'. We have*

$$\mathbb{P}[(f,x)] = \frac{1}{12} \text{ for } x\in\{1,2,3,4,5,6\},$$
$$\mathbb{P}[(m,6)] = \frac{1}{2} \ \text{ and } \ \mathbb{P}[(m,x)] = 0 \ \text{ for } x\in\{1,2,3,4,5\}.$$

*It is now straightforward to see that the distribution of $X$ is given by Table 2.1.*

**Table 2.1** Distribution of $X$ in Example 2.2.6

| $x$ | 1 | 2 | 3 | 4 | 5 | 6 |
|---|---|---|---|---|---|---|
| $p_X(x)$ | $\frac{1}{12}$ | $\frac{1}{12}$ | $\frac{1}{12}$ | $\frac{1}{12}$ | $\frac{1}{12}$ | $\frac{7}{12}$ |

*We therefore expect on average*

$$\mathbb{E}[X] = 1 \cdot \frac{1}{12} + 2 \cdot \frac{1}{12} + 3 \cdot \frac{1}{12} + 4 \cdot \frac{1}{12} + 5 \cdot \frac{1}{12} + 6 \cdot \frac{7}{12} = \frac{19}{4},$$

*which of course agrees with the result in* (2.2.3).

The quantity $\mathbb{E}[X|B]$ also has many properties similar to $\mathbb{E}[X]$. For instance it is linear in $X$. We have

**Theorem 2.2.7.** *Let $a, b \in \mathbb{R}$ and $X$ and $Y$ be random variables on the same probability space $(\Omega, \mathbb{P})$ with $\mathbb{E}[|X|] < \infty$ and $\mathbb{E}[|Y|] < \infty$ and $B$ be an event with $\mathbb{P}[B] > 0$. We then have*

$$\mathbb{E}[aX + bY|B] = a\mathbb{E}[X|B] + b\mathbb{E}[Y|B].$$

Furthermore, we have

**Theorem 2.2.8.** *Let $X$ be a random variable with $\mathbb{E}[|X|] < \infty$ and $B$ be an event with $\mathbb{P}[B] > 0$. We then have*

$$\mathbb{E}[X|B] = \sum_{\substack{x \in \mathbb{R} \\ p_X(x) \neq 0}} x \cdot \mathbb{P}[X = x|B]. \tag{2.2.4}$$

*Further, if $f : \mathbb{R} \to \mathbb{R}$ is an arbitrary function with $\mathbb{E}[|f(X)|] < \infty$, then*

$$\mathbb{E}[f(X)|B] = \sum_{\substack{x \in \mathbb{R} \\ p_X(x) \neq 0}} f(x) \cdot \mathbb{P}[X = x|B]. \tag{2.2.5}$$

The proofs of Theorems 2.2.7 and 2.2.8 are almost the same as for the unconditional ones and are therefore omitted.

### 2.2.2 Conditioning on Partitions and Random Variables

Suppose we have an experiment whose outcome is given by a random variable $X$ and we repeat this experiment $n$ times with $n$ large. The *law of large numbers* now states that the average value of $X$ is close to $\mathbb{E}[X]$. When we know nothing about the outcome of the experiment, this is the best we can predict. However, this estimate can of course be a little bit crude. The question at this point is whether we can improve this estimate if we know something about the outcome of the experiment. Let us consider some examples.

**Example 2.2.9.** *We have a room containing* 100 *people living in the UK. Our job is to guess the salary of everybody in the room. We could of course ask everybody about their salary, but most people don't want to speak about their salary. We therefore have to find another way. Using Google,* *it turns out that the average salary in the UK is* £30,000. *We could therefore guess that everybody has a salary of* £30,000. *This would lead to*

$$\textit{Average salary of these } 100 \textit{ people} \approx £30{,}000.$$

*However, there are large salary differences in the UK. Thus this estimate might not be very good. We can try to improve this estimate by asking everybody for their job. It turns out that in this room there are 30 interns, 40 train drivers and 30 photographers. The average salaries of these jobs are listed in Table 2.2.*

**Table 2.2** Salary table for Example 2.2.9

| *Job* | *Interns* | *Train drivers* | *Photographers* |
|---|---|---|---|
| *Average salary* | £12,000 | £45,000 | £24,000 |

*We still don't know the salaries of the people in the room, but now we can guess it a little bit better. Instead of guessing the average salary in the UK for each person, we guess that each intern has the salary* £12,000, *each train driver the salary* £45,000 *and each photographer the salary* £24,000. *This is definitely an improvement, but we can of course not expect it to be 100% exact. However, we can expect that it is on average a good approximation. Also, we can improve our guess for the average salary of these 100 people by*

$$\frac{30 \cdot 12{,}000 + 40 \cdot 45{,}000 + 30 \cdot 24{,}000}{100} \approx £39{,}000.$$

**Example 2.2.10.** *Suppose we would like to guess the life expectancy of somebody. We clearly cannot determine this exactly and thus have to estimate it in a different way. Since the average life expectancy in the UK is 79 years, we could just guess it as 79 years. This will not be exactly right, but it will be at least on average a good estimate. As in Example 2.2.9, we can try to improve this estimate. Looking at a table, we see that the average life expectancy in the UK for females is 81 years and for males is 76 years. We can therefore improve our guess by assigning to each male a life expectancy of 76 years and to each female a life expectancy of 81 years.*

When we study Examples 2.2.9 and 2.2.10 more carefully, we see that in both cases we have an 'interesting' random variable, which we cannot determine exactly. However, in both examples there are other properties, which we can easily determine and which have an influence on this random variable. Using these properties, we assign to each sample point an estimate for $X$ depending on these properties. More precisely:

(1) We start with an 'interesting' random variable $X$.
(2) We then split the sample space $\Omega$ into different cases, like different jobs or gender. In other words, we introduce a partition $(B_j)_{j=1}^n$ of $\Omega$, see Definition 2.1.3.
(3) For all $j$, we assign to each $\omega \in B_j$ the average behaviour of $X$ on $B_j$.

Of course, we would like to translate this into a mathematically rigorous definition. One way to do this is to use *indicator functions*. Recall that the *indicator function* for $A \subset \Omega$ is defined as

$$\mathbb{1}_A(\omega) := \begin{cases} 1, & \text{if } \omega \in A, \\ 0, & \text{if } \omega \notin A. \end{cases} \tag{2.2.6}$$

If we now have a partition $(B_j)_{j=1}^n$ and a sequence of real numbers $(a_j)_{j=1}^n$, then the function

$$\sum_{j=1,\dots,n} a_j \cdot \mathbb{1}_{B_j}(\omega) = \begin{cases} a_1, & \text{if } \omega \in B_1, \\ a_2, & \text{if } \omega \in B_2, \\ \vdots & \vdots \\ a_n, & \text{if } \omega \in B_n \end{cases} \tag{2.2.7}$$

assigns to each $\omega \in B_j$ the value $a_j$. Thus the above motivation combined with the function in (2.2.7) leads to the following definition.

**Definition 2.2.11** (Conditional expectation with respect to a partition). Let $(\Omega, \mathbb{P})$ be a discrete probability space, $X$ be a random variable with $\mathbb{E}[|X|] < \infty$ and $(B_j)_{j=1}^n$ with $n \in \mathbb{N} \cup \{\infty\}$ be a partition of $\Omega$. The *conditional expectation* $\mathbb{E}\left[X|(B_j)_{j=1}^n\right]$ *of* $X$ *given* $(B_j)_{j=1}^n$ is defined for $\omega \in \Omega$ as

$$\mathbb{E}\left[X|(B_j)_{j=1}^n\right](\omega) := \sum_{\substack{j=1,\dots,n \\ \mathbb{P}[B_j]\neq 0}} \mathbb{E}\left[X|B_j\right] \cdot \mathbb{1}_{B_j}(\omega). \tag{2.2.8}$$

As in (2.2.7), we can rewrite (2.2.8) as

$$\mathbb{E}\left[X|(B_j)_{j=1}^n\right](\omega) = \begin{cases} \mathbb{E}[X|B_1], & \text{if } \omega \in B_1, \\ \mathbb{E}[X|B_2], & \text{if } \omega \in B_2, \\ \vdots & \vdots, \\ \mathbb{E}[X|B_n], & \text{if } \omega \in B_n. \end{cases} \tag{2.2.9}$$

It is important to point out that

$\mathbb{E}\left[X|(B_j)_{j=1}^n\right](\omega)$ **is a random variable and not a constant!**

It is constant on each $B_j$, but the values on different $B_j$s are in general not equal. This is an important difference with $\mathbb{E}[X]$ and $\mathbb{E}\left[X|B_j\right]$, which are both constants. The fact that $\mathbb{E}\left[X|(B_j)_{j=1}^n\right](\omega)$ is not a constant can be a little confusing at the beginning, especially because we are using the symbol $\mathbb{E}$. The reason one uses $\mathbb{E}$ is that we assign to each $B_j$ the expectation $\mathbb{E}\left[X|B_j\right]$. In other words, $\mathbb{E}\left[X|(B_j)_{j=1}^n\right](\omega)$ is an expectation that distinguishes between different cases. As is usual for random variables, we drop the dependence on $\omega$ and just write $\mathbb{E}\left[X|(B_j)_{j=1}^n\right]$.

**Example 2.2.12.** *Consider again the situation in Example 2.2.6. Recall, X is the rolled number and*

$$F = \text{'We have selected the fair die'}.$$
$$M = \text{'We have selected the manipulated die'}.$$

*We then have*

$$\mathbb{E}[X|(F,M)] = \mathbb{E}[X|F] \cdot \mathbb{1}_F(\omega) + \mathbb{E}[X|M] \cdot \mathbb{1}_M(\omega) = 3.5 \cdot \mathbb{1}_F(\omega) + 6 \cdot \mathbb{1}_M(\omega). \tag{2.2.10}$$

*We can rewrite* (2.2.10) *as in* (2.2.9), *as*

$$\mathbb{E}[X|(F,M)] = \begin{cases} 3.5, & \text{if } \omega \in F, \\ 6, & \text{if } \omega \in M. \end{cases}$$

*Let us now explicitly evaluate X and $\mathbb{E}[X|(F,M)]$, using the same sample space as in Example 2.2.6. This then gives Table 2.3 and we clearly see that $\mathbb{E}[X|(F,M)]$ is a better approximation of X than $\mathbb{E}[X]$.*

**Table 2.3** Values of $X$ and $\mathbb{E}[X|(F,M)]$ in Example 2.2.12

| $\omega \in \Omega$ | $(f,1)$ | $(f,2)$ | $(f,3)$ | $(f,4)$ | $(f,5)$ | $(f,6)$ |
|---|---|---|---|---|---|---|
| $X$ | 1 | 2 | 3 | 4 | 5 | 6 |
| $\mathbb{E}[X\|(F,M)]$ | 3.5 | 3.5 | 3.5 | 3.5 | 3.5 | 3.5 |
| | | | | | | |
| $\omega \in \Omega$ | $(m,1)$ | $(m,2)$ | $(m,3)$ | $(m,4)$ | $(m,5)$ | $(m,6)$ |
| $X$ | 6 | 6 | 6 | 6 | 6 | 6 |
| $\mathbb{E}[X\|(F,M)]$ | 6 | 6 | 6 | 6 | 6 | 6 |

**Example 2.2.13.** *Let us consider the Brexit referendum. There are many statistical surveys about it. Some data are listed in Table 2.4.*

**Table 2.4** Results in the Brexit referendum

| | England | Scotland | Wales | Ireland | Total |
|---|---|---|---|---|---|
| Yes | 53.2% | 38.0% | 52.5% | 44.2% | 51.9% |
| No | 46.8% | 62.0% | 47.5% | 55.8% | 48.1% |

*Let us now pick somebody, purely at random, who has voted. Let us further consider the random variable*

$$Y(\omega) := \begin{cases} 1, & \text{if } \omega \text{ has voted 'Yes'}, \\ 0, & \text{if } \omega \text{ has voted 'No'}. \end{cases}$$

*It follows from Table 2.4 that*

$$\mathbb{E}[Y] = 0.519.$$

*Let us now consider the events*

$$E := \text{'}\omega \text{ voted in England'}, \qquad S := \text{'}\omega \text{ voted in Scotland'},$$
$$W := \text{'}\omega \text{ voted in Wales'}, \qquad N := \text{'}\omega \text{ voted in Ireland'}.$$

*These events clearly form a partition of the sample space and from Table 2.4 we get*

$$\mathbb{E}[Y|(E,S,W,N)] = 0.532 \cdot \mathbb{1}_E(\omega) + 0.38 \cdot \mathbb{1}_S(\omega) + 0.525 \cdot \mathbb{1}_W(\omega) + 0.442 \cdot \mathbb{1}_N(\omega).$$

In most situations we will condition on random variables instead of partitions. In other words, what can we say (on average) about a random variable $X$ if we know the random variable $Y$? For instance, if $X$ is the value of a certain stock tomorrow and $Y$ is the value of the stock today, what can we say about $X$ when we know $Y$? The question is how to define this. Note that

$$\{Y = y_1\} \cap \{Y = y_2\} = \emptyset \text{ for } y_1 \neq y_2 \qquad \text{and} \qquad \bigcup_{\substack{y \in \mathbb{R} \\ \{Y=y\} \neq \emptyset}} \{Y = y\} = \Omega.$$

We thus see that a discrete random variable induces a finite or countable partition of $\Omega$ in a natural way. Therefore, we can use Definition 2.2.11 to define conditioning on a random variable $Y$.

**Definition 2.2.14** (Conditional expectation with respect to a random variable). Let $(\Omega, \mathbb{P})$ be a discrete probability space, $X$ and $Y$ be random variables with $\mathbb{E}[|X|] < \infty$. The *conditional expectation* $\mathbb{E}[X|Y]$ *of* $X$ *given* $Y$ is defined as

$$\begin{aligned} \mathbb{E}[X|Y] &: \Omega \to \mathbb{R} \\ \omega &\to \sum_{\substack{y \in \mathbb{R} \\ \mathbb{P}[Y=y]>0}} \mathbb{E}[X|Y=y] \cdot \mathbb{1}_{\{Y=y\}}(\omega). \end{aligned} \tag{2.2.11}$$

In other words, $\mathbb{E}[X|Y]$ assigns to each element $\omega$ in the event $\{Y = y\}$ the average behaviour of $X$ on $\{Y = y\}$. We can thus rewrite (2.2.11) as

$$\mathbb{E}[X|Y](\omega) = \mathbb{E}[X|Y=y] \quad \text{if } Y(\omega) = y. \tag{2.2.12}$$

As with $\mathbb{E}\left[X|(B_j)_{j=1}^n\right](\omega)$, it is important to point out that

**$\mathbb{E}[X|Y](\omega)$ is a random variable and not a constant!**

As usual, we drop the dependence on $\omega$ and just write $\mathbb{E}[X|Y]$. Heuristically, $\mathbb{E}[X|Y]$ can be interpreted as follows:

*$\mathbb{E}[X|Y]$ is the average of the random variable $X$*
*when we know the value of the random variable $Y$.*

We recommend keeping this heuristic interpretation in mind as it allows us, in many cases, to correctly guess the value for $\mathbb{E}[X|Y]$. We will see how to do this in Section 2.2.3. Let us consider an example.

**Example 2.2.15.** *Let $X$ and $Y$ be discrete random variables on some $(\Omega, \mathbb{P})$ and suppose that the sample space $\Omega$ is split by the events $\{X = i,\ Y = j\}$ as in Figure 2.3. Let us now randomly select a sample point $\omega \in \Omega$ and determine $\mathbb{E}[X|Y]$.*

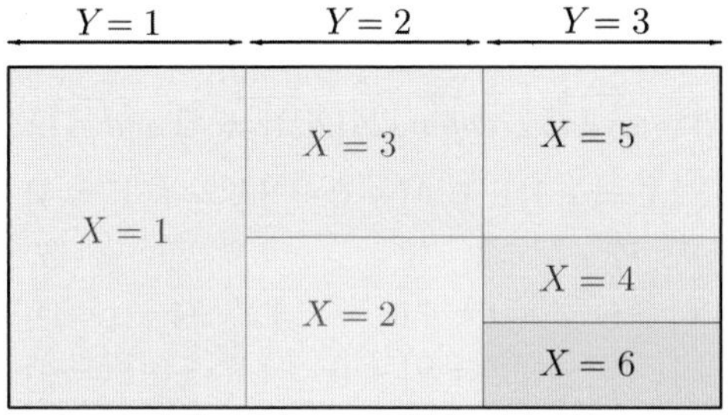

**Figure 2.3** Illustration of the splitting of the sample space by $X$ and $Y$ in Example 2.2.15.

*For this we have to determine the average behaviour of $X$ on the different events $\{Y = j\}$.*

- *If we have selected $\omega \in \{Y = 1\}$ then we always have $X = 1$, no matter which $\omega \in \{Y = 1\}$ we have selected. Thus the best prediction for $X$ on $\{Y = 1\}$ is 1.*
- *The situation looks different if we select $\omega \in \{Y = 2\}$. Here $X$ can be 2 or 3. Since we cannot distinguish the elements in $\{Y = 2\}$, we cannot determine $X$ exactly. However, in around $1/2$ of the cases we will select an $\omega$ with $X = 2$ and in around $1/2$ of the cases an $\omega$ with $X = 3$. If we repeat this many times and each time randomly choose an element of $\{Y = 2\}$, we can expect the average value of $X$ to be $2 \cdot \frac{1}{2} + 3 \cdot \frac{1}{2} = 2.5$.*
- *Similarly, if we select $\omega \in \{Y = 3\}$ then we would expect the average value of $X$ to be $5 \cdot \frac{1}{2} + 4 \cdot \frac{1}{4} + 6 \cdot \frac{1}{4} = \frac{20}{4} = 5$.*

*Combining these observations, we get*

$$\mathbb{E}[X|Y] = 1 \cdot \mathbb{1}_{\{Y=1\}}(\omega) + 2.5 \cdot \mathbb{1}_{\{Y=2\}} + 5 \cdot \mathbb{1}_{\{Y=3\}}.$$

*We can rewrite this as*

$$\mathbb{E}[X|Y] = \begin{cases} 1, & \text{if } Y = 1, \\ 2.5, & \text{if } Y = 2, \\ 5, & \text{if } Y = 3. \end{cases}$$

Consider a further example.

**Example 2.2.16.** *Let $X$ and $Y$ be discrete random variables on some $(\Omega, \mathbb{P})$ and suppose that the joint distribution of $X$ and $Y$ is given in Table 2.5.*

**Table 2.5** Joint distribution of $X$ and $Y$ in Example 2.2.16

| $X \setminus Y$ | 4 | 9 | Row sum |
|---|---|---|---|
| $-1$ | $\frac{1}{8}$ | $\frac{3}{8}$ | $\frac{1}{2}$ |
| 0 | $\frac{2}{9}$ | $\frac{1}{9}$ | $\frac{1}{3}$ |
| 2 | $\frac{1}{12}$ | $\frac{1}{12}$ | $\frac{1}{6}$ |
| Column sum | $\frac{31}{72}$ | $\frac{41}{72}$ | |

*We now determine* $\mathbb{E}[X|Y]$. *We have to compute* $\mathbb{E}[X|Y=4]$ *and* $\mathbb{E}[X|Y=9]$.

$$\begin{aligned}
\mathbb{E}[X|Y=4] &= (-1)\cdot\mathbb{P}[X=-1|Y=4]+0\cdot\mathbb{P}[X=0|Y=4]+2\cdot\mathbb{P}[X=2|Y=4]\\
&= (-1)\cdot\frac{1}{8}\cdot\frac{72}{31}+0\cdot\frac{2}{9}\cdot\frac{72}{31}+2\cdot\frac{1}{12}\cdot\frac{72}{31}=-\frac{3}{31},\\
\mathbb{E}[X|Y=9] &= (-1)\cdot\mathbb{P}[X=-1|Y=9]+0\cdot\mathbb{P}[X=0|Y=4]+2\cdot\mathbb{P}[X=2|Y=9]\\
&= (-1)\cdot\frac{3}{8}\cdot\frac{72}{41}+0\cdot\frac{1}{9}\cdot\frac{72}{41}+2\cdot\frac{1}{12}\cdot\frac{72}{41}=-\frac{15}{41}.
\end{aligned}$$

*It follows that*

$$\mathbb{E}[X|Y] = -\frac{3}{31}\cdot\mathbb{1}_{\{Y=4\}}-\frac{15}{41}\cdot\mathbb{1}_{\{Y=9\}}.$$

***Remark* 2.2.17.** Note that many people call $\mathbb{E}[X|Y]$ a prediction for $X$ when one knows $Y$. This is not really correct as in general

$$X \not\approx \mathbb{E}[X|Y].$$

This can be seen, for instance, in Example 2.2.15. However, as for the expectation $\mathbb{E}[X]$, $\mathbb{E}[X|Y]$ is a good prediction for the average value of $X$ over a large number of experiments when we know $Y$.

We can of course condition on more than one random variable. This can be done similarly to Definition 2.2.14.

**Definition 2.2.18** (Conditional expectation with respect to random variables)**.** Let $(\Omega, \mathbb{P})$ be a discrete probability space, $X$ and $Y_1, \ldots, Y_T$ be random variables with $\mathbb{E}[|X|] < \infty$. The *conditional expectation* $\mathbb{E}\left[X|(Y_j)_{j=1}^T\right]$ *of $X$ given* $Y_1, \ldots, Y_T$ is defined as

$$\mathbb{E}\left[X|(Y_j)_{j=1}^T\right](\omega) := \sum_{\substack{(y_1,\ldots,y_T)\in\mathbb{R}^T\\ \mathbb{P}[Y_1=y_1,\ldots,Y_T=y_T]>0}} \mathbb{E}[X|Y_1=y_1,\ldots,Y_T=y_T]\cdot\mathbb{1}_{\{Y_1=y_1,\ldots,Y_T=y_T\}}(\omega).$$

This definition can be described in words as: We assign to each element in the event $\{Y_1 = y_1, \ldots, Y_T = y_T\}$ the average behaviour of $X$ on the event $\{Y_1 = y_1, \ldots, Y_T = y_T\}$.

Similar to the one-variable case, $\mathbb{E}\left[X|(Y_j)_{j=1}^T\right]$ can be interpreted heuristically as:

$\mathbb{E}\left[X|(Y_j)_{j=1}^T\right]$ *is the average of the random variable* $X$
*when we know the values of the random variables* $Y_1, \ldots, Y_T$.

### 2.2.3 Properties of Conditional Expectation

We will see that conditioning is a very important tool in financial mathematics and is used very often. We therefore need to be able to compute the conditional expectation of a random variable in the most efficient way. In order to do this, we study some important properties of conditional expectations.

For simplicity of notation, in this section we only state the results for conditioning on one random variable. However, all results in this section are also true for conditioning on several random variables. Let us begin with some simple properties.

**Lemma 2.2.19.** Let $(\Omega, \mathbb{P})$ be a discrete probability space, and $X$, $Y$ and $Z$ be random variables with $\mathbb{E}[|X|] < \infty$ and $\mathbb{E}[|Z|] < \infty$. Then

(a) $\mathbb{E}[a|Y] = a$ for any fixed $a \in \mathbb{R}$.
(b) $X \geq 0 \implies \mathbb{E}[X|Y] \geq 0$.
(c) $Z \geq X \implies \mathbb{E}[Z|Y] \geq \mathbb{E}[X|Y]$.
(d) For $a, b \in \mathbb{R}$

$$\mathbb{E}[aX + bZ|Y] = a\mathbb{E}[X|Y] + b\mathbb{E}[Z|Y].$$

*Proof.* We start with (a). Let $y \in \mathbb{R}$ with $\mathbb{P}[Y = y] > 0$ be given. Then

$$\mathbb{E}[a|Y = y] = \sum_{\omega \in \Omega} a \cdot \mathbb{P}[\omega|Y = y] = a \sum_{\omega \in \{Y=y\}} \mathbb{P}[\omega|Y = y] = a.$$

(b) follows immediately from (c) by considering 0 as a degenerate random variable. Next consider (c). Let $y \in \mathbb{R}$ with $\mathbb{P}[Y = y] > 0$ be given. Then

$$\mathbb{E}[X|Y = y] = \sum_{\omega \in \Omega} X(\omega) \cdot \mathbb{P}[\omega|Y = y] \leq \sum_{\omega \in \Omega} Z(\omega) \cdot \mathbb{P}[\omega|Y = y] = \mathbb{E}[Z|Y = y].$$

Using this and the definition of conditional expectation, see Definition 2.2.14, we get

$$\begin{aligned}\mathbb{E}[X|Y] &= \sum_{\substack{y \in \mathbb{R} \\ \mathbb{P}[Y=y]>0}} \mathbb{E}[X|Y = y] \cdot \mathbb{1}_{\{Y=y\}}(\omega) \\ &\leq \sum_{\substack{y \in \mathbb{R} \\ \mathbb{P}[Y=y]>0}} \mathbb{E}[Z|Y = y] \cdot \mathbb{1}_{\{Y=y\}}(\omega) \\ &= \mathbb{E}[Z|Y].\end{aligned}$$

(d) follows immediately from Definition 2.2.14 and Theorem 2.2.7. □

As we said earlier, $\mathbb{E}[X|Y]$ can be viewed as the average of $X$ when we know $Y$. A natural question at this point is: What is $\mathbb{E}[X|Y]$ when $X$ and $Y$ are independent? Let us consider an example.

**Example 2.2.20.** *Independently throw two coins. The first coin shows 'Head' with probability $p$ and the second coin shows 'Head' with probability $q$. Define*

$$X_1 = \begin{cases} 1, & \text{if the first coin shows 'Head'}, \\ 0, & \text{if the first coin shows 'Tail'}, \end{cases}$$

*and*

$$X_2 = \begin{cases} 1, & \text{if the second coin shows 'Head'}, \\ 0, & \text{if the second coin shows 'Tail'}. \end{cases}$$

*In this situation, $X_1$ and $X_2$ are clearly independent. Let us determine $\mathbb{E}[X_1|X_2]$. We have to compute $\mathbb{E}[X_1|X_2 = 0]$ and $\mathbb{E}[X_1|X_2 = 1]$. Since $X_1$ and $X_2$ are independent, by Theorem 2.2.8,*

$$\begin{aligned} \mathbb{E}[X_1|X_2 = 0] &= 0 \cdot \mathbb{P}[X_1 = 0|X_2 = 0] + 1 \cdot \mathbb{P}[X_1 = 1|X_2 = 0] \\ &= 0 \cdot \mathbb{P}[X_1 = 0] + 1 \cdot \mathbb{P}[X_1 = 1] = 0 \cdot (1-p) + 1 \cdot p = p. \end{aligned}$$

*Similarly, $\mathbb{E}[X_1|X_2 = 1] = p$. So*

$$\begin{aligned} \mathbb{E}[X_1|X_2] &= \mathbb{E}[X_1|X_2 = 0] \cdot \mathbb{1}_{\{X_2=0\}} + \mathbb{E}[X_1|X_2 = 1] \cdot \mathbb{1}_{\{X_2=1\}} \\ &= p \cdot \mathbb{1}_{\{X_2=0\}} + p \cdot \mathbb{1}_{\{X_2=1\}} = p = \mathbb{E}[X_1]. \end{aligned}$$

*Therefore $\mathbb{E}[X_1|X_2] = \mathbb{E}[X_1]$. One can show with a similar computation that*

$$\mathbb{E}[X_2|X_1] = q = \mathbb{E}[X_2].$$

*Thus the conditional expectation is just the usual expectation in both cases.*

Let us now try to understand what happens in general when $X$ and $Y$ are independent. Intuitively, independence means that $Y$ has no influence on $X$. In other words, when we know $Y$, we still know nothing about $X$. Therefore, $Y$ cannot have an influence on the average behaviour of $X$. Thus, the average of $X$ always has to be $\mathbb{E}[X]$, regardless of whether we know the value of $Y$ or not. Therefore, we should have $\mathbb{E}[X|Y] = \mathbb{E}[X]$. This is indeed true.

**Lemma 2.2.21.** Let $(\Omega, \mathbb{P})$ be a discrete probability space, $X$ and $Y$ be independent random variables with $\mathbb{E}[|X|] < \infty$. Then

$$\mathbb{E}[X|Y] = \mathbb{E}[X].$$

*Proof.* Let $y \in \mathbb{R}$ be given with $\mathbb{P}[Y = y] > 0$. Since $X$ and $Y$ are independent, we get immediately that $\mathbb{P}[X = x|Y = y] = \mathbb{P}[X = x]$ for all $x \in \mathbb{R}$. Thus, Theorem 2.2.8 implies

$$\mathbb{E}[X|Y=y] = \sum_{\substack{x\in\mathbb{R} \\ p_X(x)\neq 0}} x\cdot\mathbb{P}[X=x|Y=y] = \sum_{\substack{x\in\mathbb{R} \\ p_X(x)\neq 0}} x\cdot\mathbb{P}[X=x] = \mathbb{E}[X].$$

Since $y$ was arbitrary, the lemma follows. □

Let us now consider the opposite situation: What can we say about $\mathbb{E}[X|Y]$ if $Y$ determines $X$ completely? We start again with an example.

**Example 2.2.22.** *Let $Y$ be a discrete random variable and define $X = Y^2$. In this situation, $Y$ clearly determines $X$ completely. Let us now determine $\mathbb{E}[X|Y]$. We have to compute $\mathbb{E}[X|Y=y]$ for all $y \in \mathbb{R}$ with $\mathbb{P}[Y=y] > 0$. Now $\mathbb{E}[X|Y=y]$ is the average behaviour of $X$ on the event $\{Y=y\}$. Obviously, we have $X = y^2$ on $\{Y=y\}$. Thus the average behaviour of $X$ on the event $\{Y=y\}$ is trivially $y^2$. Therefore*

$$\mathbb{E}[X|Y=y] = y^2$$

*and, on the event $\{Y=y\}$, $\mathbb{E}[X|Y]$ agrees with $Y^2$. Since $y$ was arbitrary, we get immediately that $\mathbb{E}[X|Y] = Y^2$.*

The observations made in Example 2.2.22 can now be generalised.

**Lemma 2.2.23.** Let $(\Omega, \mathbb{P})$ be a discrete probability space and $Y$ be a random variable on this space.

(a) If $\mathbb{E}[|Y|] < \infty$ then
$$\mathbb{E}[Y|Y] = Y.$$

(b) Let $f : \mathbb{R} \to \mathbb{R}$ be a function and suppose that $\mathbb{E}[|f(Y)|] < \infty$. Then
$$\mathbb{E}[f(Y)|Y] = f(Y).$$

(c) Let $f : \mathbb{R} \to \mathbb{R}$ be a function and $X$ be a further random variable on $(\Omega, \mathbb{P})$. If $\mathbb{E}[|f(Y)X|] < \infty$ and $\mathbb{E}[|X|] < \infty$ then
$$\mathbb{E}[f(Y)X|Y] = f(Y)\,\mathbb{E}[X|Y].$$

*Proof.* (a) follows from (b) with $f(y) = y$. (b) follows from (c) with $X \equiv 1$ since $\mathbb{E}[1|Y] = 1$ by Lemma 2.2.19. It is therefore enough to prove (c).

Let $y \in \mathbb{R}$ be given with $\mathbb{P}[Y=y] > 0$. We clearly have $f\big(Y(\omega)\big) = f(y)$ on $\{Y=y\}$. With this, we obtain immediately

$$\mathbb{E}[f(Y)X|Y=y] = \sum_{\omega\in\{Y=y\}} f\big(Y(\omega)\big)X(\omega)\cdot\mathbb{P}[\omega|Y=y] = \sum_{\omega\in\{Y=y\}} f(y)X(\omega)\cdot\mathbb{P}[\omega|Y=y]$$
$$= f(y)\sum_{\omega\in\{Y=y\}} X(\omega)\cdot\mathbb{P}[\omega|Y=y] = f(y)\,\mathbb{E}[X|Y=y].$$

Therefore $\mathbb{E}[f(Y)X|Y] = f(Y)\mathbb{E}[X|Y]$ on $\{Y=y\}$. This completes the proof since $y$ was arbitrary. □

Let us now consider some examples.

**Example 2.2.24.** *Let $X$ and $Y$ be random variables on a discrete probability space $(\Omega, \mathbb{P})$. Then*

*(a)* $\mathbb{E}\left[XY^2|Y\right] = Y^2 \cdot \mathbb{E}[X|Y]$.
*(b)* $\mathbb{E}[X+Y|Y] = \mathbb{E}[X|Y] + \mathbb{E}[Y|Y] = \mathbb{E}[X|Y] + Y$.

We can now make an important observation.

**Lemma 2.2.25.** Let $(\Omega, \mathbb{P})$ be a discrete probability space and $X$ and $Y$ be random variables on this space with $\mathbb{E}[|X|] < \infty$. Then there exists an $f : \mathbb{R} \to \mathbb{R}$ such that

$$\mathbb{E}[X|Y] = f(Y).$$

We would like to point out here that in most cases one only needs to know the existence of the function $f$ in Lemma 2.2.25, and does not need to know the function $f$ explicitly. However, this function has an intuitive interpretation: the value $f(y)$ is the average behaviour of $X$ on the event $\{Y=y\}$.

*Proof.* Let us consider the event $\{Y=y\}$ for $y \in \mathbb{R}$ with $\mathbb{P}[Y=y] > 0$. By the definition of $\mathbb{E}[X|Y]$,

$$\mathbb{E}[X|Y](\omega) = \mathbb{E}[X|Y=y] \text{ for all } \omega \in \{Y=y\}.$$

In view of this equation, we define

$$f(y) = \begin{cases} \mathbb{E}[X|Y=y], & \text{if } \mathbb{P}[Y=y] > 0, \\ 0, & \text{otherwise.} \end{cases}$$

Therefore, on $\{Y=y\}$ (with $\mathbb{P}[Y=y] > 0$)

$$\mathbb{E}[X|Y](\omega) = \mathbb{E}[X|Y=y] = f(y) = f(Y).$$

This completes the proof since $y$ is arbitrary. □

Let us now consider some examples in which we can determine the function $f$ in Lemma 2.2.25 explicitly.

**Example 2.2.26.** *Let $X$ and $Y$ be independent. By Lemma 2.2.21, $\mathbb{E}[X|Y] = \mathbb{E}[X]$. Define $f(y) = \mathbb{E}[X]$ for all $y \in \mathbb{R}$. (Observe that $f$ is a constant function.) Then we clearly have*

$$\mathbb{E}[X|Y] = \mathbb{E}[X] = f(Y).$$

**Example 2.2.27.** *Suppose that* $X = Y^2$. *Then* $\mathbb{E}[X|Y] = \mathbb{E}\left[Y^2|Y\right] = Y^2$, *so* $\mathbb{E}[X|Y] = f(Y)$ *with* $f(y) = y^2$.

An important quantity for a random variable is its expectation. As conditional expectation is a random variable, we can ask: What is the expectation of $\mathbb{E}[X|Y]$?

**Theorem 2.2.28** (Tower property, simple version). *Let* $(\Omega, \mathbb{P})$ *be a discrete probability space and* $X$ *and* $Y$ *be random variables on this space with* $\mathbb{E}[|X|] < \infty$. *Then* $\mathbb{E}\Big[|\mathbb{E}[X|Y]|\Big] < \infty$ *and*

$$\mathbb{E}\Big[\mathbb{E}[X|Y]\Big] = \mathbb{E}[X].$$

*Proof.* Observe that for an event $A \subset \Omega$,

$$\mathbb{E}[\mathbb{1}_A(\omega)] = \sum_{\omega\in\Omega} \mathbb{1}_A(\omega)\cdot\mathbb{P}[\omega] = \sum_{\omega\in A}\mathbb{P}[\omega] = \mathbb{P}[A].$$

Using Definition 2.2.14, linearity of expectation and Lemma 2.2.23

$$\begin{aligned}
\mathbb{E}\Big[\mathbb{E}[X|Y]\Big] &= \mathbb{E}\left[\sum_{\substack{y\in\mathbb{R}\\ \mathbb{P}[Y=y]>0}} \mathbb{E}[X|Y=y]\cdot\mathbb{1}_{\{Y=y\}}(\omega)\right] = \sum_{\substack{y\in\mathbb{R}\\ \mathbb{P}[Y=y]>0}} \mathbb{E}[X|Y=y]\cdot\mathbb{E}\left[\mathbb{1}_{\{Y=y\}}(\omega)\right] \\
&= \sum_{\substack{y\in\mathbb{R}\\ \mathbb{P}[Y=y]>0}} \mathbb{E}[X|Y=y]\cdot\mathbb{P}[Y=y] \\
&= \sum_{\substack{y\in\mathbb{R}\\ \mathbb{P}[Y=y]>0}} \left(\sum_{\omega\in\{Y=y\}} X(\omega)\cdot\mathbb{P}[\omega|Y=y]\right)\cdot\mathbb{P}[Y=y] \\
&= \sum_{\substack{y\in\mathbb{R}\\ \mathbb{P}[Y=y]>0}} \left(\sum_{\omega\in\{Y=y\}} X(\omega)\cdot\mathbb{P}[\omega]\right) = \sum_{\omega\in\Omega} X(\omega)\cdot\mathbb{P}[\omega] = \mathbb{E}[X].
\end{aligned}$$

It remains to justify that $\mathbb{E}[|\mathbb{E}[X|Y]|] < \infty$. By assumption, $\mathbb{E}[|X|] < \infty$ and so the sum $\sum_{\omega\in\Omega} X(\omega)\cdot\mathbb{P}[\omega]$ is absolutely convergent. Checking the above computations, we see that $\mathbb{E}[\mathbb{E}[X|Y]]$ can be obtained from $\mathbb{E}[X]$ by reordering the sum $\sum_{\omega\in\Omega} X(\omega)\cdot\mathbb{P}[\omega]$. Thus $\mathbb{E}[|\mathbb{E}[X|Y]|] < \infty$ and this completes the proof. □

**Example 2.2.29.** *Let us consider a betting game with several identical and independent rounds, where the number of rounds is random. What total gain can we expect on average when the number of rounds is independent of the gain or loss in each round? Denote by* $N$ *the number of rounds and* *by $X_j$ the gain (or loss) in round $j$. The $X_j$ are i.i.d. and $N$ is independent of the $X_j$. Assume at this point that $\mathbb{E}[|N|] < \infty$ and that $\mathbb{E}\left[|X_j|\right] < \infty$ for all $j$. The gain of this betting game is $\sum_{j=1}^{N} X_j$. To compute the expectation of this expression, use Theorem 2.2.28 to get*

$$\mathbb{E}\left[\sum_{j=1}^{N} X_j\right] = \mathbb{E}\left[\mathbb{E}\left[\sum_{j=1}^{N} X_j \middle| N\right]\right].$$

*To determine $\mathbb{E}\left[\sum_{j=1}^{N} X_j | N\right]$ we compute $\mathbb{E}\left[\sum_{j=1}^{N} X_j | N = n\right]$ for all $n \in \mathbb{N}$ with $\mathbb{P}[N = n] > 0$. We have*

$$\mathbb{E}\left[\sum_{j=1}^{N} X_j \middle| N = n\right] = \mathbb{E}\left[\sum_{j=1}^{n} X_j \middle| N = n\right] = \sum_{j=1}^{n} \mathbb{E}\left[X_j | N = n\right].$$

*Since the $X_j$ are i.i.d. and independent of $N$, we have $\mathbb{P}\left[X_j = x | N = n\right] = \mathbb{P}\left[X_j = x\right]$ for all $j \in \mathbb{N}$ and thus $\mathbb{E}\left[X_j | N = n\right] = \mathbb{E}\left[X_j\right] = \mathbb{E}[X_1]$. We therefore have*

$$\mathbb{E}\left[\sum_{j=1}^{N} X_j \middle| N = n\right] = \sum_{j=1}^{n} \mathbb{E}\left[X_j\right] = \sum_{j=1}^{n} \mathbb{E}[X_1] = n\,\mathbb{E}[X_1].$$

*So $\mathbb{E}\left[\sum_{j=1}^{N} X_j | N\right] = N\,\mathbb{E}[X_1]$ and therefore*

$$\mathbb{E}\left[\sum_{j=1}^{N} X_j\right] = \mathbb{E}[N\,\mathbb{E}[X_1]] = \mathbb{E}[N]\,\mathbb{E}[X_1].$$

Let us now consider an important generalisation of the tower property. As motivation, let $Y_1, \ldots, Y_T$ be the prices of one stock at times 1 to $T$ and $X$ the gain of a certain investment strategy. Then $Z := \mathbb{E}\left[X | (Y_j)_{j=1}^{t}\right]$ is the average gain when we know the prices of the stock up to time $t$. We can now interpret $Z$ as the gain of another investment strategy. We can ask at this point what is the average gain of this second strategy when we only know the prices of the stock up to time $s$ (for $s \leq t$), that is what is $\mathbb{E}\left[Z | (Y_j)_{j=1}^{s}\right]$?

**Theorem 2.2.30** (Tower property, general version). *Let $(\Omega, \mathbb{P})$ be a discrete probability space and $X$ and $Y_1, \ldots, Y_T$ be discrete random variables with $\mathbb{E}[|X|] < \infty$. Then for $1 \leq s \leq t \leq T$*

$$\mathbb{E}\left[\mathbb{E}\left[X | (Y_j)_{j=1}^{t}\right] \middle| (Y_j)_{j=1}^{s}\right] = \mathbb{E}\left[X | (Y_j)_{j=1}^{s}\right].$$

*Proof.* For simplicity of notation, we prove the theorem only for $s = 1$ and $t = 2$. We thus have to show that

$$\mathbb{E}\left[\mathbb{E}[X | Y_1, Y_2] \middle| Y_1\right] = \mathbb{E}[X | Y_1].$$

To prove this identity, we have to show for all $y \in \mathbb{R}$ with $\mathbb{P}[Y_1 = y] > 0$ that

$$\mathbb{E}\Big[\mathbb{E}[X|Y_1, Y_2]\Big|Y_1 = y\Big] = \mathbb{E}[X|Y_1 = y].$$

Recall, by Definition 2.2.18 that

$$\mathbb{E}[X|Y_1, Y_2] = \sum_{\substack{(y_1,y_2)\in\mathbb{R}^2 \\ \mathbb{P}[Y_1=y_1,Y_2=y_2]>0}} \mathbb{E}[X|Y_1 = y_1, Y_2 = y_2] \cdot \mathbb{1}_{\{Y_1=y_1,Y_2=y_2\}}(\omega).$$

So

$$\begin{aligned}
\mathbb{E}\Big[\mathbb{E}[X|Y_1, Y_2]\Big|Y_1 = y\Big] &= \sum_{\substack{y_1,y_2\in\mathbb{R} \\ \mathbb{P}[Y_1=y_1,Y_2=y_2]>0}} \mathbb{E}[X|Y_1 = y_1, Y_2 = y_2]\,\mathbb{E}\Big[\mathbb{1}_{\{Y_1=y_1,Y_2=y_2\}}\Big|Y_1 = y\Big] \\
&= \sum_{\substack{y_2\in\mathbb{R} \\ \mathbb{P}[Y_1=y,Y_2=y_2]>0}} \mathbb{E}[X|Y_1 = y, Y_2 = y_2]\,\mathbb{E}\Big[\mathbb{1}_{\{Y_1=y,Y_2=y_2\}}\Big|Y_1 = y\Big]
\end{aligned}$$

since $\mathbb{1}_{\{Y_1=y_1,Y_2=y_2\}}$ vanishes on the event $\{Y_1 = y\}$ unless $y_1 = y$. Now

$$\begin{aligned}
\mathbb{E}\Big[\mathbb{1}_{\{Y_1=y,Y_2=y_2\}}\Big|Y_1 = y\Big] &= \sum_{\omega\in\Omega} \mathbb{1}_{\{Y_1=y,Y_2=y_2\}} \cdot \mathbb{P}[\omega|Y_1 = y] = \sum_{\omega\in\{Y_1=y\}} \frac{\mathbb{1}_{\{Y_1=y,Y_2=y_2\}} \cdot \mathbb{P}[\omega]}{\mathbb{P}[Y_1 = y]} \\
&= \sum_{\omega\in\{Y_1=y,Y_2=y_2\}} \frac{\mathbb{P}[\omega]}{\mathbb{P}[Y_1 = y]} = \frac{\mathbb{P}[Y_1 = y,\ Y_2 = y_2]}{\mathbb{P}[Y_1 = y]}.
\end{aligned}$$

Inserting this into the above equation gives

$$\begin{aligned}
\mathbb{E}\Big[\mathbb{E}[X|Y_1, Y_2]\Big|Y_1 = y\Big] &= \sum_{\substack{y_2\in\mathbb{R} \\ \mathbb{P}[Y_1=y,Y_2=y_2]>0}} \mathbb{E}[X|Y_1 = y, Y_2 = y_2]\,\frac{\mathbb{P}[Y_1 = y,\ Y_2 = y_2]}{\mathbb{P}[Y_1 = y]} \\
&= \sum_{\substack{y_2\in\mathbb{R} \\ \mathbb{P}[Y_1=y,Y_2=y_2]>0}} \left(\sum_{\omega\in\{Y_1=y,\ Y_2=y_2\}} X(\omega)\frac{\mathbb{P}[\omega]}{\mathbb{P}[Y_1 = y,\ Y_2 = y_2]}\right)\frac{\mathbb{P}[Y_1 = y,\ Y_2 = y_2]}{\mathbb{P}[Y_1 = y]} \\
&= \sum_{\substack{y_2\in\mathbb{R} \\ \mathbb{P}[Y_1=y,Y_2=y_2]>0}} \sum_{\omega\in\{Y_1=y,\ Y_2=y_2\}} X(\omega) \cdot \mathbb{P}[\omega|Y_1 = y] = \sum_{\omega\in\{Y_1=y\}} X(\omega) \cdot \mathbb{P}[\omega|Y_1 = y] \\
&= \mathbb{E}[X|Y_1 = y].
\end{aligned}$$

This completes the proof. □

## 2.3 Modelling the Information Available in the Future

The topic of this section is how to mathematically model the information we have at a given time $t$. We begin with the classical approach.

### 2.3.1 Motivation

Consider this situation.

*Suppose that Victor and Peter play a game:*
*Victor can toss a fair coin up to 7 times, but he may stop whenever he chooses.*
*Each time Victor tosses 'Head', he gets £1 from Peter. Each time Victor tosses 'Tail', he has to pay £1 to Peter. What strategy should Victor use?*

If Victor knows before the game starts that he will toss

'Head', 'Head', 'Head', 'Tail', 'Tail', 'Tail', 'Head',

then Victor will toss the coin three times and would win £3. However, Victor has the same problem as us all: He cannot predict the future. Therefore his decisions at time $t$ can only take into account the information available at time $t$. The question at this point is how we can formulate this mathematically.

Tossing the coin can be modelled mathematically as a sequence $(Y_t)_{t=1}^7$ of i.i.d. Bernoulli$(1/2)$-distributed random variables, where we interpret 1 as 'Head' and 0 as 'Tail'. We now have to determine what we know after tossing the coin $t$ times. Clearly, we know the values of $Y_1, \dots, Y_t$. Victor's strategy at time $t$ on whether to stop or continue can depend only on the information available at time $t$. The classical way to model the information available is to say

*After tossing the coin $t$ times, we know the values of $Y_1, \dots, Y_t$.*

If we know the values of $Y_1, \dots, Y_t$ then we know also the values of $(Y_1)^2$, $Y_1 + Y_t$ and $Y_1 - 5Y_t$. In fact, if we know the values of $Y_1, \dots, Y_t$ then we know the values of all random variables of the form

$$X = f(Y_1, \dots, Y_t) \tag{2.3.1}$$

with $f : \mathbb{R}^t \to \mathbb{R}$ a fixed function. All random variables $X$ of the form given by (2.3.1) have the following property: if we know the values of $Y_1, \dots, Y_t$ then we also know the value of $X$. In other words, all such random variables are uniquely determined by $Y_1, \dots, Y_t$. On the other hand, there are random variables $X$, which are not determined by $Y_1, \dots, Y_t$. For those random variables, there are cases where even if we know the values of $Y_1, \dots, Y_t$, there are still several values that $X$ could take. Therefore we don't know the value of $X$ in this situation and $X$ cannot be written as in (2.3.1). Summarising the above observations, we obtain:

*After tossing the coin $t$ times, we know the value of a random variable $X$*
*if and only if $X$ can be written as $X = f(Y_1, \dots, Y_t)$ for some function $f : \mathbb{R}^t \to \mathbb{R}$.*

Suppose now that we have tossed the coin $t$ times. Victor now has to decide whether to stop the game or toss the coin again. This decision can be modelled with the help of a random variable $E_t$ with

$$E_t = \begin{cases} 1, & \text{Victor tosses the coin again,} \\ 0, & \text{Victor stops.} \end{cases} \tag{2.3.2}$$

The important point here is that $E_t$ can only depend on the information we have after $t$ times tossing the coin. So $E_t$ has to be determined by the values of $Y_1, \ldots, Y_t$. In view of the above considerations, this means $E_t$ must have the form

$$E_t = f_t(Y_1, \ldots, Y_t), \tag{2.3.3}$$

where $f_t$ is a fixed function. Therefore Victor's strategy must be expressible in the form $(E_t)_{t=1}^T$ with each $E_t$ as in (2.3.3). We will not discuss how to explicitly choose $(E_t)_{t=1}^T$ here (although the optional stopping theorem implies that the expected profit of any such strategy is 0). The important point at this stage was to determine what such a strategy has to look like.

The situation in the game above occurs in many other places, for instance with trading strategies in a financial market. Traders have the same problem as Victor: they cannot predict the future. Thus one can only allow trading strategies that use, at time $t$, the information available at time $t$. The above approach can easily be generalised as long as the time $t$ in the model is discrete, that is $t \in \{0, \ldots, T\}$ or $t \in \mathbb{N}_0$. However, it no longer works well if the time $t$ in the model is continuous, that is $t \in [0, T]$ or $t \in \mathbb{R}$. In these situations, one has to work with $\sigma$-algebras and filtrations. In the first half of this book, we will work mostly with discrete time, that is $t \in \{0, \ldots, T\}$, and so could just use the approach above. However, within financial mathematics, $\sigma$-algebras and filtrations are commonly used everywhere, regardless of whether the time is discrete or continuous. It is therefore important to have some basic knowledge of $\sigma$-algebras and filtrations. We will give only an overview here and mention the most important results we need later.

## 2.3.2 Definition of a $\sigma$-Algebra

In the betting game at the beginning of Section 2.3, we specified which random variables we know at time $t$. The modern approach is slightly different and one specifies instead the events we know at time $t$. To get an idea of how this works, let us consider some examples and see what influence the time has on events.

**Example 2.3.1.** *Suppose we roll a fair die and consider the event $A = \{2, 3, 5\}$. Before we roll the die, we don't know the outcome and so don't know whether $A$ occurs or not. We can therefore say that $A$ is random. However, after we roll the die, we know the outcome and can say whether $A$ has occurred or not. So $A$ is not random anymore and is now deterministic.*

Let us now consider an example with more than one time step.

**Example 2.3.2.** *Suppose we roll a fair die twice. We can model this experiment with the sample space*

$$\Omega = \{(i,j);\ 1 \le i,j \le 6\},$$

*where $(i,j)$ denotes the outcome in which we first roll the number i and then the number j. Consider the events*

$$A := \{(1,1), (2,2), (3,3), (4,4), (5,5), (6,6)\},$$
$$B := \{(6,1), (6,2), (6,3), (6,4), (6,5), (6,6)\}.$$

*Before we roll the die we don't know whether A and B occur and so both events are random. On the other hand, once we have rolled the die twice then we know the outcome and we can say with certainty whether A and B have occurred. Therefore A and B are deterministic in this case. Suppose now that we have rolled the die once.*

- *Can we say whether the event A has occurred or not?*
  *Suppose we rolled a 1. Then the sample point has the form $(1,*)$. Now $(1,1) \in A$ and $(1,2) \notin A$. Since $(1,1)$ and $(1,2)$ can both still occur, we cannot say whether A has occurred or not. So A is still random.*
- *Can we say whether the event B has occurred or not?*
  *Suppose we rolled a 1. Then B has not occurred as all sample points in B have the form $(6,*)$. Similarly, B has not occurred if we rolled a 2, a 3, a 4 or a 5 in the first roll. The situation is different if we rolled a 6. In this case B can still occur. Since B contains all sample points of the form $(6,*)$, we know already that B has occurred. So, after rolling the die once, we know whether B has occurred or not. Therefore B is no longer random after rolling the die once.*

We see at this point that the time has an influence on whether an event is random or deterministic. In other words, at time $t$ some events are no longer random since we can say with certainty whether they have occurred or not. We use the formulation

*We know an event A at time t (or an event A can be observed at time t)*
*if we can say with certainty at time t whether A has occurred or has not occurred.*

As we have seen in Examples 2.3.1 and 2.3.2, some events can be observed at time $t$ and some others cannot. Let us now consider the general situation. Denote by $\mathcal{F}_t$ the family of all events that can be observed up to time $t$, that is

$$\mathcal{F}_t = \{A \subset \Omega;\ \text{we know the event } A \text{ at time } t\}.$$

What properties does the family $\mathcal{F}_t$ have?

- The trivial event '*Something will happen*' has to be contained in $\mathcal{F}_t$.
- If we know the event $A$ at time $t$ then we also know the event $A^c$ at time $t$.
- If we know the events $A$ and $B$ at time $t$ then we also know $A \cup B$ and $A \cap B$.
- If we know the events $A_1, A_2, \dots$ at time $t$ then we also know the event $\bigcup_{j=1}^{\infty} A_j$.

Any set of events satisfying these desired properties is called a $\sigma$-algebra. We give the formal definition below.

**Definition 2.3.3.** A family $\mathcal{F}$ of subsets of $\Omega$ is called a *$\sigma$-algebra* (or *$\sigma$-field*) over $\Omega$, pronounced sigma-algebra, if

- $\Omega \in \mathcal{F}$.
- $A \in \mathcal{F} \Longrightarrow A^c \in \mathcal{F}$, where $A^c = \Omega \setminus A = \{\omega \in \Omega; \omega \notin A\}$.
- $A_i \in \mathcal{F}$ for all $i \in \mathbb{N} \Longrightarrow \bigcup_{i=1}^{\infty} A_i \in \mathcal{F}$.

Note that the definition only specifies three of the four desirable properties. The choice of the three axioms in the definition was made to minimise the overlap in the four properties. We will see below that the fourth follows from these three. In most situations the set $\Omega$ is clear from the context, so we typically drop the dependence on $\Omega$ and just say $\mathcal{F}$ is a $\sigma$-algebra. Let us now consider some examples.

### Example 2.3.4.

(*a*) *Let $\Omega$ be an arbitrary, non-empty set. Then $\mathcal{F}_0 = \{\emptyset, \Omega\}$ and $\mathcal{P}\Omega$ are $\sigma$-algebras, where $\mathcal{P}\Omega$ is the power set of $\Omega$ (the set of all subsets).*

(*b*) *Suppose that $\Omega = \{0, 1\}$. Consider the family*

$$\mathcal{G} = \big\{\emptyset, \{0\}, \Omega\big\}.$$

*Is $\mathcal{G}$ a $\sigma$-algebra? The subset $\{0\} \in \mathcal{G}$, but $\{0\}^c = \{1\} \notin \mathcal{G}$. So $\mathcal{G}$ is not a $\sigma$-algebra.*

(*c*) *Let $\Omega$ be $\{0, 1\}$. How many $\sigma$-algebras exist over $\Omega$? We have*

$$\mathcal{P}\Omega = \big\{\emptyset, \{0\}, \{1\}, \Omega\big\}.$$

*Since the power set $\mathcal{P}\Omega$ has 4 elements, there exist $2^4 = 16$ families of sets over $\Omega$. If a family of sets is a $\sigma$-algebra then it has to contain $\emptyset$ and $\Omega$. Thus the first candidate for a $\sigma$-algebra is $\mathcal{F}_0 = \{\emptyset, \Omega\}$. This is indeed a $\sigma$-algebra. Let $\mathcal{F}$ be another $\sigma$-algebra. Then $\emptyset \in \mathcal{F}$ and $\Omega \in \mathcal{F}$. Since $\mathcal{F} \neq \mathcal{F}_0$, there have to be more sets in $\mathcal{F}$ than $\emptyset$ and $\Omega$. However, there are only two candidates left: $\{0\}$ and $\{1\}$. If $\{0\} \in \mathcal{F}$ then $\{0\}^c = \{1\} \in \mathcal{F}$. Similarly, if $\{1\} \in \mathcal{F}$ then $\{1\}^c = \{0\} \in \mathcal{F}$. Thus $\mathcal{F} = \mathcal{P}\Omega$. Therefore there only exist two $\sigma$-algebras over $\{0, 1\}$, namely*

$$\mathcal{F}_0 = \{\emptyset, \Omega\},$$
$$\mathcal{P}\Omega = \{\emptyset, \{0\}, \{1\}, \Omega\}.$$

Let us now take a look at some basic properties of $\sigma$-algebras.

**Lemma 2.3.5.** Let $\mathcal{F}$ be a *$\sigma$-algebra* over a set $\Omega$. Then

- $\emptyset \in \mathcal{F}$.
- If $A_i \in \mathcal{F}$ for all $i \in \mathbb{N}$ then $\bigcap_{i \in \mathbb{N}} A_i \in \mathcal{F}$.
- If $A_1, A_2, \ldots, A_n \in \mathcal{F}$ then $\bigcup_{i=1}^{n} A_i \in \mathcal{F}$ and $\bigcap_{i=1}^{n} A_i \in \mathcal{F}$.

*Proof.* The first point holds because $\emptyset = \Omega^c$. For the second point, observe that since $A_i^c \in \mathcal{F}$ for all $i \in \mathbb{N}$, $\bigcup_{i=1}^{\infty}(A_i)^c \in \mathcal{F}$. But by de Morgan's rule

$$\left(\bigcap_{i=1}^{\infty} A_i\right)^c = \bigcup_{i=1}^{\infty}(A_i)^c.$$

Taking complements of both sides proves the assertion. The third point follows by using $A_i = \emptyset$, respectively $A_i = \Omega$ for $i \geq n+1$. □

If a family $\mathcal{F}$ of subsets is finite then the following proposition is helpful to check whether $\mathcal{F}$ is a $\sigma$-algebra.

**Proposition 2.3.6.** Let $\Omega$ be a non-empty set and $\mathcal{F}$ be a family of subsets of $\Omega$, consisting only of finitely many subsets, that is

$$\mathcal{F} = \{F_1, \ldots, F_n\} \text{ with } F_j \subset \Omega.$$

Suppose further that

$$\Omega \in \mathcal{F}, \quad A \in \mathcal{F} \Longrightarrow A^c \in \mathcal{F} \quad \text{and} \quad A, B \in \mathcal{F} \Longrightarrow A \cup B \in \mathcal{F}.$$

Then the family $\mathcal{F}$ is a $\sigma$-algebra.

*Proof.* The first two properties are the same as in the definition of a $\sigma$-algebra. Therefore we have only to show that

$$A_j \in \mathcal{F} \; \forall j \in \mathbb{N} \Longrightarrow \bigcup_{j=1}^{\infty} A_j \in \mathcal{F}.$$

However, $\mathcal{F}$ is a finite family. This implies that in the sequence $A_1, A_2, \ldots$, there are only finitely many distinct $A_j$. Therefore we can write

$$\bigcup_{j=1}^{\infty} A_j = \bigcup_{k \in K} F_k \text{ with } K \subset \{1, 2, \ldots, n\}.$$

It follows immediately by induction that a finite union of sets in $\mathcal{F}$ is again contained in $\mathcal{F}$. This completes the proof. □

## 2.3.3 Visualisation of $\sigma$-Algebras

A natural question is: What picture should one keep in mind when working with a $\sigma$-algebra? By definition, a $\sigma$-algebra is a set of sets. The problem is that in most situations, one cannot really illustrate a set of sets. A better approach is to ask what the members in a set of sets look like. Let us consider an example.

**Example 2.3.7.** *Let $\Omega = \mathbb{R}^2$ and let $x_0$ be an arbitrary point in $\Omega$. Let $\mathcal{S}_{x_0}$ be the family of all sets containing $x_0$, that is $\mathcal{S}_{x_0} = \{A \subset \Omega;\, x_0 \in A\}$. Clearly, the family $\mathcal{S}_{x_0}$ contains uncountably many sets $A$ as $\Omega = \mathbb{R}^2$ is not countable. However, we can easily illustrate the sets in $\mathcal{S}_{x_0}$, see Figure 2.4.*

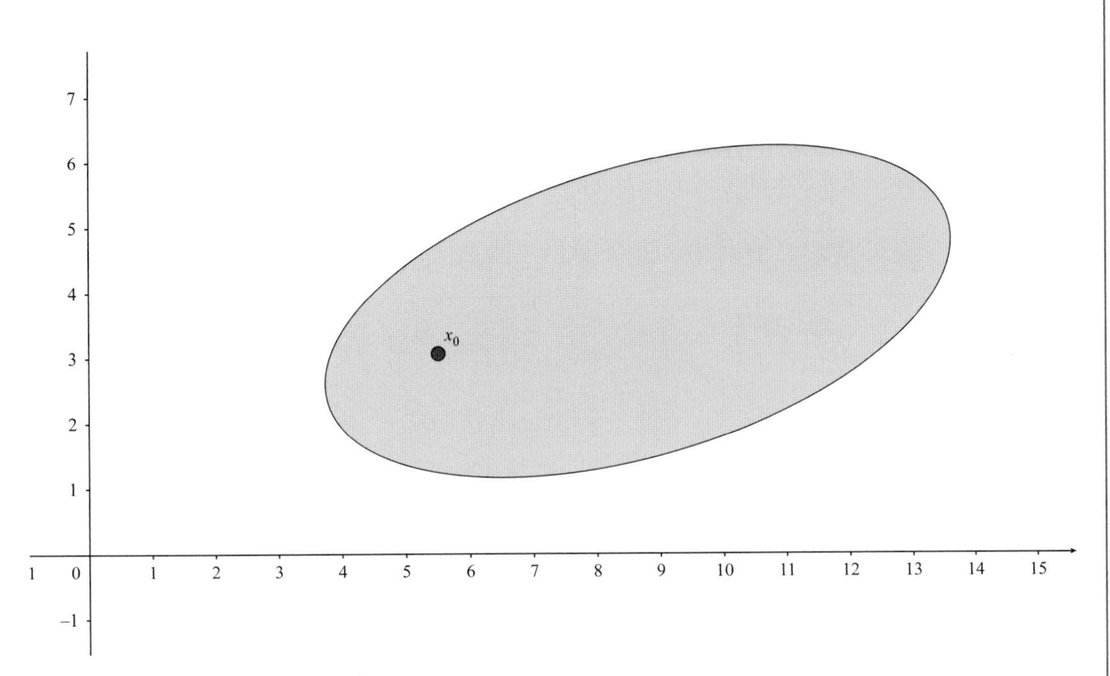


**Figure 2.4** Illustration of a set in $\mathcal{S}_{x_0}$.

*For the avoidance of confusion, note that the family $\mathcal{S}_{x_0}$ is not a $\sigma$-algebra. Indeed, if $A \in \mathcal{S}_{x_0}$ then $A^c \notin \mathcal{S}_{x_0}$.*

Using Example 2.3.7 as a reference, we would like to answer the following question:

*What do the sets in the $\sigma$-algebra $\mathcal{F}_t$ look like?*

Recall, $\mathcal{F}_t$ is the family of events that we know at time $t$. We will work here primarily with stochastic models for stock markets. At time $t$, we know the stock prices up to time $t$. The stock prices in a stochastic model are of course random variables. Therefore the question is: Which events do we know if we know the value of a random variable? Let us look at a simple example.

**Example 2.3.8.** *Consider a random variable $Y$ taking only the three values 2, 3 and 5. Which events do we know if we know the value of $Y$? We will denote the family of these events by $\sigma(Y)$. Let us start with the observation that the random variable $Y$ partitions the sample space $\Omega$ into three events. These events are $\{Y = 2\}$, $\{Y = 3\}$ and $\{Y = 5\}$, see Figure 2.5(a). Let us now consider the events $A$ and $B$ in Figure 2.5.*

- *Do we know $A$ if we (only) know the value of $Y$? Suppose we carry out an experiment and this experiment realises the value $Y = 2$. Then we know that an $\omega \in \{Y = 2\}$ has been selected, but we do not know which one has been selected. Now, there are some $\omega \in \{Y = 2\}$ which are contained in $A$ while some other $\omega \in \{Y = 2\}$ are not contained in $A$. Since we do not know which $\omega$ has been selected, we cannot say if $A$ has occurred. Thus $A \notin \sigma(Y)$.*
- *Do we know $B$ if we know the value of $Y$? Suppose we carry out an experiment and this experiment gives the value $Y = 2$. Then an $\omega \in \{Y = 2\}$ has been selected. Since $\{Y = 2\}$ is contained in $B$, we know that $B$ has occurred. Similarly for the case $Y = 3$. Finally, if we have $Y = 5$ then we know that $B$ did not occur since no $\omega \in \{Y = 5\}$ is contained in $B$. Thus, we have $B \in \sigma(Y)$ since we always know if $B$ has occurred if we know the value of $Y$.*

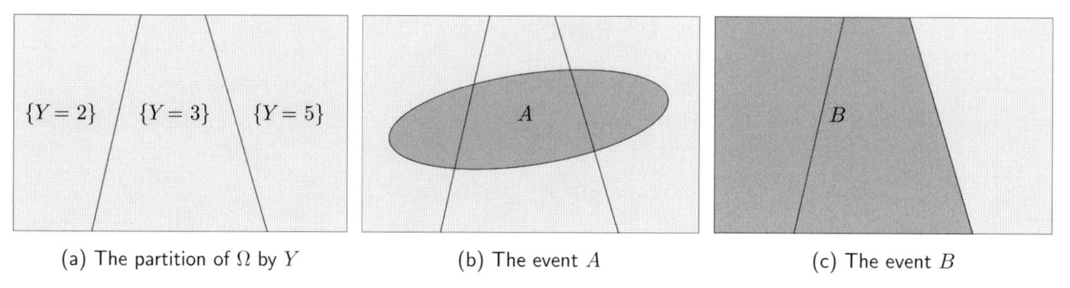


(a) The partition of $\Omega$ by $Y$ (b) The event $A$ (c) The event $B$

**Figure 2.5** Venn diagrams in Example 2.3.8.

The considerations in Example 2.3.8 can easily be generalised. Suppose that we have a random variable $Y$ taking values $y_1, y_2, \ldots, y_K$. As in Example 2.3.8, we denote by $\sigma(Y)$ the family of events where we know the value of $Y$. Then an event $A \in \sigma(Y)$ if and only if we have for each $j$

$$A \cap \{Y = y_j\} = \{Y = y_j\} \text{ or } A \cap \{Y = y_j\} = \emptyset. \tag{2.3.4}$$

Indeed, suppose there exists a $j$ such that $\emptyset \neq A \cap \{Y = y_j\} \subsetneq \{Y = y_j\}$. Then there is some $\omega \in \{Y = y_j\}$ which is contained in $A$, while some other $\omega \in \{Y = y_j\}$ is not contained in $A$. If in an experiment $Y = y_j$ then we know that an $\omega \in \{Y = y_j\}$ has occurred, but we do not know which one. We therefore cannot decide whether $\omega \in A$ or $\omega \notin A$. In other words, we do not know whether $A$ has occurred and hence $A \notin \sigma(Y)$.

The observation in (2.3.4) can be used to visualise the events in $\sigma(Y)$. Let us illustrate this with a random variable $Y$ taking the values 1, 2, 3, 4, 5 and 6.

This random variable $Y$ leads to a partition of $\Omega$ as in Figure 2.6(a). If we now would like to construct an event $A \in \sigma(Y)$ then we have to decide for each $j$ separately if $\{Y = j\}$ is contained in $A$ or if $\{Y = j\}$ and $A$ are disjoint. Graphically, this process corresponds to colouring each area in the Venn diagram in Figure 2.6(a) either black or white, without taking into account the colour of the other areas. This can lead, for instance, to the Venn diagram in Figure 2.6(b). An advantage of this illustration is that we can determine the number of events contained in $\sigma(Y)$. Indeed, if the random variable $Y$ can take the values $y_1, y_2, \ldots, y_K$ then $\sigma(Y)$ contains $2^K$ events.

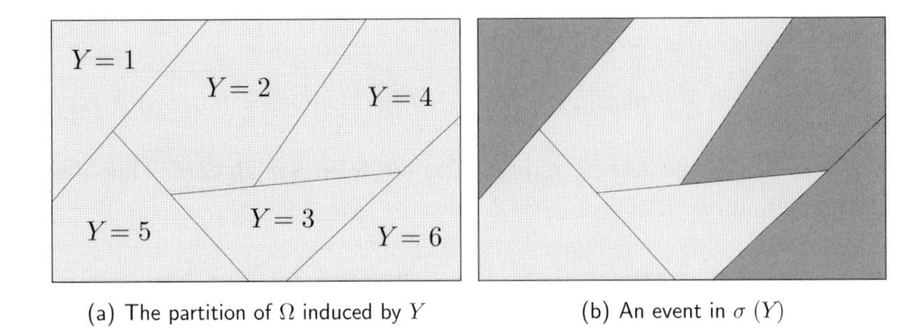


(a) The partition of $\Omega$ induced by $Y$ (b) An event in $\sigma(Y)$

**Figure 2.6** Illustration of events $A \in \sigma(Y)$.

### 2.3.4 $\sigma$-Algebras Generated by Random Variables

In Section 2.3.3 we gave an heuristic definition of $\sigma(Y)$. The aim of this section is to give a mathematically precise definition of $\sigma(Y)$. We start again with the following question:

*Which events do we know if we know the value of a random variable Y?*

We denote the family of these events as $\sigma(Y)$. In view of the considerations in Section 2.3.2, we see that $\sigma(Y)$ has to be a $\sigma$-algebra. Let us now consider the events

$$\{Y \in [0,2]\},\ \{Y \in [-\infty, 0.5]\} \text{ and } \{Y \in [1.5,2]\}. \tag{2.3.5}$$

Do we know these events if we know the value of $Y$? Suppose we have $Y = y$. Then $\{Y \in [0,2]\}$ has occurred if $y \in [0,2]$ and $\{Y \in [0,2]\}$ hasn't if $y \notin [0,2]$. A similar argument also holds for $\{Y \in [-\infty, 0.5]\}$ and $\{Y \in [1.5,2]\}$. In view of these observations, it is natural to request that

$$\{Y \in [a,b]\} \in \sigma(Y) \text{ for all } a,b \in \mathbb{R} \text{ with } a \le b. \tag{2.3.6}$$

The question at this point is: Is the family

$$\mathcal{H}_Y := \big\{\{Y \in [a,b]\};\ a,b \in \mathbb{R} \text{ with } a \le b\big\} \subset \mathcal{P}\Omega \tag{2.3.7}$$

already a $\sigma$-algebra? In order to answer this question, we take a look at an example.

**Example 2.3.9.** *Let us consider* $\Omega = \{\omega_1, \omega_2, \omega_3\}$, $\mathbb{P}\left[\omega_j\right] = \frac{1}{3}$ *for all* $j \in \{1,2,3\}$ *and*

$$Y(\omega_1) = 2,\ Y(\omega_2) = 3,\ Y(\omega_3) = 5.$$

*Let us now determine the family* $\mathcal{H}_Y$ *in* (2.3.7) *explicitly. Note that* $\mathcal{H}_Y$ *is a family of subsets. This is equivalent to saying that* $\mathcal{H}_Y \subset \mathcal{P}\Omega$. *We now have, for instance,*

$$\{Y \in [1,2]\} = \{\omega_1\} \text{ and } \{Y \in [1,3]\} = \{\omega_1, \omega_2\}.$$

*Going through all a and b, one arrives at*

$$\mathcal{H}_Y = \big\{\emptyset,\ \{\omega_1\},\ \{\omega_2\},\ \{\omega_3\},\ \{\omega_1,\omega_2\},\ \{\omega_2,\omega_3\},\ \{\omega_1,\omega_2,\omega_3\}\big\}.$$

*We now see that $\mathcal{H}_Y$ is not a $\sigma$-algebra since*

$$\{\omega_1\} \in \mathcal{H}_Y \text{ and } \{\omega_3\} \in \mathcal{H}_Y, \quad \text{but } \{\omega_1\} \cup \{\omega_3\} = \{\omega_1, \omega_3\} \notin \mathcal{H}_Y.$$

This shows us that the sets $\{Y \in [a,b]\}$ do not form a $\sigma$-algebra. This observation leads to the following definition.

**Definition 2.3.10.** Let $\Omega$ be a set and $\mathcal{H}$ be a non-empty family of subsets of $\Omega$, that is $\mathcal{H} \subset \mathcal{P}\Omega$. We then define $\sigma(\mathcal{H})$ to be the smallest $\sigma$-algebra containing $\mathcal{H}$. This means in formulas: If $\mathcal{F}$ is a $\sigma$-algebra with $\mathcal{H} \subset \mathcal{F}$ then $\sigma(\mathcal{H}) \subset \mathcal{F}$.

At this point one should check that this definition is well-defined as it is not immediately clear that a $\sigma$-algebra with the properties in Definition 2.3.10 exists. However, we will omit the details.

The formulation of $\sigma(\mathcal{H})$ in Definition 2.3.10 is not particularly intuitive. A much more intuitive interpretation of $\sigma(\mathcal{H})$ is to view $\sigma(\mathcal{H})$ as the family of sets that can be obtained by forming all possible (countable) unions, intersections and complements with the sets in $\mathcal{H}$. We will consider some examples, see Example 2.3.13. Before we do this, we give the definition of $\sigma(Y)$.

**Definition 2.3.11.** Let $(\Omega, \mathbb{P})$ be a (discrete) probability space, $Y$ be a random variable on this space and $\mathcal{H}_Y$ be as in (2.3.7). Define

$$\sigma(Y) := \sigma(\mathcal{H}_Y).$$

In words: $\sigma(Y)$ is the smallest $\sigma$-algebra containing all events

$$\{Y \in [a,b]\} \text{ with } a \le b \text{ and } a, b \in \mathbb{R}. \tag{2.3.8}$$

It is straightforward to see that the events

$$\{Y \in ]a,b[\},\ \{Y \in [a,b[\},\ \{Y \in ]a,b]\},\ \{Y \in [a,\infty]\},\ \{Y \in [-\infty,b[\}$$

are also contained in $\sigma(Y)$. We recommend keeping in mind the intuitive interpretation

*$\sigma(Y)$ is the family of the events we know when we know the value of $Y$*

together with the visualisation in Section 2.3.3. We will see in Lemma 2.3.15 that the heuristic and mathematical approaches give the same result when $Y$ is discrete. Also, we would like to mention that Definition 2.3.11 can be used for arbitrary random variables and that the above heuristic interpretation is still valid. However, the visualisation in Section 2.3.3 unfortunately works only for discrete random variables.

When one first sees Definition 2.3.11, a typical approach is to try and determine $\sigma(Y)$ explicitly. In order to do this, one has to first determine the family $\mathcal{H}_Y$ and then $\sigma(\mathcal{H}_Y)$. Let us consider an example.

**Example 2.3.12.** *Let $\Omega$ and $Y$ be as in Example 2.3.9. We have shown in this case that*

$$\begin{aligned}\mathcal{H}_Y &= \big\{\{Y \in [a,b]\};\ a,b \in \mathbb{R} \text{ and } a \le b\big\} \\ &= \big\{\emptyset,\ \{\omega_1\},\ \{\omega_2\},\ \{\omega_3\},\ \{\omega_1,\omega_2\},\ \{\omega_2,\omega_3\},\ \{\omega_1,\omega_2,\omega_3\}\big\}.\end{aligned}$$

*Next we have to determine $\sigma(\mathcal{H}_Y)$. We have to determine all events which can be obtained by forming all possible (countable) unions, intersections and complements with the sets in $\mathcal{H}_Y$. We get immediately that*

$$\mathcal{H}_Y \subset \sigma(Y) \text{ and } \{\omega_1, \omega_3\} \in \sigma(Y).$$

*Since $\Omega = \{\omega_1, \omega_2, \omega_3\}$, we must have $\sigma(Y) = \mathcal{P}\Omega$.*

In Example 2.3.12, it was easy to determine $\sigma(\mathcal{H}_Y)$. However, this is not always the case. We look at two small examples of $\sigma$-algebras generated by some families $\mathcal{H}$.

**Example 2.3.13.** *Let $A$ be a subset of $\Omega$. Consider the family $\mathcal{H}$ consisting only of $A$, that is $\mathcal{H} = \{A\}$. Let us now determine $\sigma(\mathcal{H})$. We write here $\sigma(A)$ instead of $\sigma(\mathcal{H})$ as there is no danger of confusion. Using Definition 2.3.3 and Lemma 2.3.5, we get that*

$$A \in \sigma(A),\ A^c \in \sigma(A),\ \emptyset \in \sigma(A) \text{ and } \Omega \in \sigma(A).$$

*We claim now that $\mathcal{F} = \{A, A^c, \emptyset, \Omega\}$ is a $\sigma$-algebra. Clearly, we have $\Omega \in \mathcal{F}$ and if $B \in \mathcal{F}$ then also $B^c \in \mathcal{F}$. Since this family is finite, Proposition 2.3.6 shows that it is sufficient to check that the union of two arbitrary sets in $\mathcal{F}$ is again contained in $\mathcal{F}$. This condition is clearly also fulfilled. Thus $\mathcal{F}$ is a $\sigma$-algebra and therefore $\mathcal{F} = \sigma(A)$. An illustration of the events in $\sigma(A)$ is given in Figure 2.7.*

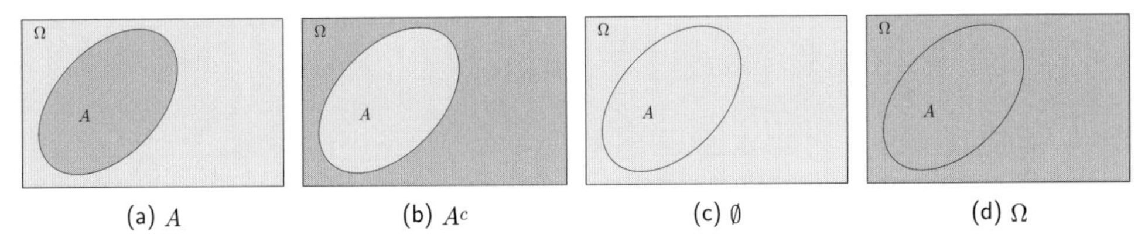


(a) $A$ (b) $A^c$ (c) $\emptyset$ (d) $\Omega$

**Figure 2.7** Illustration of the events in $\sigma(A)$.

**Example 2.3.14.** *Let $A$ and $B$ be subsets of $\Omega$ and let us consider the family $\mathcal{H}$ consisting of $A$ and $B$, that is $\mathcal{H} = \{A, B\}$. Let us determine $\sigma(\mathcal{H})$. We write here $\sigma(A, B)$ instead of $\sigma(\mathcal{H})$. If we draw a Venn diagram for this situation, see Figure 2.8, then we see that $A$ and $B$ partition the sample space $\Omega$ into four disjoint events. These events are*

$$A \setminus B,\ B \setminus A,\ A \cap B \text{ and } (A \cup B)^c. \tag{2.3.9}$$

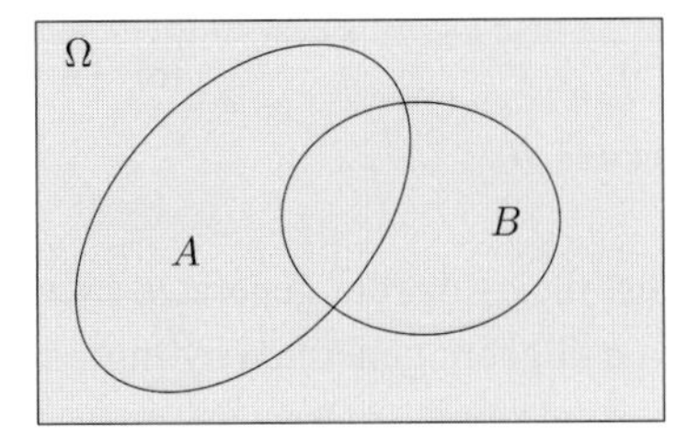


**Figure 2.8** Venn diagram for the events $A$ and $B$ in Example 2.3.14.

*Since $\sigma(A, B)$ is a $\sigma$-algebra, we see that the events in* (2.3.9) *also have to be contained in* $\sigma(A, B)$. *We are now in a similar situation to that in Section 2.3.3, see Figure 2.6, so can try to use a similar argument. More precisely, we look for all events which can be constructed by taking unions of the events in* (2.3.9), *see Figure 2.9. Since we can decide separately for each event in* (2.3.9) *whether the event is contained in the union or not, we have* $2^4 = 16$ *possibilities. This procedure leads to the events*

$$A,\ B,\ A \cap B,\ A \cup B,\ A \setminus B,\ B \setminus A,\ (A \cup B) \setminus (A \cap B),\ \emptyset,$$
$$(A \cup B)^c,\ A^c,\ B^c,\ \big((A \cup B) \setminus (A \cap B)\big)^c,\ A^c \cup B,\ A \cup B^c,\ (A \cap B)^c,\ \Omega. \tag{2.3.10}$$

*All these have to be contained in* $\sigma(A, B)$. *As in Example 2.3.14, one can now show that the events in* (2.3.10) *form a $\sigma$-algebra and therefore $\sigma(A, B)$ consists of the 16 events in* (2.3.10). *These events are illustrated in Figure 2.9.*

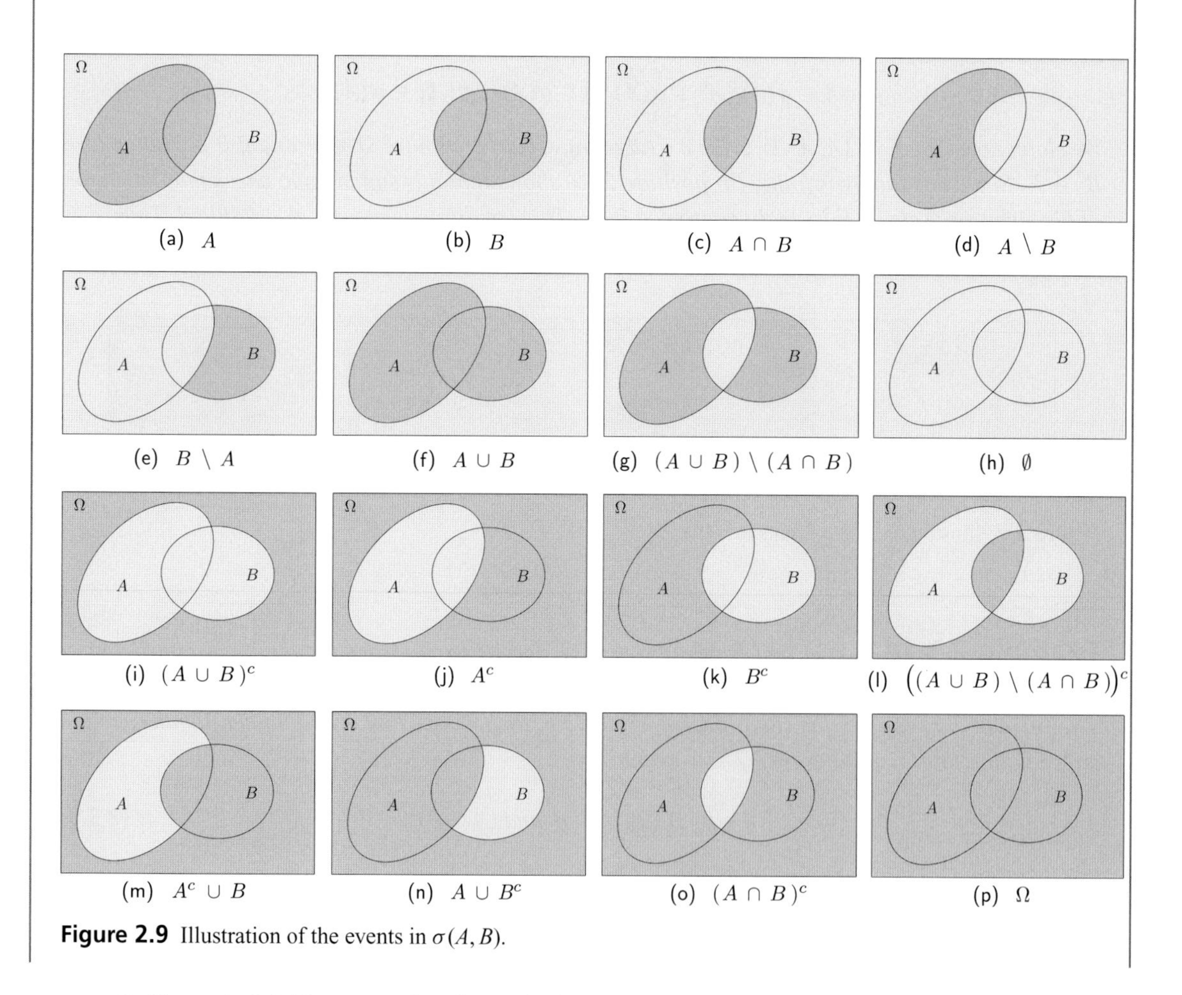


(a) $A$ (b) $B$ (c) $A \cap B$ (d) $A \setminus B$

(e) $B \setminus A$ (f) $A \cup B$ (g) $(A \cup B) \setminus (A \cap B)$ (h) $\emptyset$

(i) $(A \cup B)^c$ (j) $A^c$ (k) $B^c$ (l) $\big((A \cup B) \setminus (A \cap B)\big)^c$

(m) $A^c \cup B$ (n) $A \cup B^c$ (o) $(A \cap B)^c$ (p) $\Omega$

**Figure 2.9** Illustration of the events in $\sigma(A, B)$.

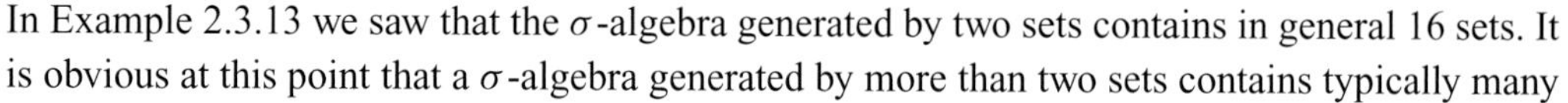

In Example 2.3.13 we saw that the $\sigma$-algebra generated by two sets contains in general 16 sets. It is obvious at this point that a $\sigma$-algebra generated by more than two sets contains typically many more sets. An illustration as in Figure 2.9 is therefore not practical in general. Instead one has to use the trick we used in Section 2.3.3. More precisely, we just draw the Venn diagram as in Figures 2.6(a) and 2.8 and (implicitly) assume that the reader is familiar with the fact that each event in the $\sigma$-algebra is a union of some of the disjoint events in the Venn diagram.

We are mainly interested in $\sigma$-algebras generated by random variables. In the case when we have a discrete random variable, we can adapt the arguments used in Section 2.3.3. This leads to the following lemma.

**Lemma 2.3.15.** Let $(\Omega, \mathbb{P})$ be a (discrete) probability space. Let $Y$ be a random variable defined on this probability space and let $\sigma(Y)$ be defined as in Definition 2.3.11. Suppose that $Y$ takes only the values $y_1, y_2, \ldots, y_n$ with $y_i \neq y_j$ for $i \neq j$. Then each $A \in \sigma(Y)$ can be written as a union of some $\{Y = y_j\}$. Explicitly, we have

$$A \in \sigma(Y) \Longrightarrow A = \bigcup_{j\in J}\{Y = y_j\} \text{ for some } J \subset \{1,2,3,\ldots,n\}. \tag{2.3.11}$$

*Proof.* We first observe that $\{Y = y_j\} \in \sigma(Y)$. This follows immediately by using $a = b = y_j$ in (2.3.8). Thus $\sigma(Y)$ has to contain all events $\bigcup_{j\in J}\{Y = y_j\}$ with $J \subset \{1,2,3,\ldots,n\}$. Furthermore, for all $a \leq b$

$$\{Y \in [a,b]\} = \bigcup_{\substack{1\leq j\leq n\\ y_j\in[a,b]}} \{Y = y_j\}.$$

Therefore it is sufficient to show that the family of events $\bigcup_{j\in J}\{Y = y_j\}$ is a $\sigma$-algebra. Now

$$\Omega = \bigcup_{j=1}^{n}\{Y = y_j\}, \qquad \left(\bigcup_{j\in J}\{Y = y_j\}\right) \cup \left(\bigcup_{j\in J'}\{Y = y_j\}\right) = \bigcup_{j\in J\cup J'} \{Y = y_j\},$$
$$\left(\bigcup_{j\in J}\{Y = y_j\}\right)^c = \bigcup_{j\in J^c}\{Y = y_j\} \text{ with } J^c = \{1,2,3,\ldots,n\} \setminus J.$$

Since this family is finite, we can use Proposition 2.3.6 to complete the proof. □

We see at this point that Lemma 2.3.15 shows that heuristic considerations about $\sigma(Y)$ in Section 2.3.3 lead to the same result as the mathematical approach in this section. In particular, Lemma 2.3.15 shows that the visualisation in Section 2.3.3 gives the correct result.

We would like to mention at this point that it is almost impossible to determine $\sigma(Y)$ explicitly for general random variables $Y$. This includes in particular continuous random variables. However, in most situations it is neither useful nor necessary to determine $\sigma(Y)$ explicitly. Normally it is better to work with the properties that define $\sigma(Y)$. In other words, one can use that $\sigma(Y)$ is the smallest $\sigma$-algebra containing all events $\{Y \in [a,b]\}$ and manipulate the events $\{Y \in [a,b]\}$

accordingly. For instance, suppose we would like to show that $\sigma(Y) \subset \mathcal{F}$ where $\mathcal{F}$ is another $\sigma$-algebra. In this case it is enough to show that $\{Y \in [a,b]\} \in \mathcal{F}$ for each $a, b$. Let us illustrate this with an example.

**Example 2.3.16.** *Let $(\Omega, \mathbb{P})$ be a (discrete) probability space, $Y$ be a random variable on this space and define $X = Y^2 + 1$. Let $A$ be an event. We make the following observation: if we know an event $A$ when we know the value of $X$ then we also know $A$ when we know the value of $Y$. In other words, if $A \in \sigma(X)$ then we must/should have $A \in \sigma(Y)$. Hence, we expect in this situation*

$$\sigma(X) \subset \sigma(Y). \tag{2.3.12}$$

*One way to verify* (2.3.12) *is to first determine $\sigma(X)$ and $\sigma(Y)$ explicitly and then check that* (2.3.12) *holds. However, this is inefficient and not even doable in most cases. Using the characterisation of $\sigma(X)$ in Definition 2.3.11, it is enough to show that $\{X \in [a,b]\} \in \sigma(Y)$ for all $a$, $b$. First consider the case $1 \le a \le b$. Then*

$$\begin{aligned}\{X \in [a,b]\} = \{Y^2+1 \in [a,b]\} &= \left\{Y \in [\sqrt{a-1}, \sqrt{b-1}] \text{ or } Y \in [-\sqrt{b-1}, -\sqrt{a-1}]\right\} \\ &= \left\{Y \in [\sqrt{a-1}, \sqrt{b-1}]\right\} \cup \left\{Y \in [-\sqrt{b-1}, -\sqrt{a-1}]\right\} \in \sigma(Y).\end{aligned}$$

*The remaining cases for $a$, $b$ are handled similarly. Therefore $\{X \in [a,b]\} \in \sigma(Y)$ for all $a$, $b$. The definition of $\sigma(X)$ then implies* (2.3.12).

We will normally work with more than one random variable. The $\sigma$-algebra generated by several random variables is defined similarly.

**Definition 2.3.17.** Let $(\Omega, \mathbb{P})$ be a (discrete) probability space and $Y_1, \ldots, Y_T$ be random variables on this space. Define $\sigma(Y_1, \ldots, Y_T)$ to be the smallest $\sigma$-algebra containing the events

$$\{Y_1 \in [a_1, b_1], \ldots, Y_T \in [a_T, b_T]\} \text{ with } a_j \le b_j \text{ for all } 1 \le j \le T.$$

Note that for $s \le t$

$$\sigma(Y_1, \ldots, Y_s) \subset \sigma(Y_1, \ldots, Y_t).$$

Intuitively, if $Y_1, \ldots, Y_T$ are the prices of a stock at times 1 to $T$, then we can interpret $\mathcal{F}_t$ as the information available up to time $t$. It thus makes sense to have $\mathcal{F}_s \subset \mathcal{F}_t$ for $s \le t$ since we know more at time $t$ than at time $s$. In order to model the available information at different times, we introduce the following definition.

**Definition 2.3.18.** Let $(\Omega, \mathbb{P})$ be a (discrete) probability space and $T \in \mathbb{N} \cup \{\infty\}$ be given. A *filtration* $\mathbb{F}$ is a sequence $\{\mathcal{F}_t\}_{0 \le t \le T}$ of $\sigma$-algebras with

$$\mathcal{F}_0 \subset \mathcal{F}_1 \subset \mathcal{F}_2 \subset \cdots \subset \mathcal{F}_T. \tag{2.3.13}$$

Note that for discrete-time models in finance, all the filtrations we consider are generated by random variables. In other words, we consider only filtrations $\mathbb{F} = (\mathcal{F}_t)_{t=1}^T$ with $\mathcal{F}_t := \sigma(Y_1, \ldots, Y_t)$.

### 2.3.5 Measurable Random Variables and Conditioning

We have seen that events can be random or deterministic, depending on the time. A similar observation is true for random variables. Let us consider an example.

**Example 2.3.19.** *Let $Y_1, \ldots, Y_T$ be the prices of a stock at times 1 to $T$ and consider the random variables*

$$Y_2, \quad \frac{Y_1 + Y_t}{2} \quad \text{and } M := \max_{1 \le t \le T} Y_t.$$

*We know the value of $Y_2$ at time 2 and later, but at time 1 we don't know the value of $Y_2$. Similarly, we know the value of $\frac{Y_1+Y_t}{2}$ only at time $t$ and later. Finally, we know the value of $M$ only at time $T$.*

We therefore see that a random variable can also be random or deterministic (depending on the time). The question at this point is: *For which random variables is the value determined by time $t$?* Let us formulate this more precisely. Suppose a $\sigma$-algebra $\mathcal{F}_t$ and a random variable $X$ are given. As before, we interpret $\mathcal{F}_t$ as the family of events which we know up to time $t$. The main question is now:

*Under which conditions do we know the value of $X$ if we know the events in $\mathcal{F}_t$?*

We have two cases to consider.

- Suppose we know the value of $X$ at time $t$. We can then say whether $\{X \in [a,b]\}$ has occurred or not. This means that $\{X \in [a,b]\} \in \mathcal{F}_t$ since $\mathcal{F}_t$ consists of all events, for which we can decide whether they have occurred or not.
- On the other hand, if we can say whether or not the event $\{X \in [a,b]\}$ has occurred, we also know whether $X \in [a,b]$ or $X \notin [a,b]$. If this is true for all $a$ and $b$ then the value of $X$ is clearly determined.

We therefore see that

$$\text{The value of } X \text{ is determined by time } t \iff \{X \in [a,b]\} \in \mathcal{F}_t \text{ for all } a \le b.$$

This observation leads to the following definition.

**Definition 2.3.20.** Let $(\Omega, \mathbb{P})$ be a discrete probability space, $X$ be a random variable on this space and $\mathcal{F}$ be a $\sigma$-algebra over $\Omega$.

The random variable $X$ is called *$\mathcal{F}$-measurable* if

$$\{X \in [a,b]\} \in \mathcal{F} \text{ for all } a \leq b \text{ with } a,b \in \mathbb{R}.$$

Let us consider some examples.

### Example 2.3.21.

(a) *Let $Y$ be a random variable and define $X := Y + 3$. Is $X$ $\sigma(Y)$-measurable? Intuitively, $\sigma(Y)$ consists of all events one knows when $Y$ is given. When we know the value of $Y$, we clearly also know the value of $X$ and thus $X$ should be $\sigma(Y)$-measurable. Indeed, we have*

$$\{X \in [a,b]\} = \{Y + 3 \in [a,b]\} = \{Y \in [a-3, b-3]\} \in \sigma(Y).$$

(b) *Let $Y$ be a random variable and define $X := Y^2$. Is $X$ $\sigma(Y)$-measurable? We argue similarly as in the previous case. We get for $0 \leq a \leq b$*

$$\{X \in [a,b]\} = \{Y^2 \in [a,b]\} = \{Y \in [-\sqrt{b}, -\sqrt{a}] \cup [\sqrt{a}, \sqrt{b}]\} \in \sigma(Y).$$

*Further, if $a \leq 0$ then $\{X \in [a,b]\} = \{X \in [0,b]\} \in \sigma(Y)$. Finally, if $b < 0$ then $\{X \in [a,b]\} = \emptyset \in \sigma(Y)$.*

In Example 2.3.21 we studied random variables $X$ of the form $f(Y)$, where $f$ is a fixed function. The question at this point is: Do there exist random variables $X$, which are $\sigma(Y)$-measurable and are not of the form $f(Y)$? The following lemma shows that this is not the case.

**Lemma 2.3.22** (Dynkin's lemma, discrete version). Let $(\Omega, \mathbb{P})$ be a discrete probability space and $X$, $Y_1$, ..., $Y_T$ be random variables on this space. Then the following are equivalent:

- $X$ is $\sigma(Y_1, \ldots, Y_T)$-measurable.
- There exists a function $f$ such that $X = f(Y_1, \ldots, Y_T)$.

*Proof.* We assume for simplicity that $T = 1$, write $Y = Y_1$ and assume that $Y$ only takes values in $\mathbb{N}$. These simplifications don't have an influence on the proof itself, but make the notation easier. We know from Lemma 2.3.15 that

$$A \in \sigma(Y) \implies A = \bigcup_{j \in J} \{Y = j\} \text{ for some } J \subset \mathbb{N}.$$

The case $J = \emptyset$ corresponds to $A = \emptyset$ and $J = \mathbb{N}$ corresponds to $A = \Omega$.

Suppose $X$ is $\sigma(Y)$-measurable. We have to show that there is a function $f$ with $X = f(Y)$. Consider the event $\{X = x\}$ for $x \in \mathbb{R}$ fixed with $\{X = x\} \neq 0$. Since $X$ is $\sigma(Y)$-measurable,

$$\{X = x\} \in \sigma(Y) \implies \{X = x\} = \bigcup_{j \in J} \{Y = j\} \text{ for some } J \subset \mathbb{N}. \tag{2.3.14}$$

Since $\{X = x\} \neq \emptyset$, we have $J \neq \emptyset$. Now (2.3.14) implies that

$$Y = j \text{ with } j \in J \implies X = x.$$

Define $f(j) := x$ for $j \in J$. With this choice, $X = f(Y)$ on $\{X = x\}$. Since $\{X = x_1\} \cap \{X = x_2\} = \emptyset$ if $x_1 \neq x_2$, $f(j)$ is well-defined. Further, the events $\{X = x\}$ with $x \in \mathbb{R}$ arbitrary cover $\Omega$ and this completes the first implication.

Conversely, suppose $X = f(Y)$. Clearly, $X$ is constant on each event $\{Y = j\}$ with $j \in \mathbb{N}$. If $\omega_1$, $\omega_2 \in \{Y = j\}$ are given then

$$X(\omega_1) \in [a,b] \iff X(\omega_2) \in [a,b] \iff f(j) \in [a,b].$$

We therefore see that either $\{Y = j\} \subset \{X \in [a,b]\}$ or $\{Y = j\} \cap \{X \in [a,b]\} = \emptyset$. This implies that $\{X \in [a,b]\} = \bigcup_{j \in J} \{Y = j\}$ for some $J \subset \mathbb{N}$, which completes the proof. □

**Example 2.3.23.** *Let $\Omega = \{a,b,c,d,m,n,u,v\}$ and $\mathbb{P}[\omega] = \frac{1}{8}$ for all $\omega \in \Omega$. Further, let $X, Y, Z$ be random variables on the probability space $(\Omega, \mathbb{P})$, where the values of $X$, $Y$ and $Z$ are given in Table 2.6.*

**Table 2.6** Table of $X$, $Y$ and $Z$ values

| $\omega \in \Omega$ | $a$ | $b$ | $c$ | $d$ | $m$ | $n$ | $u$ | $v$ |
|---|---|---|---|---|---|---|---|---|
| $Y(\omega)$ | 1 | 1 | 1 | 1 | 2 | 2 | 3 | 3 |
| $X(\omega)$ | 1 | 2 | 2 | 1 | 3 | 2 | 1 | 3 |
| $Z(\omega)$ | 4 | 4 | 4 | 4 | $\pi$ | $\pi$ | $-3$ | $-3$ |

- *Is $X$ a $\sigma(Y)$-measurable random variable? Let us first argue heuristically: do we know the value of $X$ if we know the value of $Y$? The answer is no. For instance, if $Y = 2$ then we know that $m$ or $n$ has been selected, but we don't know which one. Since $3 = X(m) \neq X(n) = 2$, we do not know which value $X$ has. So we expect that $X$ is not $\sigma(Y)$-measurable. Dynkin's lemma justifies this since $X \neq f(Y)$. Indeed, $X = f(Y)$ would imply $f(2) = 3$ and $f(2) = 2$, which is clearly not possible.*
- *Is $Z$ a $\sigma(Y)$-measurable random variable? We see immediately that if $Y = 1$ then $Z = 4$, if $Y = 2$ then $Z = \pi$ and if $Y = 3$ then $Z = -3$. So we know the value of $Z$ when we know the value of $Y$ and therefore expect that $Z$ is $\sigma(Y)$-measurable. Indeed, we can write $Z = f(Y)$, where $f$ is a function with $f(1) = 4$, $f(2) = \pi$ and $f(3) = -3$. Dynkin's lemma implies that $Z$ is $\sigma(Y)$-measurable.*

Let us now briefly compare the statement in Dynkin's lemma with the considerations at the beginning of Section 2.3. Recall, we tossed a coin several times and modelled this experiment with a sequence $(Y_t)_{t=1}^T$ of i.i.d. Bernoulli(1/2)-distributed random variables. We then arrived at the conclusion that

*After tossing the coin t times, we know the value of a random variable X if and only if X can be written as* $X = f(Y_1, \ldots, Y_t)$.

Using Dynkin's lemma, we see that this is equivalent to saying that we know the value of a random variable $X$ if and only if $X$ is $\sigma(Y_1, \ldots, Y_T)$-measurable. In other words, saying *we know the values of* $Y_1, \ldots, Y_t$ gives the same information as *specifying the events we know when we know* $Y_1, \ldots, Y_t$. Although this shows that, in the discrete-time setting, the classical approach described at the start of Section 2.3 is equivalent to using $\sigma$-algebras, we will use the notation of $\sigma$-algebras from now on as this is standard today.

Another reason to use $\sigma$-algebras is that the classical approach fails in the general situation. Suppose that we have a continuous model for a financial market. In this model, the price of one stock is modelled by a process $(S_t)_{t\in[0,1]}$, where $S_t \in \mathbb{R}_+$ is the price of the stock at time $t$. In other words, for each $t \in [0, 1]$ we have a random variable $S_t$, so there are infinitely many of them. As in the discrete case, we cannot predict the future. The classical approach to model the available information at time $t$ would be to say

*At time t, we know the values of* $S_s$ *for* $s \leq t$.

The question at this point is: Which random variables do we know when we know the values of $S_s$ for $s \leq t$? If one follows the approach at the beginning of Section 2.3, then one would say that one knows all random variables $X$, which are functions of $S_s$ with $s \leq t$. However, how do we define a function with uncountably many arguments, that is what is $f\big((S_s)_{s\in[0,t]}\big)$? This approach therefore leads to large technical difficulties. On the other hand, to use the $\sigma$-algebra approach, define

$$\mathcal{F}_t := \sigma\left(\{S_s \in [a, b]\} \text{ with } s \leq t,\ a \leq b\right). \tag{2.3.15}$$

We can interpret $\mathcal{F}_t$ as the family of all events we know up to time $t$. The $\sigma$-algebra $\mathcal{F}_t$ is well-defined since $\sigma(\mathcal{H})$ is well-defined for each family $\mathcal{H}$ of subsets, see Definition 2.3.10. Furthermore, we know the value of a random variable $X$ at time $t$ if and only if $X$ is $\mathcal{F}_t$-measurable. The argument to see why these statements are equivalent is the same as in the discrete case and can be seen at the beginning of Section 2.3.5. Therefore, the approach with $\sigma$-algebras still works in this situation while the classical approach fails.

Next, we will consider some simple consequences of Dynkin's lemma.

**Corollary 2.3.24.** Let $(\Omega, \mathbb{P})$ be a discrete probability space and $Y_1, \ldots, Y_T$ be random variables on this space. Suppose that $X$ and $Z$ are $\sigma(Y_1, \ldots, Y_T)$-measurable. Then

(a) $aX + bZ$ is $\sigma(Y_1, \ldots, Y_T)$-measurable for all $a, b \in \mathbb{R}$.
(b) $X \cdot Z$ is $\sigma(Y_1, \ldots, Y_T)$-measurable.
(c) $\frac{X}{Z}$ is $\sigma(Y_1, \ldots, Y_T)$-measurable if $\mathbb{P}[Z = 0] = 0$.

*Proof.* We prove only the first point as the proof of the others is similar. Since $X$ and $Z$ are $\sigma(Y_1,\dots,Y_T)$-measurable, there exist functions $f$, $g$ such that $X = f(Y_1,\dots,Y_T)$ and $Z = g(Y_1,\dots,Y_T)$. Therefore

$$aX + bZ = af(Y_1,\dots,Y_T) + bg(Y_1,\dots,Y_T) = h(Y_1,\dots,Y_T)$$

with $h = af + bg$. Dynkin's lemma implies that $aX + bZ$ is $\sigma(Y_1,\dots,Y_T)$-measurable. □

**Corollary 2.3.25.** Let $(\Omega,\mathbb{P})$ be a discrete probability space and $X$, $Y_1$, …, $Y_T$ be random variables on this space. If $X$ is $\sigma(Y_1,\dots,Y_T)$-measurable then $X$ is a constant on each event $\{Y_1 = y_1,\dots,Y_T = y_T\}$ with $\mathbb{P}[Y_1 = y_1,\dots,Y_T = y_T] > 0$.

*Proof.* Since $X$ is $\sigma(Y_1,\dots,Y_T)$-measurable, it follows from Dynkin's lemma that there exists a function $f$ such that $X = f(Y_1,\dots,Y_T)$. The random variables $Y_1,\dots,Y_T$ are clearly constant on the event $\{Y_1 = y_1,\dots,Y_T = y_T\}$ and therefore $X = f(Y_1,\dots,Y_T)$ also has to be constant on the event $\{Y_1 = y_1,\dots,Y_T = y_T\}$. This completes the proof. □

At this point, we would like to look at the special case $\mathcal{F}_0 = \{\emptyset,\Omega\}$.

**Lemma 2.3.26.** Let $(\Omega,\mathbb{P})$ be a discrete probability space, $X$ be a random variable on this space and $\mathcal{F}_0 = \{\emptyset,\Omega\}$. Then the following are equivalent:

- $X$ is $\mathcal{F}_0$-measurable.
- $X$ is constant.

*Proof.* If $X$ is constant (i.e. $X = c$ for some $c \in \mathbb{R}$), then

$$\{X = x\} = \begin{cases} \Omega, & \text{if } x = c, \\ \emptyset, & \text{if } x \neq c. \end{cases}$$

Therefore $X$ is $\mathcal{F}_0$-measurable.

Suppose now that $X$ is $\mathcal{F}_0$-measurable. Choose an arbitrary, but fixed, $\omega \in \Omega$ and define $x := X(\omega)$. Then $\{X = x\} \neq \emptyset$ since $\omega \in \{X = x\}$. Since $X$ is $\mathcal{F}_0$-measurable, we must have $\{X = x\} = \Omega$ and so $X$ is constant. □

Another special case we would like to mention is the following.

**Lemma 2.3.27.** Let $(\Omega,\mathbb{P})$ be a (discrete) probability space and let $X$ and $Y$ be independent random variables on this space. Then

$$X \text{ is } \sigma(Y)\text{-measurable} \iff X \text{ is a constant.}$$

Let us explain heuristically why this lemma has to be true. Intuitively, $\sigma(Y)$ consists of all events which we know when we know the value of $Y$. Now $X$ and $Y$ are independent, which means knowing the value of $Y$ gives us no information about $X$. In other words, even when we know the value of $Y$, we don't know whether the event $\{X = x\}$ has occurred or not, except when $\{X = x\} = \emptyset$ or $\{X = x\} = \Omega$.

*Proof.*
'$\Longleftarrow$' If $X$ is a constant then, using Lemma 2.3.26, $X$ is trivially $\sigma(Y)$-measurable.
'$\Longrightarrow$' We argue by contradiction. Assume that $X$ is not a constant and is $\sigma(Y)$-measurable. Dynkin's lemma tells us that there exists a function $f$ with $X = f(Y)$. Since $X$ is not a constant, we can find an $x$ such that $0 < \mathbb{P}[X = x] < 1$. Choose a $y \in \mathbb{R}$ such that $\mathbb{P}[Y = y] > 0$. Then

$$0 < \mathbb{P}[X = x] \cdot \mathbb{P}[Y = y] = \mathbb{P}[X = x, Y = y] = \mathbb{P}[f(Y) = x, Y = y] = \mathbb{P}[f(y) = x, Y = y].$$

If $f(y) \neq x$ then $\mathbb{P}[f(y) = x, Y = y] = 0$, which is a contradiction. If $f(y) = x$ then $\mathbb{P}[f(y) = x, Y = y] = \mathbb{P}[Y = y]$. This implies that $\mathbb{P}[X = x] \cdot \mathbb{P}[Y = y] = \mathbb{P}[X = x]$ and thus $\mathbb{P}[X = x] = 1$, which is also a contradiction. This completes the proof. □

For completeness we would like to mention conditioning on $\sigma$-algebras. We saw in Section 2.2 that we can interpret $\mathbb{E}[X|Y_1, \ldots, Y_t]$ heuristically as

$\mathbb{E}[X|Y_1, \ldots, Y_t]$ *is the average of the random variable* $X$
*when we know the random variables* $Y_1, \ldots, Y_t$.

On the other hand, we have seen that a $\sigma$-algebra can also be interpreted as expressing the information which we know. It is thus natural to ask: If a random variable $X$ and $\sigma$-algebra $\mathcal{F}_t$ are given, what can we say about $X$ if we know the events in $\mathcal{F}_t$? For instance, consider the case where $X$ is the gain of an investment strategy at time $T$ and $\mathcal{F}_t$ contains the events observable up to time $t < T$. One approach is to try to reproduce the idea behind the construction of $\mathbb{E}[X|Y_1, \ldots, Y_t]$. Therefore one is looking for a random variable $\mathbb{E}[X|\mathcal{F}_t]$ such that

- we know the value of $\mathbb{E}[X|\mathcal{F}_t]$ at time $t$ and
- $\mathbb{E}[X|\mathcal{F}_t]$ has the same average as $X$ on each $A \in \mathcal{F}_t$, that is $\mathbb{E}\big[\mathbb{E}[X|\mathcal{F}_t]\,|A\big] = \mathbb{E}\big[X|A\big]$.

This then leads to the following definition.

**Definition 2.3.28** (Conditional expectation with respect to a $\sigma$-algebra). Let $(\Omega, \mathbb{P})$ be a discrete probability space, $X$ be a random variable with $\mathbb{E}[|X|] < \infty$ and $\mathcal{F}$ be a $\sigma$-algebra over $\Omega$. Then the *conditional expectation* $\mathbb{E}[X|\mathcal{F}]$ *of* $X$ *given* $\mathcal{F}$ is a random variable which is $\mathcal{F}$-measurable and fulfils

$$\mathbb{E}[X \cdot \mathbb{1}_A(\omega)] = \mathbb{E}\Big[\mathbb{E}[X|\mathcal{F}] \cdot \mathbb{1}_A(\omega)\Big] \quad \text{for all } A \in \mathcal{F}. \tag{2.3.16}$$

At first sight, this definition is not very intuitive as it gives a property that $\mathbb{E}[X|\mathcal{F}]$ must satisfy, but no indication of how to go about computing $\mathbb{E}[X|\mathcal{F}]$. It is not even immediately clear that $\mathbb{E}[X|\mathcal{F}]$ exists at all. One can show that $\mathbb{E}[X|\mathcal{F}]$ always exists, using the Radon–Nikodym theorem, but we will not discuss this here. A disadvantage of this general formulation of $\mathbb{E}[X|\mathcal{F}]$ is that explicit formulas for $\mathbb{E}[X|\mathcal{F}]$ only exist in special cases, but there is no general formula. In situations where there is no general formula, the typical approach is to make a guess for $\mathbb{E}[X|\mathcal{F}]$ and then prove that this guess is indeed $\mathbb{E}[X|\mathcal{F}]$ by checking the conditions in Definition 2.3.16. Fortunately, conditioning on a discrete random variable $Y$ and conditioning on the $\sigma$-algebra $\sigma(Y)$ generated by $Y$ results in the same random variable.

**Lemma 2.3.29.** Let $(\Omega, \mathbb{P})$ be a discrete probability space and $X$ and $Y$ be random variables with $\mathbb{E}[|X|] < \infty$. Then

$$\mathbb{E}[X|Y] = \mathbb{E}[X|\sigma(Y)].$$

*Proof.* By Lemma 2.2.25, $\mathbb{E}[X|Y] = f(Y)$ for some function $f$. Dynkin's lemma implies that $\mathbb{E}[X|Y]$ is $\sigma(Y)$-measurable. The property (2.3.16) follows immediately by inserting the definition of $\mathbb{E}[X|Y]$ (see Definition 2.2.14) into (2.3.16). We leave the details as an exercise for the reader. □

An analogous result to that in Lemma 2.3.29 is true for conditioning on partitions. If $(B_j)_{j=1}^n$ is a partition of $\Omega$ then

$$\mathbb{E}\left[X|(B_j)_{j=1}^n\right] = \mathbb{E}\left[X|\sigma\left((B_j)_{j=1}^n\right)\right]. \tag{2.3.17}$$

In other words, conditioning on a partition or conditioning on the $\sigma$-algebra generated by this partition are the same. A consequence of this is that when conditioning on a discrete random variable, we only need the partition of $\Omega$ generated by the random variable. The actual values that the random variable takes do not matter.

Lemma 2.3.29 is also true for conditioning on more than one random variable. At certain points we will use the characterisation of $\mathbb{E}[X|Y]$ in Definition 2.3.28 as it is sometimes more convenient than Definition 2.2.14.

Finally, we would like to mention at this point that Definition 2.3.28 can be used to define conditional expectations on arbitrary probability spaces while Definition 2.2.14 only works on discrete probability spaces.

## 2.4 Exercises

**Exercise 2.1.** *An investor plans to invest £10 in an asset. Let $S_0$ denote the price of this asset today, $S_1$ the price in one year and $S_2$ the price in two years. According to the estimates of this* *investor, the prices of this asset can be realised on a probability space with* $\Omega = \{\omega_1, \omega_2, \omega_3, \omega_4\}$ *and take the values in Table 2.7.*

**Table 2.7** Prices of the asset $S_t$ in Exercise 2.1

| *Time* | $t = 0$ | $t = 1$ | $t = 2$ |
|---|---|---|---|
| *Asset* $S_t$ | $S_0 = 10$ | $S_1 = \begin{cases} 9.5, & \text{if } \omega_1 \text{ occurs,} \\ 9.5, & \text{if } \omega_2 \text{ occurs,} \\ 11, & \text{if } \omega_3 \text{ occurs,} \\ 11, & \text{if } \omega_4 \text{ occurs,} \end{cases}$ | $S_2 = \begin{cases} 9.85, & \text{if } \omega_1 \text{ occurs,} \\ 10.45, & \text{if } \omega_2 \text{ occurs,} \\ 10.45, & \text{if } \omega_3 \text{ occurs,} \\ 12.3, & \text{if } \omega_4 \text{ occurs.} \end{cases}$ |

*Furthermore, the investor estimates that*

$$\mathbb{P}[\omega_1] = \frac{3}{20}, \ \mathbb{P}[\omega_2] = \frac{9}{20}, \ \mathbb{P}[\omega_3] = \frac{6}{25} \text{ and } \mathbb{P}[\omega_4] = \frac{4}{25}.$$

*(a) Draw a tree diagram illustrating the structure of the problem.*
*(b) Determine* $\mathbb{E}[S_1]$ *and* $\mathbb{E}[S_2]$*.*
*(c) Determine* $\mathbb{E}[S_2|S_1 = 9.5]$ *and* $\mathbb{E}[S_2|S_1 = 11]$*.*
*(d) Determine* $\mathbb{E}[S_2|S_1]$*. Is* $\mathbb{E}[S_2|S_1]$ *a constant in this case?*

**Exercise 2.2.** *We toss a fair coin three times, independently, and denote by* $X$ *the number of heads. Set*

$$X_j = \begin{cases} 1, & \text{if the } j\text{th throw is 'Head',} \\ 0, & \text{if the } j\text{th throw is 'Tail'.} \end{cases}$$

*Determine the conditional expectations*

*(a)* $\mathbb{E}[X|X_1]$
*(b)* $\mathbb{E}[X|X_1 + X_2]$
*(c)* $\mathbb{E}[X_1|X]$.

**Exercise 2.3.** *Let* $(\Omega, \mathbb{P})$ *be a discrete probability space and* $X$*,* $Y$ *be random variables on this space. Suppose that* $\mathbb{E}[X|Y] = \mathbb{E}[Y|X]$*. Does this imply that* $X = Y$*? If yes, prove it. Otherwise, give a counterexample.*

**Exercise 2.4.** *Let* $X$ *and* $Y$ *be (discrete) random variables with* $\mathbb{E}\left[X^2\right] < \infty$ *and* $\mathbb{E}[|Y|] < \infty$*, and let* $f : \mathbb{R} \to \mathbb{R}$ *be a function such that* $\mathbb{E}\left[|f(Y)|^2\right] < \infty$*.*

*(a) Show that*

$$\mathbb{E}\left[(\mathbb{E}[X|Y])^2\right] = \mathbb{E}\left[X\,\mathbb{E}[X|Y]\right].$$

Hint: Use first Lemma 2.2.25, then Lemma 2.2.23 and then the tower property.

(b) *Show that*

$$\mathbb{E}\left[(X - f(Y))^2\right] = \mathbb{E}\left[(X - \mathbb{E}[X|Y])^2\right] + \mathbb{E}\left[(\mathbb{E}[X|Y] - f(Y))^2\right]. \tag{2.4.1}$$

(c) *Prove that*

$$\mathbb{E}\left[(X - \mathbb{E}[X|Y])^2\right] \leq \mathbb{E}\left[(X - f(Y))^2\right]$$

*with equality if and only if* $f(Y) = \mathbb{E}[X|Y]$ *with probability 1.*

**Exercise 2.5.** *Let* $(\Omega, \mathbb{P})$ *be a discrete probability space and let* $X$ *and* $Y$ *be two random variables on this space, taking only integer values. Show that* $\sigma(X, Y) = \sigma(X, X + Y)$.

**Exercise 2.6.** *Let* $(\Omega, \mathbb{P})$ *be a discrete probability space and* $(Y_t)_{t=1}^T$ *be a sequence of i.i.d., non-degenerate random variables on* $\Omega$. *Consider the* $\sigma$*-algebra* $\mathcal{F}_T = \sigma(Y_1, \ldots, Y_T)$. *Determine whether the random variable*

$$\exp\left(\sum_{t=1}^{T} Y_t^3\right)$$

*is* $\mathcal{F}_T$*-measurable. Justify your answer.*

# 3 Binomial or CRR Model

The **binomial** or **Cox–Ross–Rubinstein (CRR)** model is a simple discrete model for the financial market. The advantage of this model is that it is simple enough that it can be implemented and analysed without advanced mathematics like stochastic integration or Itô calculus, yet it is complicated enough to reflect a financial market. In particular, it allows us to easily understand concepts like *risk-neutral measures* and *hedging strategies*.

## 3.1 Model Specification

For the binomial model, we consider only finitely many times $t \in \{0, 1, 2, \ldots, T\}$ with $T < \infty$. Furthermore, in this model we work with a primary market consisting of only two assets. These two assets are a *bond* and a *stock*. We write $B_t$ for the price of the *bond* at time $t$ and $S_t$ for the price of the *stock* at time $t$. Recall, a *bond* is a security with a given interest rate. We make the following assumptions on $B_t$ and $S_t$.

- The *bond* is assumed to have a constant interest rate $r > -1$. This means that

$$B_t = B_0\,(1 + r)^t\,, \tag{3.1.1}$$

  where $B_0$ is the price of the bond at time $t = 0$. We normally assume that $B_0 = 1$.
- The price $S_t$ of the *stock* at time $t$ has the form

$$S_t = S_{t-1} \cdot \xi_t = S_0 \prod_{i=1}^{t} \xi_i, \tag{3.1.2}$$

  where $S_0 > 0$ is the price of the stock at $t = 0$ and $(\xi_t)_{t=1}^{T}$ is an i.i.d. sequence of random variables. Furthermore, we assume that each $\xi_t$ takes only the two values $u$ and $d$ with $u > d > 0$, and that

$$\mathbb{P}\left[\xi_t = u\right] = p \;\text{ and }\; \mathbb{P}\left[\xi_t = d\right] = 1 - p, \tag{3.1.3}$$

  for some $0 < p < 1$. Here, $u$ stands for 'up' while $d$ stands for 'down'.

We see at this point that the price of the *stock* jumps independently at each $t$ from $S_t$ to $u \cdot S_t$ or $d \cdot S_t$. The probability $p$ gives the entire dynamics of the model and determines $\mathbb{P}$ entirely. The probability $p$ is referred to as the *historical probability* or *real-world* probability.

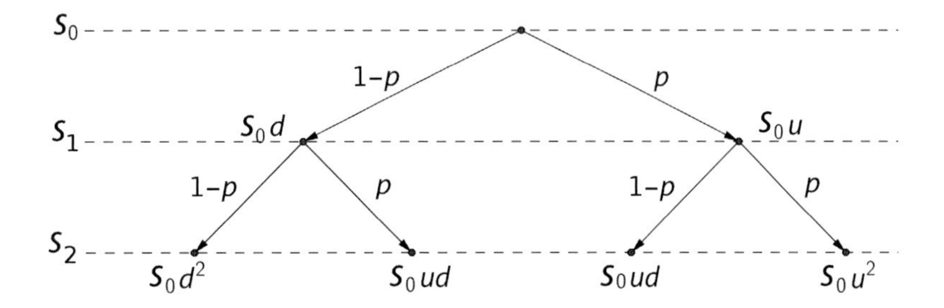


**Figure 3.1** Tree diagram for prices of the stock $S_t$.

For concreteness, let us choose a sample space for this model. Set

$$\Omega = \{(\omega_1, \dots, \omega_T); \omega_t \in \{u, d\}\} = \{u, d\}^T \tag{3.1.4}$$

with the probability measure

$$\mathbb{P}[\omega] = p^n(1-p)^{T-n}, \tag{3.1.5}$$

where $n$ is the number of $u$'s occurring in $\omega$. For instance, if $T = 4$ then

$$\mathbb{P}[(d, d, u, d)] = p(1-p)^3 \text{ and } \mathbb{P}[(u, u, u, u)] = p^4. \tag{3.1.6}$$

Finally, we define $\xi_t$. For $1 \leq t \leq T$, set

$$\xi_t(\omega_1, \dots, \omega_T) := \omega_t. \tag{3.1.7}$$

It is now straightforward to see that the $(\xi_t)_{t=1}^T$ are i.i.d. with this definition, and the distribution of each $\xi_t$ is as in (3.1.3).

The possible prices of the stock $S_t$ can be illustrated with a tree diagram, see Figure 3.1.

We see at this point that $S_2$ can have only three distinct values. We see also that there are two paths leading to the value $S_0ud$. A similar observation can be made for the remaining $S_t$. More precisely, $S_t$ can only have $t+1$ distinct values $S_0 \cdot u^n\, d^{t-n}$ with $n \in \{0, 1, 2, \dots, t\}$. Furthermore, there are $\binom{t}{n}$ paths leading to $S_0 \cdot u^n\, d^{t-n}$ and so

$$\mathbb{P}\left[S_t = S_0\, u^n\, d^{t-n}\right] = \binom{t}{n} p^n(1-p)^{t-n}. \tag{3.1.8}$$

One could at this point put all equal values of $S_t$ together and work only with the distribution of $S_t$. However, in general this is not a good idea as one normally needs knowledge of the full path leading to some value. In particular, two different paths can lead to the same value of $S_t$, but at the same time to a different investment strategy. For instance, consider

$$\omega' = (d, d, d, d, d, u, u, u, u, u) \text{ and } \omega'' = (u, u, u, u, u, d, d, d, d, d).$$

Now $S_{10}(\omega') = S_{10}(\omega'') = S_0u^5d^5$. However, if $\omega'$ occurs then it is very likely that one would sell the stock at time $t = 5$. If, on the other hand, $\omega''$ occurs then it is very likely that one would hold the stock.

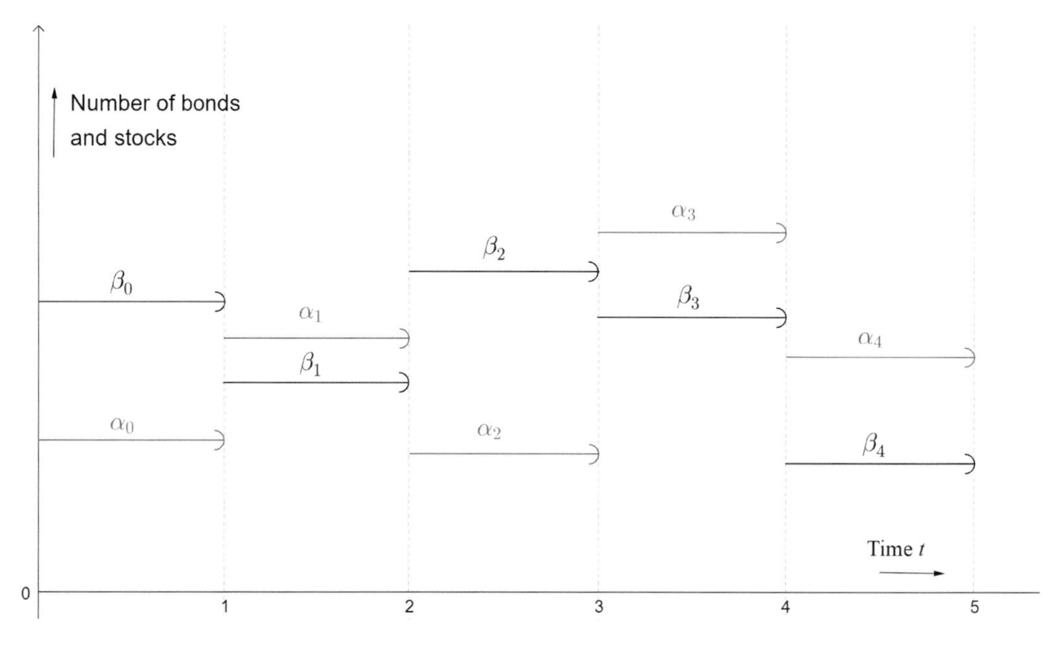


**Figure 3.2** Illustration of bonds and stocks held over the time period $[0, 5]$.

*Remark* 3.1.1. It is standard in finance to give the interest rate for a contract per annum even when the duration of the contract has a different length. The reason behind this is that interest rates are more easily comparable. However, we normally do not specify the duration of a time step in our model. Therefore the interest rate $r$ will be the interest rate for one time step (unless explicitly stated).

## 3.2 Trading Strategies

We would like to buy/sell *bonds* and *stocks*. In the binomial model, trading is only allowed at times $t \in \{0, \dots, T\}$. Therefore, the numbers of bonds and stocks are constant in each of the intervals $[t, t+1)$. Denote by

$$\alpha_t := \text{the number of shares of stock we hold in the time period } [t, t+1)$$
$$\beta_t := \text{the number of bonds we hold during that same period.}$$

We can illustrate the number of stocks and bonds we hold graphically. An example is given in Figure 3.2. An important point is that we determine the portfolio holding $(\beta_t, \alpha_t)$ at time $t$. More precisely, we start at time $t = 0$ with $(\beta_0, \alpha_0)$ and hold this till time $t = 1$. At time $t = 1$, we rebalance our portfolio to $(\beta_1, \alpha_1)$ and hold this till time $t = 2$. At time $t = 2$, we rebalance our portfolio to $(\beta_2, \alpha_2)$ and hold this till time $t = 3$ and so on. Since our model runs over the time interval $[0, T]$, we stop trading at time $T$. For simplicity, we assume that $\alpha_T = \alpha_{T-1}$ and $\beta_T = \beta_{T-1}$. This assumption is mainly to simplify the notation, see for instance (3.2.4).

At first glance, one might expect that $\alpha_t, \beta_t \in \mathbb{N}_0$. However, negative values for $\alpha_t$ and $\beta_t$ are also allowed. If $\beta_t < 0$ then this is called **borrowing** (think about taking out a loan from a bank and using this loan to purchase something) and if $\alpha_t < 0$ then this is called **short-selling**. One speaks of **short-selling** if a trader borrows a security like a stock and then sells it on. At some point the short-seller will have to buy back the borrowed security from the market, in order to return it to the lender. Between selling and then buying back the security, the *short-seller* is said to have a **short position**. For simplicity, we will allow $\alpha_t \in \mathbb{R}$, $\beta_t \in \mathbb{R}$. As mentioned in Chapter 1, we always assume that markets are *liquid*. This means that we can buy/sell/borrow as many assets as we would like and transactions happen instantly. We also assume there are no transaction fees and that our interaction with the market has no influence on the prices.

*Remark* 3.2.1. Sometimes it can cause some confusion that, for a trader starting from having no bonds, borrowing a bond leads to $\beta_t = -1$, as one might think that if someone borrows a bond then they hold that bond in their hands and therefore should have $\beta_t = 1$. At the very instant of borrowing, both of these states are true, that is the trader holds a bond in their hands, but also has a debt of one bond that will need paying back. If the trader did nothing with the borrowed bond and continued to hold it, then this would be equivalent to having $\beta_t = 0$ as the bond could be instantly returned to the lender at any point. However, typically a trader will borrow a bond in order to buy something, such as some stocks (similar to taking out a mortgage in order to buy a house). As soon as the trader sells on the bond, they are left with a debt in bonds, indicating that at some point they will need to buy back a bond to return to the lender. This state is indicated by setting $\beta_t = -1$.

A natural question at this point is: Why do we only keep track of one number for the stocks and one for the bonds? A more realistic portfolio might look like

- Stocks: 3 stocks in the safe and $-5$ stocks with Peter (i.e. we owe Peter 5 stocks).
- Bonds: 2 bonds in the safe and 2 bonds with Peter (i.e. we lent Peter 2 bonds).

Keeping track like this would be necessary if we could earn fees by lending somebody stocks or bonds. However, the market assumptions that we made exclude such things. It therefore does not make a difference whether we keep our assets all in one place or lend them to several people. As a consequence, it is sufficient to keep track only of the total number of assets we own. The scenario described above is therefore equivalent to saying $(\beta_t, \alpha_t) = (4, -2)$.

We cannot allow arbitrary trading strategies and need to make some restrictions. We have to exclude all trading strategies that look into the future. Furthermore, we also exclude any kind of random trading strategies like coin tossing. One can of course discuss whether it is worth allowing such strategies, but every trader with experience will strongly recommend against using random strategies. Therefore we will allow only trading strategies that

- do not require knowledge about the future and
- are based only on the information in the market.

Let us now look for a mathematically rigorous definition that satisfies both points. Note that we determine $(\beta_t, \alpha_t)$ at time $t$ and thus $(\beta_t, \alpha_t)$ can only take into account the information available at time $t$. Also our strategy should be based only on the information in the market. Therefore $\alpha_t$ and $\beta_t$ have to be determined by the values of $S_0, \ldots, S_t$. Mathematically, this means that there exist functions $a_t$ and $b_t$ such that

$$\alpha_t = a_t(S_0, \ldots, S_t) \text{ and } \beta_t = b_t(S_0, \ldots, S_t). \tag{3.2.1}$$

Dynkin's lemma, Lemma 2.3.22, shows that for this formulation this is equivalent to saying that $\alpha_t$ and $\beta_t$ are $\mathcal{F}_t$-measurable with $\mathcal{F}_t = \sigma(S_0, \ldots, S_t)$. In view of this, it is natural to model the time in this model with the filtration

$$\mathbb{F} = (\mathcal{F}_t)_{t=0}^T \text{ with } \mathcal{F}_t = \sigma(S_0, \ldots, S_t) \tag{3.2.2}$$

and to require that $\alpha_t$ and $\beta_t$ are $\mathcal{F}_t$-measurable for all $t \in \{0, \ldots, T\}$. Such sequences have a special name.

**Definition 3.2.2.** Let $(\Omega, \mathbb{P})$ be a (discrete) probability space and $\mathbb{F} = \{\mathcal{F}_t\}_{0 \leq t \leq T}$ be a filtration, see Definition 2.3.18. A sequence of random variables $(X_t)_{0 \leq t \leq T}$ is called $\mathbb{F}$-*adapted* or $\mathcal{F}_t$-*adapted* if $X_t$ is $\mathcal{F}_t$-measurable for all $0 \leq t \leq T$.

With this definition, we now define

**Definition 3.2.3.** Consider the binomial model and endow it with the filtration $\mathbb{F}$ in (3.2.2). An *(admissible) trading strategy* in the *primary market* is an *adapted* process $\varphi = (\varphi_t)_{0 \leq t \leq T}$ with $\varphi_t = (\beta_t, \alpha_t)$ with respect to the filtration $\mathbb{F}$.

We call all trading strategies not fulfilling Definition 3.2.3 invalid. We will work here only with *admissible trading strategies* and so for simplicity will just call them *trading strategies*.

A natural question at this point is:

*Why do $\alpha_t$ and $\beta_t$ in* (3.2.1) *not depend on the bonds?*

At first glance, it looks much more logical to replace (3.2.1) by

$$\alpha_t = a_t(S_0, \ldots, S_t, B_0, \ldots, B_t) \text{ and } \beta_t = b_t(S_0, \ldots, S_t, B_0, \ldots, B_t). \tag{3.2.3}$$

However, the prices of the bonds $B_t$ are not random and we know them already at time $t = 0$. We can therefore incorporate $B_0, \ldots, B_t$ into the functions $a_t$ and $b_t$. Furthermore, it is straightforward to see that

$$\sigma(S_0, \ldots, S_t, B_0, \ldots, B_t) = \sigma(S_0, \ldots, S_t)$$

since $B_0, \ldots, B_t$ are constants. It therefore makes no difference whether we work with (3.2.1) or with (3.2.3). We prefer to highlight only the dependence on $S_0, \ldots, S_t$ as it simplifies the notation.

Note that $S_0$ is the initial price of the stock and is not random. Therefore, $\mathcal{F}_0 = \sigma(S_0) = \{\Omega, \emptyset\}$. Since $\alpha_0$ and $\beta_0$ are $\mathcal{F}_0$-measurable, it immediately follows from Proposition 2.3.26 that $\alpha_0$ and $\beta_0$ are constants. In other words, the first step of a trading strategy is always deterministic.

**Example 3.2.4.** *Let us consider some examples of trading strategies.*

(*a*) *At time $t = 0$ we buy one stock and hold it until the first time that the stock's price goes down. Then we sell the stock. Is this a* valid trading strategy*? Intuitively, the answer is yes as this strategy does not require us to know the future and only depends on the prices in the market. Let us now check whether this strategy fulfils Definition 3.2.3. For this, we have to determine $\alpha_t$ and $\beta_t$ explicitly. We always have $\beta_t = 0$ and therefore $\beta_t$ is $\mathcal{F}_t$-measurable. Furthermore,*

$$\alpha_t = \begin{cases} 1, & \text{if } S_0 \leq \cdots \leq S_t, \\ 0, & \text{otherwise.} \end{cases}$$

*As $\alpha_t$ depends only on $S_1, \ldots, S_t$, it is $\mathcal{F}_t$-measurable and this is a valid* trading strategy.

(*b*) *At time $t = 0$ we buy one stock and borrow one bond. We hold the stock until the price of the stock is strictly less than $S_0/2$. If this happens, we sell the stock and invest all the money in bonds. Is this a valid* trading strategy*? This is a valid* trading strategy. *At time t, we only use $S_0$, $\ldots$, $S_t$ to determine the values $\alpha_t$ and $\beta_t$. We omit here the explicit expression for $\alpha_t$ and $\beta_t$.*

(*c*) *At time $t = 0$ we buy one stock and borrow one bond. We sell the stock at that time t when $S_t$ is maximal in the time interval $0 \leq t \leq T$. Is this a valid* trading strategy*? The answer is clearly no as we would need to know all $S_0$, $\ldots$, $S_T$ to know at which time t the stock $S_t$ is maximal.*

(*d*) *At time $t = 0$ we buy one stock and borrow one bond. Furthermore, for each $t \geq 1$ we independently toss a fair coin. If we toss 'Head', we buy a further stock and if we toss 'Tail', we sell a stock. Is this a valid* trading strategy*? For this strategy, one does not need to know the future, but it depends on more than just the prices in the market. It is therefore not a valid* trading strategy. *Indeed,*

$$\alpha_1 = \alpha_0 + X_1 = 1 + X_1,$$

*where $X_1$ is a random variable independent of $S_0$ and $S_1$, taking the values 1 and $-1$ with probability $1/2$. It follows from Lemma 2.3.27 that $X_1$ is not $\mathcal{F}_1$-measurable.*

A natural question is: How much is my portfolio worth at time $t$? We denote this value by $V_t(\varphi)$. For $t = 0$ we clearly have that

$$V_0(\varphi) = \beta_0 B_0 + \alpha_0 S_0.$$

For $t > 0$, we have to be a little more careful. The reason is that we are allowed to buy and sell bonds and stocks at time $t \in \{0, 1, \ldots, T\}$. Buying and selling bonds and stocks, in other words adjusting the portfolio, is typically called rebalancing. Before we rebalance the portfolio at time $t$, we own $\alpha_{t-1}$ stocks and $\beta_{t-1}$ bonds. In real life, we would look at the stock price, think about what we should do and then contact the stockbroker. After our order has been executed, our portfolio consists of $\alpha_t$ stocks and $\beta_t$ bonds. In real life, this rebalancing of the portfolio needs some time. This can take a few hours or just a fraction of a second if a computer is trading. In the considered model, on the other hand, we assume that the rebalancing of the portfolio needs no time at all, that is it happens instantly. Therefore, at time $t$ our portfolio has two values: the value before and the value after rebalancing. Denote by $V_{t-}(\varphi)$ the value before and by $V_{t+}(\varphi)$ the value after rebalancing. Now

$$V_{t-}(\varphi) = \beta_{t-1}B_t + \alpha_{t-1}S_t \quad \text{and} \quad V_{t+}(\varphi) = \beta_t B_t + \alpha_t S_t.$$

If one rebalances a portfolio then one can remove or add further money into the portfolio. For instance, one can sell all bonds and stocks and use this money to buy a new car. Therefore, there is a priori no need to have $V_{t-}(\varphi) = V_{t+}(\varphi)$. However, here we will only allow *self-financing strategies*. This means that we are neither allowed to remove nor to add money to our portfolio. In other words, when we rebalance our portfolio, the value before and after rebalancing must be equal. We therefore must have $V_{t-}(\varphi) = V_{t+}(\varphi)$. This leads to the following definition.

**Definition 3.2.5** (Self-financing condition). A *self-financing strategy* is a *trading strategy* $\varphi$ such that

$$\beta_{t-1}B_t + \alpha_{t-1}S_t = \beta_t B_t + \alpha_t S_t \text{ for all } t \in \{1, 2, \dots, T\}.$$

We work only with self-financing trading strategies, so we can write $V_t(\varphi)$ instead of $V_{t-}(\varphi)$ or $V_{t+}(\varphi)$. Therefore, the value of the portfolio of a self-financing trading strategy is

$$V_t(\varphi) = \beta_t B_t + \alpha_t S_t \text{ for all } t \in \{0, \dots, T\}. \tag{3.2.4}$$

Note that for $t = T$ we use in (3.2.4) the assumption $\alpha_T = \alpha_{T-1}$ and $\beta_T = \beta_{T-1}$. The trading strategy in Example 3.2.4(a) is a valid trading strategy, but it is not *self-financing*. It will thus be excluded from now on. On the other hand, the trading strategy in Example 3.2.4(b) is a valid trading strategy and is also *self-financing*. It is thus the kind of strategy we are interested in.

Recall, an *arbitrage* is an opportunity to make money out of nothing or to make money without risk. The above definitions now enable us to give a formal definition of arbitrage opportunity in the binomial model.

**Definition 3.2.6** (Arbitrage opportunity). An *arbitrage opportunity* (in the primary market) is a self-financing trading strategy $\varphi$ with the following properties.

- No initial cost: $V_0(\varphi) = 0$.
- Non-negative final value: $V_T(\varphi) \geq 0$.
- The possibility of a positive final gain:

$$\mathbb{E}[V_T(\varphi)] > 0.$$

Note that under the condition $V_T(\varphi) \geq 0$,

$$\mathbb{E}[V_T(\varphi)] > 0 \iff \mathbb{P}[V_T(\varphi) > 0] > 0. \tag{3.2.5}$$

In our arguments, we will thus use whichever condition in (3.2.5) is more suitable.

We will show that the *binomial model* is arbitrage-free if and only if $d < 1 + r < u$, where $r$, $d$ and $u$ are as in the model specifications on page 62. We will prove this later using the tool of *risk-neutral probability*, see Definition 3.3.4. However, we can easily show that

**Proposition 3.2.7.** If the condition $d < 1 + r < u$ is not fulfilled, then the *binomial model* is not arbitrage-free.

*Proof.* Assume for simplicity that $S_0 = B_0 = 1$. This has no influence on the proof, but makes the expressions for $\alpha_t$ and $\beta_t$ simpler.

Start with the case $r + 1 \leq d < u$. This means that the increase in the value of the stock is greater than the increase in the value of the bond (due to the interest rate), regardless of whether the stock price increases or decreases. Consider a trading strategy in which, at time $t = 0$, we borrow one bond and buy one stock and hold them until time $T$. In formulas, we have $\beta_t = -1$ and $\alpha_t = 1$ for $t \geq 0$. This is clearly a *self-financing trading strategy*, since $\alpha_t$ and $\beta_t$ are constants. Furthermore, $V_0(\varphi) = 0$. The assumption $r + 1 \leq d < u$ immediately implies that $(1 + r) \leq \xi_t$ and so

$$B_t \leq S_t \quad \text{for all } t.$$

The value of our portfolio at time $T$ is

$$V_T(\varphi) = \alpha_{T-1} S_T + \beta_{T-1} B_T = S_T - B_T \geq 0.$$

Since $0 < p < 1$, $\mathbb{P}[(1 + r) < \xi_t] > 0$ and hence $\mathbb{P}[S_T - B_T > 0] > 0$. In other words, the strategy $\beta_t = -1$ and $\alpha_t = 1$ is an arbitrage opportunity. The argument for $d < u \leq r + 1$ is similar, but in this case we choose $\beta_t = 1$ and $\alpha_t = -1$. This completes the proof. □

In the proof of Proposition 3.2.7 we used that the interest (increase in value) of one security is always larger than the interest of another security. Many arbitrage opportunities take this form.

## 3.3 Pricing of European Contingent Claims

In this section we study the pricing of *European contingent claims* such as *call* and *put options*. We assume that the *maturity* of all *contingent claims* is $T$ (i.e. they will or can be exercised at time $T$). Recall, the *payoff* of a *contingent claim* denotes the money we earn with the *contingent claim* at *maturity*, see Section 1.2. Denote the payoff of a given *contingent claim* by $\Phi_T$. The payoff $\Phi_T$ should clearly depend only on the prices of the *stock* $S_t$ and therefore $\Phi_T$ has to be $\mathcal{F}_T$-measurable. We have seen some payoffs of *European contingent claims* in Example 1.2.3. We repeat the most important examples.

### Example 3.3.1.

- Stocks*: The payoff of one stock is* $\Phi_T = S_T$.
- Bonds*: The payoff of one bond is* $\Phi_T = B_T$. *Note that* $B_T$ *is a constant and is therefore measurable for all* $\sigma$*-algebras, including* $\mathcal{F}_T$.
- European call option*: When the* strike price *is* $K \in (0, \infty)$ *then the payoff is*

$$\Phi_T = \max\{S_T - K, 0\}.$$

- European put option*: When the* strike price *is* $K \in (0, \infty)$ *then the payoff is*

$$\Phi_T = \max\{K - S_T, 0\}. \tag{3.3.1}$$

In order to simplify the terminology, we identity a *contingent claim* with its payoff $\Phi_T$. Thus, if two *contingent claims* have the same payoff, we will not distinguish between them.

An important question is: *What is the arbitrage-free price P for a given* European contingent claim $\Phi_T$ *(if it exists)*? To answer this question, we have to change our viewpoint a bit. Suppose you are working in a bank and your boss comes to you and says:

> 'Please find a self-financing *trading strategy* $\varphi$ to generate a payoff $\Phi_T$ and determine how much money we have to invest at the start.'

In formulas: Find a self-financing *trading strategy* $\varphi$ such that

$$\Phi_T = V_T(\varphi) \text{ and determine } V_0(\varphi).$$

If such a strategy $\varphi$ exists, we call the *European contingent claim* $\Phi_T$ **replicable** and such a strategy is called a **replicating** or **hedging** strategy. Let us consider some simple examples.

### Example 3.3.2.

(*a*) *Suppose that* $\Phi_T = S_T$. *Then* $\Phi_T$ *is replicable. The corresponding trading strategy is*

$$\alpha_t = 1 \text{ and } \beta_t = 0 \text{ for all } t \in \{0, \ldots, T\}.$$

*In words: At time* $t = 0$ *buy one stock and hold it for the rest of the time. The initial wealth needed for this trading strategy is* $V_0(\varphi) = S_0$.

(*b*) *Suppose that* $\Phi_T = B_T$. *Then* $\Phi_T$ *is replicable. The corresponding trading strategy is*

$$\alpha_t = 0 \text{ and } \beta_t = 1 \text{ for all } t \in \{0, \ldots, T\}.$$

*In words: At time* $t = 0$ *buy one bond and hold it for the rest of the time. The initial wealth needed for this trading strategy is* $V_0(\varphi) = B_0$.

(*c*) *Suppose that* $\Phi_T = 3S_T - B_T$. *Then* $\Phi_T$ *is replicable. The corresponding trading strategy is*

$$\alpha_t = 3 \text{ and } \beta_t = -1 \text{ for all } t \in \{0, \ldots, T\}.$$

*In words: At time $t = 0$ buy three stocks and borrow one bond and hold it for the rest of the time. The initial wealth needed for this trading strategy is $V_0(\varphi) = 3S_0 - B_0$.*

Let us now see how the arbitrage-free price $P$ of a *contingent claim* $\Phi_T$ and a replicating strategy $\varphi$ are related. At first glance, these seem to have nothing to do with each other. However, they are related by a simple trick. To illustrate this, let us consider an example.

**Example 3.3.3.** *Suppose that you can buy or sell a pen for £1 in store A. Suppose further that you can buy or sell the same pen for £2 in store B. A natural thing to do at this point is to buy a pen in store A and then sell it in store B. This gives a risk-free profit of £1. This is clearly an arbitrage opportunity.*

Let us apply this trick to the *European contingent claim* $\Phi_T$. Suppose that $\Phi_T$ is replicable, so there exists a *trading strategy* $\varphi$ with $\Phi_T = V_T(\varphi)$, with initial price $V_0(\varphi)$. We have two ways to produce the payoff $\Phi_T$: The first way is with the *contingent claim* itself and the second way is with the replicating strategy $\varphi$. If the price $P$ of the *contingent claim* $\Phi_T$ is not equal to $V_0(\varphi)$, then we can argue as Example 3.3.3:

Produce $\Phi_T$ the cheap way and sell it the expensive way.

Let us make this more explicit. Suppose that the price $P$ of the *European contingent claim* is larger than the initial value needed for the trading strategy, that is $P > V_0(\varphi)$. In this case, at time $t = 0$ we sell one contingent claim for the price $P$, invest $V_0(\varphi)$ into the replicating strategy and buy bonds with the left-over income $P - V_0(\phi)$. The initial cost of this portfolio is zero. At time $T$ we have to pay the owner of the *contingent claim* a payoff $\Phi_T$. On the other hand, the *trading strategy* $\varphi$ generates a payoff $V_T(\varphi) = \Phi_T$, and the money invested in bonds is now worth $(1+r)^T(P - V_0(\varphi))$. As both payoffs cancel each other, our total profit/loss is

$$\begin{aligned} -\Phi_T + V_T(\varphi) + (1+r)^T\left(P - V_0(\varphi)\right) &= -\Phi_T + \Phi_T + (1+r)^T\left(P - V_0(\varphi)\right) \\ &= (1+r)^T(P - V_0(\varphi)) > 0. \end{aligned}$$

We therefore indeed have an arbitrage opportunity. Similarly, one can show that we also have an arbitrage opportunity if we can buy the *contingent claim* for a price $P < V_0(\varphi)$. Note that in finance one uses the terminology '*holding a short position*' for selling and the terminology '*holding a long position*' for buying.

The two aims of this section are now to show that

- In the binomial model, under the assumption $d < 1 + r < u$, every *European contingent claim* is replicable.
- The unique arbitrage-free price of a *European contingent claim* $\Phi_T$ is $V_0(\varphi)$, where $\varphi$ is the corresponding replicating strategy.

## 3.3.1 One Period ($T = 1$)

### Replicating Strategies for European Contingent Claims

We begin with the case $T = 1$. Suppose that we have a *European contingent claim* $\Phi_1$. Then $\Phi_1$ is $\mathcal{F}_1$-measurable. In fact, our argument works for all $\mathcal{F}_1$-measurable functions, not only for *contingent claims*. Our aim is to find a *trading strategy* $\varphi$ with a payoff $\Phi_1$. As we have only one time period, each trading strategy $\varphi$ has the form $\varphi = \varphi_0 = (\beta_0, \alpha_0)$. Since $\alpha_0$, $\beta_0$ are $\mathcal{F}_0$-measurable, they are just constants; see the lines below Definition 3.2.3. By the self-financing property, Definition 3.2.5, we are looking for $\alpha_0$, $\beta_0 \in \mathbb{R}$ with

$$\Phi_1 \stackrel{!}{=} V_1(\varphi) = \beta_0 B_1 + \alpha_0 S_1. \tag{3.3.2}$$

In words, we have to determine $\alpha_0$, $\beta_0$ at $t = 0$ so that (3.3.2) holds at $t = 1$ regardless of how the stock price develops. Therefore, we must consider (3.3.2) in all possible cases. Since $\Phi_1$ is an $\mathcal{F}_1$-measurable random variable and $\mathcal{F}_1 = \sigma(S_1)$, it follows from Corollary 2.3.25 that $\Phi_1$ is constant on the events $\{S_1 = uS_0\}$ and $\{S_1 = dS_0\}$. Therefore, we have to consider only two cases. We denote by $V_1^u$ the value of $\Phi_1$ on the event $\{S_1 = uS_0\}$, and by $V_1^d$ the value on $\{S_1 = dS_0\}$. Note that $V_1^d$ and $V_1^u$ are constants. An illustration of our situation is given in Figure 3.3. We can thus write $\Phi_1$ as

$$\Phi_1 = V_1^u \cdot \mathbb{1}_{\{S_1 = uS_0\}} + V_1^d \cdot \mathbb{1}_{\{S_1 = dS_0\}}. \tag{3.3.3}$$

Similarly, we can also distinguish the values of $V_1(\varphi)$ in the two cases $\{S_1 = uS_0\}$ and $\{S_1 = dS_0\}$. If $S_1 = uS_0$ then we have $V_1(\varphi) = \beta_0 B_1 + \alpha_0 u S_0$ and if $S_1 = dS_0$ then we have $V_1(\varphi) = \beta_0 B_1 + \alpha_0 d S_0$. Using indicator functions, we can write

$$V_1(\varphi) = \left(\beta_0 B_1 + \alpha_0 u S_0\right) \cdot \mathbb{1}_{\{S_1 = uS_0\}} + \left(\beta_0 B_1 + \alpha_0 d S_0\right) \cdot \mathbb{1}_{\{S_1 = dS_0\}}. \tag{3.3.4}$$

Inserting (3.3.3) and (3.3.4) into (3.3.2) immediately leads to the linear system

$$V_1^u = \beta_0 B_0 (1 + r) + \alpha_0 S_0 u, \tag{3.3.5}$$
$$V_1^d = \beta_0 B_0 (1 + r) + \alpha_0 S_0 d. \tag{3.3.6}$$

Here, (3.3.5) corresponds to the case $S_1 = uS_0$ and (3.3.6) to $S_1 = uS_0$. The equations (3.3.5) and (3.3.6) are linearly independent since $u > d$ and thus $\alpha_0$ and $\beta_0$ are unique. Solving the linear system, one obtains

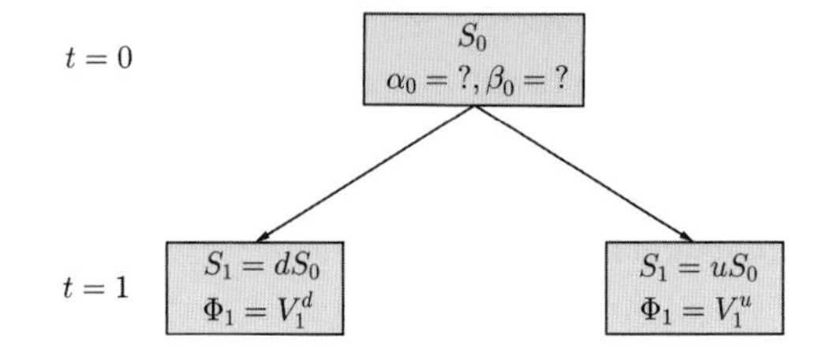


**Figure 3.3** Probability tree for determining a replicating strategy for $\Phi_1$.

$$\alpha_0 = \frac{V_1^u - V_1^d}{S_0\,(u-d)} \quad \text{and} \quad \beta_0 = \frac{uV_1^d - dV_1^u}{B_0(1+r)\,(u-d)}. \tag{3.3.7}$$

We therefore see that there exists a *replicating strategy* $\varphi$ for $\Phi_1$ and that it is unique. Furthermore, we can now also determine the initial wealth needed for the *trading strategy* $\varphi$. We have

$$\begin{aligned}
V_0(\varphi) = \alpha_0 S_0 + \beta_0 B_0 &= \frac{V_1^u - V_1^d}{S_0\,(u-d)} \cdot S_0 + \frac{uV_1^d - dV_1^u}{B_0(1+r)\,(u-d)} \cdot B_0 \\
&= \frac{V_1^u - V_1^d}{u-d} + \frac{uV_1^d - dV_1^u}{(1+r)\,(u-d)} = \frac{1}{1+r}\left(\frac{(1+r)-d}{u-d}V_1^u + \frac{u-(1+r)}{u-d}V_1^d\right) \\
&= \frac{1}{1+r}\left(p^*V_1^u + (1-p^*)V_1^d\right) \quad \text{with } p^* = \frac{(1+r)-d}{u-d}.
\end{aligned} \tag{3.3.8}$$

Now $0 < p^* < 1 \iff d < 1+r < u$ and so, under our assumptions, $p^*$ is a non-degenerate probability. This probability $p^*$ is important and thus has a name.

**Definition 3.3.4.** Consider the binomial model with $d$, $u$ and $r$ as on page 62.

$$\text{The probability } p^* = \frac{(1+r)-d}{u-d}$$

is called the *risk-neutral probability*. We write $\mathbb{P}_{p^*}[\,.\,]$ and $\mathbb{E}_{p^*}[\,.\,]$ for the probability measure and expectation obtained by replacing $p$ with $p^*$.

We now have for instance

$$\mathbb{P}_{p^*}[\xi_1 = u] = \mathbb{P}_{p^*}[S_1 = uS_0] = p^* \quad \text{and } \mathbb{P}_{p^*}[\xi_1 = d] = \mathbb{P}_{p^*}[S_1 = dS_0] = 1 - p^*.$$

Further, we have

$$\begin{aligned}
\mathbb{E}_{p^*}[S_1] &= S_0 u \cdot p^* + S_0 d \cdot (1-p^*) \\
&= S_0\left(u \cdot \frac{(1+r)-d}{u-d} + d \cdot \frac{u-(1+r)}{u-d}\right) = (1+r)\,S_0.
\end{aligned} \tag{3.3.9}$$

We see at this point that both $S_1$ and $B_1$ have the same expected payoff under $p^*$ if $S_0 = B_0$. This observation is the origin of the name *risk-neutral probability*. Let us explain this briefly. If $p > p^*$ then $\mathbb{E}_p[S_1] > B_1$ and it is on average better to invest in 'risky' stocks. On the other hand, if $p < p^*$ then $\mathbb{E}_p[S_1] < B_1$ and it is on average better to invest in bonds. However, if $p = p^*$ then $\mathbb{E}_p[S_1] = B_1$ and it makes on average no difference whether we invest in 'risky' stocks or in 'risk-free' bonds.

We have to mention an important point about $p^*$: **The probability $p^*$ is not in any way related to $p$ and is only used as a mathematical device.** It does not yield any predictions regarding real-world spot dynamics. The dynamics of the model are determined by $p$ and not by $p^*$! For instance, if one has to compute the expected payoff of a call option or the probability that this call option is exercised, then one has to use $p$.

We now go back to (3.3.8). Recall that $\Phi_1$ is constant on the events $\{S_1 = uS_0\}$ and $\{S_1 = dS_0\}$. With the help of $p^*$ and (3.3.3) we can rewrite (3.3.8) as

$$V_0(\varphi) = \frac{1}{1+r}\left(p^* V_1^u + (1-p^*)V_1^d\right) = \frac{1}{1+r}\mathbb{E}_{p^*}[\Phi_1]. \qquad (3.3.10)$$

We see that, using $p^*$, we can compute the initial wealth needed for a *trading strategy* to produce a payoff $\Phi_1$. Let us now summarise our computations.

**Theorem 3.3.5.** *Consider the binomial model with $T = 1$ and $d < 1 + r < u$. Then*

- *All $\mathcal{F}_1$-measurable $\Phi_1$ (and thus all* European contingent claims*) are* replicable*, that is there exists a self-financing* trading strategy $\varphi = (\beta_0, \alpha_0)$ *with payoff* $\Phi_1$.
- *This* trading strategy $\varphi$ *is unique and $\alpha_0$ and $\beta_0$ are given by* (3.3.7).
- *The initial wealth needed for this* replicating strategy $\varphi$ *is*

$$V_0(\varphi) = \mathbb{E}_{p^*}\left[\frac{\Phi_1}{1+r}\right].$$

*Remark* 3.3.6. Theorem 3.3.5 shows that in the binomial model all $\mathcal{F}_1$-measurable $\Phi_1$ are replicable and that the replicating strategy is unique. This is not automatically true for more general models. In Chapter 4 we will see some examples of contingent claims that are not replicable or have more than one replicating strategy.

Let us now consider some examples.

**Example 3.3.7.** *We consider the binomial model with $B_0 = S_0 = £10$, $u = 1.03$, $d = 1.01$, $r = 0.02$ and $p = 0.75$. Then*

$$p^* = \frac{(1+r)-d}{u-d} = \frac{(1+0.02)-1.01}{1.03-1.01} = \frac{0.01}{0.02} = \frac{1}{2}.$$

(*a*) *Suppose we would like to get a payoff of* £102 *regardless of whether the stock goes up or down. This means that $\Phi_1 \equiv £102$. Recall, $V_1^u$ is the value of $\Phi_1$ on the event $\{S_1 = uS_0\}$ and $V_1^d$ is the value of $\Phi_1$ on $\{S_1 = dS_0\}$. Therefore $V_1^u = V_1^d = £102$. Inserting this into* (3.3.7) *gives*

$$\alpha_0 = \frac{V_1^u - V_1^d}{S_0\,(u-d)} = 0 \ \text{and} \ \beta_0 = \frac{uV_1^d - dV_1^u}{B_0(1+r)\,(u-d)} = \frac{(1.03-1.01)102}{10(1+0.02)\,(0.02)} = 10.$$

*The initial wealth needed to obtain this payoff is*

$$V_0(\varphi) = \mathbb{E}_{p^*}\left[\frac{\Phi_1}{1+r}\right] = \frac{1}{1+r}\mathbb{E}_{p^*}[\Phi_1] = \frac{1}{1.02}\cdot 102 = 100.$$

*Let us now interpret this. This strategy means that we have to invest* £100 *and buy* 10 *bonds and no stocks. This makes perfect sense as we would like to have a completely risk-free payoff and* *the natural way to achieve this is to invest everything in bonds. Note that payoff $\neq$ profit. The profit of this strategy is £2.*

(*b*) *Suppose we would like to get a payoff of £102 if the stock goes up and £51 if the stock goes down. Then $V_1^u = £102$ and $V_1^d = £51$, so*

$$\alpha_0 = \frac{V_1^u - V_1^d}{S_0(u-d)} = \frac{102-51}{10(1.03-1.01)} = 255 \text{ and}$$
$$\beta_0 = \frac{uV_1^d - dV_1^u}{B_0(1+r)(u-d)} = \frac{1.03 \cdot 51 - 1.01 \cdot 102}{10(1+0.02)(0.02)} = -247.5.$$

*The initial wealth needed to obtain this payoff is*

$$V_0(\varphi) = \frac{1}{1+r}\mathbb{E}_{p^*}[\Phi_1] = \frac{1}{1+r}\left(V_1^u \cdot p^* + V_1^d \cdot (1-p^*)\right)$$
$$= \frac{1}{1.02} \cdot \left(102 \cdot \frac{1}{2} + 51 \cdot \frac{1}{2}\right) = 75.$$

### The Risk-Neutral Measure and the Primary Market

Apart from determining the initial wealth needed, one can also use $p^*$ to show that there are no arbitrage opportunities in the primary market in the sense of Definition 3.2.6 (for $T = 1$). Suppose there is an arbitrage opportunity in the primary market. This means there is a trading strategy $\varphi$ such that

$$V_0(\varphi) = 0,\ V_1(\varphi) \geq 0 \text{ and } \mathbb{E}_p[V_1(\varphi)] > 0.$$

Explicitly, this means

$$V_0(\varphi) = \beta_0 B_0 + \alpha_0 S_0 = 0,$$
$$V_1(\varphi) = \beta_0 B_0(1+r) + \alpha_0 S_1 \geq 0 \text{ and}$$
$$\mathbb{E}_p[V_1(\varphi)] = \beta_0 B_0(1+r) + \alpha_0 \mathbb{E}_p[S_1] > 0.$$

We now compute the expectation of $V_1(\varphi)$ with respect to $p^*$. Using that $\alpha_0$, $\beta_0$ are constants, together with (3.3.9),

$$\mathbb{E}_{p^*}[V_1(\varphi)] = \beta_0 B_0(1+r) + \alpha_0 \mathbb{E}_{p^*}[S_1] = \beta_0 B_0(1+r) + \alpha_0(1+r)\mathbb{E}_{p^*}[S_0]$$
$$= (1+r)\big(\beta_0 B_0 + \alpha_0 S_0\big) = (1+r)V_0(\varphi) = 0.$$

Therefore $V_1(\varphi) \geq 0$ and $\mathbb{E}_{p^*}[V_1(\varphi)] = 0$ and so

$$\mathbb{P}_{p^*}[V_1(\varphi) > 0] = 0.$$

Note that $0 < p^* < 1$, so $\mathbb{P}_{p^*}[\omega] \neq 0$ is equivalent to $\mathbb{P}_p[\omega] \neq 0$ (see (3.1.4)). It follows that

$$\mathbb{P}_{p^*}[V_1(\varphi) > 0] = 0 \iff \mathbb{P}_p[V_1(\varphi) > 0] = 0. \tag{3.3.11}$$

This implies $\mathbb{E}_p [V_1 (\varphi)] = 0$, which is a contradiction. Therefore our model is arbitrage-free. Note that this proof is remarkably simple for a result as strong as '*there are no arbitrage opportunities*'. To summarise, the main ingredients of this proof are

- $\mathbb{P}_p [A] = 0 \iff \mathbb{P}_{p^*} [A] = 0$ for $A \subset \Omega$ and
- $\mathbb{E}_{p^*} [V_1(\varphi)]$ is easily computable.

Note that in general $\mathbb{E}_{p^*} [X] \neq \mathbb{E}_p [X]$ and in most situations one cannot say anything about $\mathbb{E}_p [X]$ if one knows $\mathbb{E}_{p^*} [X]$.

### The Arbitrage-Free Price of a Contingent Claim

We are interested in an arbitrage-free price of a *European contingent claim*. To establish this, we need an extended market consisting of a stock $S_t$, a bond $B_t$ and a *contingent claim* $\Phi_1$. We assume at this point

- the dynamics of the stock $S_t$ and the bond $B_t$ are given by the binomial model,
- the *contingent claim* has no influence on the prices of the stock $S_t$ and the bond $B_t$,
- the *maturity* of the *contingent claim* is 1 and
- the *contingent claim* $\Phi_1$ is $\mathcal{F}_1$-measurable.

A *trading strategy* $\psi$ in this extended market is a triple $(\beta_0, \alpha_0, \gamma_0)$, where $\alpha_0$ and $\beta_0$ are as before and $\gamma_0$ denotes the number of *contingent claims*. As for $\alpha_0$ and $\beta_0$, we allow $\gamma_0 \in \mathbb{R}$. Also, we determine $\gamma_0$ at time $t = 0$ so $\gamma_0$ is $\mathcal{F}_0$-measurable. The initial value of the portfolio at time $t = 0$ is

$$V_0(\psi) = \beta_0 B_0 + \alpha_0 S_0 + \gamma_0 P,$$

where $P$ is the price of the *contingent claim*. The value of the portfolio at $t = 1$ is

$$V_1(\psi) = \beta_0 B_1 + \alpha_0 S_1 + \gamma_0 \Phi_1,$$

where $\Phi_1$ is the payoff for the contingent claim. Similar to Definition 3.2.6, we call a *trading strategy* $\psi$ an arbitrage opportunity if

$$V_0 (\psi) = 0, \; V_1 (\psi) \geq 0 \text{ and } \mathbb{E}_p [V_1 (\psi)] > 0. \tag{3.3.12}$$

**Theorem 3.3.8.** *Consider the extended market described above and let $P$ be the initial price of the* European contingent claim $\Phi_1$. *This market is arbitrage-free if and only if*

$$P = \mathbb{E}_{p^*} \left[ \frac{\Phi_1}{1 + r} \right].$$

*In other words, the unique arbitrage-free initial price of $\Phi_1$ is $\mathbb{E}_{p^*} \left[ \frac{\Phi_1}{1+r} \right]$.*

*Proof.* By Theorem 3.3.5, the *contingent claim* $\Phi_1$ is replicable so there exists a trading strategy $\varphi = (\beta_0, \alpha_0)$ with $\Phi_1 = V_1(\varphi)$. The initial wealth needed for this trading strategy is $V_0(\varphi) = \mathbb{E}_{p^*}\left[\frac{\Phi_1}{1+r}\right]$.

We first show that there is an arbitrage opportunity if $P \neq \mathbb{E}_{p^*}\left[\frac{\Phi_1}{1+r}\right]$. For this part of the proof, we use the argument we have already used on page 70.

Suppose that $P > V_0(\varphi)$. We construct an arbitrage opportunity $\psi$. Note that we have two ways to produce a payoff $\Phi_1$: with the replicating strategy $\varphi$ and the contingent claim itself. However, $P > V_0(\varphi)$ so the contingent claim is more expensive than the replicating strategy $\varphi$. We use the usual '*buy cheap and sell expensive*' to construct an arbitrage opportunity. Explicitly, an arbitrage opportunity $\psi$ can be constructed using the following three pieces.

- At time $t = 0$, sell one *contingent claim*, that is $\gamma_1 = -1$. This gives
$$\psi^{(1)} = (0, 0, -1).$$
- Invest the amount $V_0(\varphi)$ into the trading strategy $\varphi$. This gives
$$\psi^{(2)} = (\varphi, 0) = (\beta_0, \alpha_0, 0).$$
- For an arbitrage opportunity, we require $V_0(\psi) = 0$. However, buying $\psi^{(2)}$ costs less money than we earn from selling $\psi^{(1)}$. In order to obtain $V_0(\psi) = 0$, we invest the remaining money, that is $P - V_0(\varphi)$, into bonds. This gives
$$\psi^{(3)} = \left(\frac{P - V_0(\varphi)}{B_0}, 0, 0\right).$$

Combining these three pieces gives

$$\psi = \psi^{(1)} + \psi^{(2)} + \psi^{(3)} = \left(\beta_0 + \frac{P - V_0(\varphi)}{B_0}, \alpha_0, -1\right). \tag{3.3.13}$$

Now

$$V_0(\psi) = 0 \quad \text{and} \quad V_1(\psi) = \beta_0 B_1 + \alpha_0 S_1 + \frac{P - V_0(\varphi)}{B_0} B_1 - \Phi_1.$$

However, we have chosen $\varphi = (\beta_0, \alpha_0)$ such that $\Phi_1 = \beta_0 B_1 + \alpha_0 S_1$, so the value of our portfolio at time $t = 1$ is

$$V_1(\psi) = \big(P - V_0(\varphi)\big)(1 + r) > 0.$$

We therefore see that this strategy gives us an arbitrage opportunity. The argument for $P < V_0(\varphi)$ is similar. In this case we buy the contingent claim and sell the trading strategy $\varphi$. 'Selling' the trading strategy $\varphi$ has to be interpreted as using the trading strategy $-\varphi = (-\beta_0, -\alpha_0)$ to produce a payoff $-\Phi_1$. This then leads to the strategy

$$\psi = \left(-\beta_0 + \frac{V_0(\varphi) - P}{B_0}, -\alpha_0, 1\right),$$

which is an arbitrage opportunity.

We now show that there is no arbitrage opportunity when $P = V_0(\varphi)$. For this assume that $\psi = (\widehat{\beta}_0, \widehat{\alpha}_0, \widehat{\gamma}_0)$ is an arbitrage opportunity. Then

$$0 = V_0(\psi) = \widehat{\beta}_0 B_0 + \widehat{\alpha}_0 S_0 + \widehat{\gamma}_0 P \text{ and}$$
$$0 \leq V_1(\psi) = \widehat{\beta}_0 B_1 + \widehat{\alpha}_0 S_1 + \widehat{\gamma}_0 \Phi_1.$$

We compute the expectation of $V_1(\psi)$ with respect to $\mathbb{P}_{p^*}[\,.\,]$. Use (3.3.9) and that $P = \mathbb{E}_{p^*}\left[\frac{\Phi_1}{1+r}\right]$ to obtain

$$\begin{aligned}\mathbb{E}_{p^*}[V_1(\psi)] &= \widehat{\beta}_0 B_1 + \widehat{\alpha}_0 \mathbb{E}_{p^*}[S_1] + \widehat{\gamma}_0 \mathbb{E}_{p^*}[\Phi_1] \\ &= \widehat{\beta}_0(1+r)B_0 + \widehat{\alpha}_0(1+r)S_0 + \widehat{\gamma}_0(1+r)P \\ &= (1+r)V_0(\psi) = 0.\end{aligned}$$

Therefore, $V_1(\psi) \geq 0$ and $\mathbb{E}_{p^*}[V_1(\psi)] = 0$. By the same argument as on page 75 one immediately gets that $\mathbb{E}_p[V_1(\psi)] = 0$. This is a contradiction so there are no arbitrage opportunities. □

**Example 3.3.9.** *We look again at the example which we studied in the introduction. We considered there a stock $S_t$ for which $S_0 = 100$ and $S_1$ could only be £99 or £101. We assumed that*

$$\mathbb{P}[S_1 = £99] = 0.4 \text{ and } \mathbb{P}[S_1 = £101] = 0.6$$

*and that we could borrow money without interest. We then asked the question: What is the price of the contingent claim $\Phi_1$ with*

$$\Phi_1 = \begin{cases} £2, & \text{if } S_1 = £101 \text{ and} \\ £0, & \text{if } S_1 = £99. \end{cases} \tag{3.3.14}$$

*We then claimed that the arbitrage-free price of $\Phi_1$ was £1. Let us justify this. Now $r = 0$, $u = 1.01$ and $d = 0.99$ so the* risk-neutral probability *is*

$$p^* = \frac{(1+r)-d}{u-d} = \frac{0.01}{0.02} = \frac{1}{2}.$$

*Theorem 3.3.8 implies that the arbitrage-free price of $\Phi_1$ is*

$$P = \mathbb{E}_{p^*}\left[\frac{\Phi_1}{1+r}\right] = \mathbb{E}_{p^*}\left[\frac{\Phi_1}{1+r}\right] = 2 \cdot p^* + 0 \cdot (1-p^*) = 1.$$

*For completeness let us also determine the replicating strategy for $\Phi_1$. We have that the value $V_1^u$ of $\Phi_1$ on the event $\{S_1 = 101\}$ is 2. Similarly, we have $V_1^d = 0$. By (3.3.7)*

$$\alpha_0 = \frac{V_1^u - V_1^d}{S_0\,(u-d)} = \frac{2-0}{100 \cdot 0.02} = 1 \text{ and}$$
$$\beta_0 = \frac{uV_1^d - dV_1^u}{B_0(1+r)\,(u-d)} = \frac{1.01 \cdot 0 - 0.99 \cdot 2}{0.02} = -99,$$

*where we have assumed that $B_0 = 1$. Therefore the replicating strategy is $\varphi = (-99, 1)$. This is exactly the same strategy we used in the introduction to reproduce the payoff of $\Phi_1$.*

**Example 3.3.10.** *Let us compute the arbitrage-free price for the* contingent claims *in Example 3.3.1 (for $T = 1$).*

- Stocks*: The payoff of one stock is $\Phi_1 = S_1$, so we have to compute $\mathbb{E}_{p^*}\left[\frac{S_1}{1+r}\right]$. Using* (3.3.9), *we obtain*

$$\mathbb{E}_{p^*}\left[\frac{S_1}{1+r}\right] = \frac{1}{1+r}\mathbb{E}_{p^*}[S_1] = \frac{1}{1+r}(1+r)\mathbb{E}_{p^*}[S_0] = S_0.$$

- Bonds*: The payoff of one bond is $\Phi_1 = B_1$. Since $B_1$ is a constant, we obtain*

$$\mathbb{E}_{p^*}\left[\frac{B_1}{1+r}\right] = \frac{B_1}{1+r} = B_0.$$

- European call option*: Recall, the payoff of a* European call option *is $\Phi_1 = \max\{S_1 - K, 0\}$, where $K \in (0, \infty)$ is the* strike price. *We get*

$$\begin{aligned}\mathbb{E}_{p^*}\left[\frac{\max\{S_1 - K, 0\}}{1+r}\right] &= \frac{\max\{uS_0 - K, 0\}}{1+r}\cdot p^* + \frac{\max\{dS_0 - K, 0\}}{1+r}\cdot(1-p^*) \\ &= \begin{cases} 0, & \text{if } uS_0 \le K, \\ \frac{(uS_0 - K)p^*}{1+r}, & \text{if } dS_0 \le K \le uS_0, \\ S_0 - \frac{K}{1+r}, & \text{if } K \le dS_0. \end{cases}\end{aligned}$$

*Note that the cases $uS_0 \le K$ and $K \le dS_0$ are trivial since in these cases $\Phi_1 = 0$ and $\Phi_1 = S_1 - K$, respectively.*

- European put option*: Recall, the payoff of a* European put option *is $\Phi_T = \max\{K - S_1, 0\}$, where $K \in (0, \infty)$ is the* strike price. *We get*

$$\mathbb{E}_{p^*}\left[\frac{\max\{K - S_1, 0\}}{1+r}\right] = \begin{cases} \frac{K}{1+r} - S_0, & \text{if } uS_0 \le K, \\ \frac{(K - dS_0)(1-p^*)}{1+r}, & \text{if } dS_0 \le K \le uS_0, \\ 0, & \text{if } K \le dS_0. \end{cases}$$

A question that is often asked is: Can the arbitrage-free price of a contingent claim be negative? The answer to this question is: Yes! If we hold a call option then this call option gives us an advantage over the writer of this option. However, a contingent claim does not necessarily have to give us an advantage over the writer of this contingent claim. It is possible that a contingent claim gives us a disadvantage compared to the writer of this contingent claim. This is for instance the case if the payoff of the contingent claim is always negative. If this is the case then we clearly don't pay any money for this contingent claim and the writer of this contingent claim has to pay us for taking this disadvantage. Therefore, the arbitrage-free price of such a contingent claim has to be negative.

### 3.3.2 Multi-period Case ($T$ Arbitrary)

#### Replicating Strategies for *European Contingent Claims*

We now consider the case $T$ arbitrary and show that all $\mathcal{F}_T$-measurable random variables $\Phi_T$ are replicable. This includes of course all *European contingent claims* $\Phi_T$. We will use conditional expectations, see Section 2.2.2. Recall that we consider the filtration $\mathbb{F} = \{\mathcal{F}_t\}_{t=0}^T$ with $\mathcal{F}_t = \sigma(S_0, \ldots, S_t)$ and that Lemma 2.3.29 implies

$$\mathbb{E}\left[\Phi_T | (S_s)_{s=0}^t\right] = \mathbb{E}\left[\Phi_T | \mathcal{F}_t\right].$$

To keep the notation consistent with later chapters, we will mainly use the notation $\mathbb{E}\left[\Phi_T | \mathcal{F}_t\right]$. However, we recommend to keep the above identity in mind.

Suppose that we have an $\mathcal{F}_T$-measurable random variable $\Phi_T$. Our aim is to show that $\Phi_T$ is replicable. We therefore have to find a self-financing *trading strategy* $\varphi = (\varphi_t)_{t=0}^{T-1}$ with $\varphi_t = (\beta_t, \alpha_t)$ and $\Phi_T = V_T(\varphi)$. Our strategy will be to use backward induction combined with the argument in Section 3.3.1. More precisely, we first construct $\alpha_{T-1}$ and $\beta_{T-1}$ and determine $V_{T-1}(\varphi)$. After this, we construct $\alpha_{T-2}$ and $\beta_{T-2}$ to replicate $V_{T-2}(\varphi)$ and determine $V_{T-3}(\varphi)$ and so on.

Let us first start with a concrete example before we look at the general case.

**Example 3.3.11.** *Consider the case $T = 2$ with $S_0 = B_0 = 10$, $u = 1.1$, $d = 0.9$, $r = 0$ and $p = \frac{3}{4}$. We can realise this model on the probability space*

$$\Omega = \{(u,u), (u,d), (d,u), (d,d)\},$$

$$\mathbb{P}[(u,u)] = \frac{9}{16}, \ \mathbb{P}[(u,d)] = \mathbb{P}[(d,u)] = \frac{3}{16} \ \textit{and} \ \mathbb{P}[(d,d)] = \frac{1}{16}.$$

*Consider the* contingent claim $\Phi_2$ *with*

$$\Phi_2\big((u,u)\big) = 10, \ \Phi_2\big((u,d)\big) = 13, \ \Phi_2\big((d,d)\big) = 11, \ \Phi_2\big((d,u)\big) = 14.$$

*We can illustrate the stock price and values of the contingent with a probability tree, see Figure 3.4.*

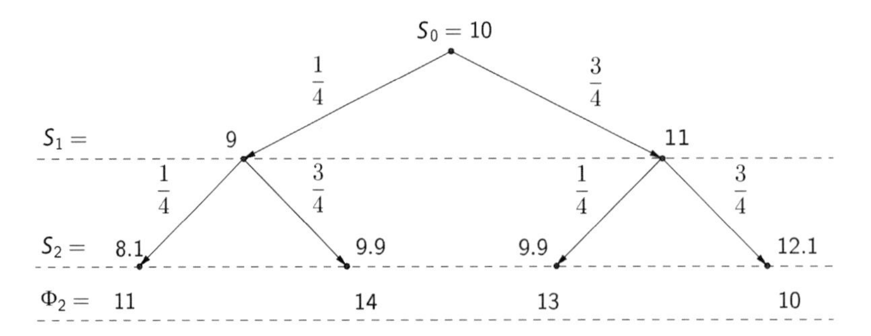


**Figure 3.4** Probability tree for the stock price and $\Phi_2$ in Example 3.3.11.

*We now determine a replicating strategy for* $\Phi_2$. *In the first step, we are looking for* $\alpha_1$ *and* $\beta_1$ *satisfying*

$$\Phi_2 \overset{!}{=} V_2(\varphi) = \beta_1 B_2 + \alpha_1 S_2 \tag{3.3.15}$$

*with* $\alpha_1$ *and* $\beta_1$ $\mathcal{F}_1$*-measurable. Therefore* $\alpha_1$ *and* $\beta_1$ *can depend on* $S_1$, *but not on* $S_2$. *It follows that* $\alpha_1$ *and* $\beta_1$ *can distinguish the events* $\{S_1 = 11\}$ *and* $\{S_1 = 9\}$, *but not* $\{S_1 = 11, S_2 = 12.1\}$ *and* $\{S_1 = 11, S_2 = 9.9\}$. *Since* $\alpha_1$ *and* $\beta_1$ *are* $\mathcal{F}_1$*-measurable, Corollary 2.3.25 implies that they are constant on* $\{S_1 = 11\}$ *and* $\{S_1 = 9\}$. *We can write* $\alpha_1$ *and* $\beta_1$ *as*

$$\alpha_1 = \alpha_1^u \cdot \mathbb{1}_{\{S_1=11\}} + \alpha_1^d \cdot \mathbb{1}_{\{S_1=9\}}, \qquad \beta_1 = \beta_1^u \cdot \mathbb{1}_{\{S_1=11\}} + \beta_1^d \cdot \mathbb{1}_{\{S_1=9\}}$$

*with* $\alpha_1^u, \alpha_1^d, \beta_1^u, \beta_1^d \in \mathbb{R}$. *Consider* (3.3.15) *separately on the events* $\{S_1 = 11\}$ *and* $\{S_1 = 9\}$. *We start with* $\{S_1 = 11\}$. *Heuristically speaking, we have to determine, in the case* $S_1 = 11$, *how many bonds* $\beta_1^u$ *and stocks* $\alpha_1^u$ *we have to hold at time* $t = 1$ *so that the value of our portfolio at time* $t = 2$ *is* 10 *if the stock goes up and* 13 *if the stock goes down. Formally, on* $\{S_1 = 11\}$

$$\Phi_2 \overset{!}{=} \beta_1^u B_2 + \alpha_1^u S_2. \tag{3.3.16}$$

*The equation* (3.3.16) *has to be fulfilled for all* $\omega \in \{S_1 = 11\}$. *Since* $\{S_1 = 11\} = \{(u, u), (u, d)\}$,

$$10 = \beta_1^u\, 10 + \alpha_1^u\, 12.1,$$
$$13 = \beta_1^u\, 10 + \alpha_1^u\, 9.9.$$

*This gives* $\alpha_1^u = -\frac{15}{11}$ *and* $\beta_1^u = \frac{53}{20}$. *In a similar way, we can determine* $\alpha_1^d$ *and* $\beta_1^d$. *This gives* $\alpha_1^d = \frac{5}{3}$ *and* $\beta_1^d = -\frac{1}{4}$, *determining* $\alpha_1$ *and* $\beta_1$. *We next determine* $V_1(\varphi)$. *Since our aim is to construct a self-financing trading strategy,*

$$\begin{aligned} V_1(\varphi) &= \beta_0 B_1 + \alpha_0 S_1 = \beta_1 B_1 + \alpha_1 S_1 \\ &= \left(\beta_1^u B_1 + \alpha_1^u S_1\right) \cdot \mathbb{1}_{\{S_1=11\}} + \left(\beta_1^d B_1 + \alpha_1^d S_1\right) \cdot \mathbb{1}_{\{S_1=9\}} \\ &= \frac{23}{2} \cdot \mathbb{1}_{\{S_1=11\}} + \frac{25}{2} \cdot \mathbb{1}_{\{S_1=9\}}. \end{aligned}$$

*We now repeat the same procedure to determine* $\alpha_0$ *and* $\beta_0$. *This computation is similar to the computation above and the computation in the case* $T = 1$, *so we just state the results*

$$\alpha_0 = -\frac{1}{2}, \ \beta_0 = \frac{17}{10} \ \text{ and } \ V_0(\varphi) = 12.$$

Let us now consider the general case. For the first step, we have to find $\mathcal{F}_{T-1}$-measurable $\alpha_{T-1}$, $\beta_{T-1}$ such that

$$\Phi_T \overset{!}{=} V_T(\varphi) = \beta_{T-1} B_T + \alpha_{T-1} S_T. \tag{3.3.17}$$

In principle we can argue as in Example 3.3.15 and study (3.3.17) on each of the events $\{S_1 = s_1, \ldots, S_{T-1} = s_{T-1}\}$ separately. However, it is easier to use a slightly different argument. First note that the price of the stock can go either up or down between time $T-1$ and time $T$, and therefore either $S_T = uS_{T-1}$ or $S_T = dS_{T-1}$. We will consider (3.3.17) separately on the events $\{S_T = uS_{T-1}\} = \{\xi_T = u\}$ and $\{S_T = dS_{T-1}\} = \{\xi_T = d\}$ (see page 69 for the definition of $\xi_T$). Also note that $\Phi_T$ is $\mathcal{F}_T$-measurable. By Dynkin's lemma, Lemma 2.3.22, there exists a function $f$ such that $\Phi_T = f(S_0, \ldots, S_T)$. We therefore have

$$\Phi_T = f(S_0, \ldots, S_{T-1}, uS_{T-1}) \text{ on the event } \{\xi_T = u\} \text{ and}$$
$$\Phi_T = f(S_0, \ldots, S_{T-1}, dS_{T-1}) \text{ on the event } \{\xi_T = d\}.$$

Introduce

$$V_T^u := f(S_0, \ldots, S_{T-1}, uS_{T-1}) \text{ and } V_T^d := f(S_0, \ldots, S_{T-1}, dS_{T-1}). \tag{3.3.18}$$

Note that at this point $V_T^u$ and $V_T^d$ are both $\mathcal{F}_{T-1}$-measurable as they only depend on $S_0, \ldots, S_{T-1}$. We can now write $\Phi_T$ as

$$\Phi_T = V_T^u \cdot \mathbb{1}_{\{\xi_T = u\}} + V_T^d \cdot \mathbb{1}_{\{\xi_T = d\}}. \tag{3.3.19}$$

Similarly, we can write

$$\beta_{T-1}B_T + \alpha_{T-1}S_T = (\beta_{T-1}B_T + \alpha_{T-1}S_{T-1}u) \cdot \mathbb{1}_{\{\xi_T = u\}} + (\beta_{T-1}B_T + \alpha_{T-1}S_{T-1}d) \cdot \mathbb{1}_{\{\xi_T = d\}}. \tag{3.3.20}$$

Using (3.3.19), (3.3.20) and the fact that the events $\{\xi_T = u\}$ and $\{\xi_T = d\}$ are disjoint, we see that $(\beta_{T-1}, \alpha_{T-1})$ satisfies (3.3.17) if and only if

$$V_T^u = \beta_{T-1}B_T + \alpha_{T-1}S_{T-1}u \text{ and } V_T^d = \beta_{T-1}B_T + \alpha_{T-1}S_{T-1}d.$$

Solving the system as in the one-period case (see (3.3.7)), gives

$$\alpha_{T-1} = \frac{V_T^u - V_T^d}{(u-d)S_{T-1}} \text{ and } \beta_{T-1} = \frac{1}{(1+r)\,B_{T-1}} \frac{uV_T^d - dV_T^u}{(u-d)}. \tag{3.3.21}$$

We see that $\alpha_{T-1}$ and $\beta_{T-1}$ are uniquely determined. Furthermore, they are $\mathcal{F}_{T-1}$-measurable, as required, since $V_T^u$ and $V_T^d$ are both $\mathcal{F}_{T-1}$-measurable.

Let us now determine $V_{T-1}(\varphi)$. Using the expressions above for $\alpha_{T-1}$ and $\beta_{T-1}$, we obtain

$$\begin{aligned}
V_{T-1}(\varphi) &= \alpha_{T-1}S_{T-1} + \beta_{T-1}B_{T-1} \\
&= \left(\frac{V_T^u - V_T^d}{(u-d)S_{T-1}}\right) S_{T-1} + \left(\frac{1}{(1+r)\,B_{T-1}} \frac{uV_T^d - dV_T^u}{(u-d)}\right) B_{T-1} \\
&= \frac{V_T^u - V_T^d}{(u-d)} + \frac{1}{(1+r)} \frac{uV_T^d - dV_T^u}{(u-d)} = \frac{1}{1+r}\left(\frac{(1+r)-d}{u-d} V_T^u + \frac{u-(1+r)}{u-d} V_T^d\right) \\
&= \frac{1}{1+r}\left(p^* V_T^u + (1-p^*)\, V_T^d\right),
\end{aligned} \tag{3.3.22}$$

where $p^* = \frac{(1+r)-d}{u-d}$ is the *risk-neutral probability*, see Definition 3.3.4. Recall, in the one-period case $T = 1$, $V_0(\varphi) = \mathbb{E}_{p^*}[\Phi_1]/(1+r)$. In view of this and taking a look at the expression for $V_{T-1}(\varphi)$, one can guess that

$$V_{T-1}(\varphi) = \frac{1}{(1+r)} \mathbb{E}_{p^*}[\Phi_T \,|\mathcal{F}_{T-1}]. \tag{3.3.23}$$

This is indeed true. Let us compute $\mathbb{E}_{p^*}[\Phi_T \,|\mathcal{F}_{T-1}]$ and justify (3.3.23). We use (3.3.19) together with Lemma 2.2.21 and Lemma 2.2.23 to get

$$\begin{aligned}\mathbb{E}_{p^*}\left[\Phi_T \,\middle|\mathcal{F}_{T-1}\right] &= \mathbb{E}_{p^*}\left[V_T^u \cdot \mathbb{1}_{\{\xi_T=u\}} + V_T^d \cdot \mathbb{1}_{\{\xi_T=d\}} \,\middle|\mathcal{F}_{T-1}\right] \\ &= V_T^u \cdot \mathbb{E}_{p^*}\left[\mathbb{1}_{\{\xi_T=u\}} \,\middle|(S_s)_{s=0}^{T-1}\right] + V_T^d \cdot \mathbb{E}_{p^*}\left[\mathbb{1}_{\{\xi_T=d\}} \,\middle|\mathcal{F}_{T-1}\right] \\ &= V_T^u \cdot \mathbb{E}_{p^*}\left[\mathbb{1}_{\{\xi_T=u\}}\right] + V_T^d \cdot \mathbb{E}_{p^*}\left[\mathbb{1}_{\{\xi_T=d\}}\right] \\ &= V_T^u \cdot p^* + V_T^d \cdot (1-p^*). \end{aligned} \tag{3.3.24}$$

We therefore see that (3.3.22) and (3.3.24) imply (3.3.23). This completes the proof and the construction for $T \to T-1$.

The construction for $T-1 \to T-2$ (and from $t \to t-1$) is similar, so we only compute $V_{T-2}(\varphi)$. Using (3.3.23), we get

$$\begin{aligned}V_{T-2}(\varphi) &= \frac{1}{1+r}\mathbb{E}_{p^*}\left[V_{T-1}(\varphi)\middle|\mathcal{F}_{T-2}\right] = \frac{1}{(1+r)^2}\mathbb{E}_{p^*}\left[\mathbb{E}_{p^*}\left[\Phi_T|\mathcal{F}_{T-1}\right]\middle|\mathcal{F}_{T-2}\right] \\ &= \frac{1}{(1+r)^2}\mathbb{E}_{p^*}\left[\Phi_T\middle|\mathcal{F}_{T-2}\right], \end{aligned} \tag{3.3.25}$$

where in the last equality we used the tower property, see Theorem 2.2.30. By induction, for $0 \le t \le T$,

$$V_t(\varphi) = \frac{1}{(1+r)^{T-t}}\mathbb{E}_{p^*}\left[\Phi_T|\mathcal{F}_t\right] \tag{3.3.26}$$

and for $0 \le t \le T-1$,

$$\alpha_t = \frac{V_{t+1}^u - V_{t+1}^d}{(u-d)S_t} \quad \text{and} \quad \beta_t = \frac{1}{(1+r)\,B_t}\,\frac{uV_{t+1}^d - dV_{t+1}^u}{(u-d)}, \tag{3.3.27}$$

where $V_t^u$ is the value of $V_t(\varphi)$ on the event $\{\xi_t = u\}$ and $V_t^d$ is the value of $V_t(\varphi)$ on the event $\{\xi_t = d\}$. Note that (3.3.26) is also true for $t = T$ since $\Phi_T$ is $\mathcal{F}_T$-measurable and therefore Lemma 2.2.23 implies

$$\mathbb{E}_{p^*}\left[\Phi_T|\mathcal{F}_T\right] = \Phi_T.$$

Summarising our computations, we get the following result.

**Theorem 3.3.12.** *Consider the binomial model with $T < \infty$ and $d < 1+r < u$. Then*

- *All $\mathcal{F}_T$-measurable random variables $\Phi_T$ (and thus all* European contingent claims*) are* replicable*, that is there exists a self-financing* trading strategy $\varphi = (\varphi_t)_{t=0}^{T-1}$ *with* $\varphi_t = (\beta_t, \alpha_t)$ *with payoff* $\Phi_T$.
- *This* trading strategy $\varphi$ *is unique and $\alpha_t$ and $\beta_t$ are given by* (3.3.27). *The value of the portfolio at time t is*

$$V_t(\varphi) = \frac{1}{(1+r)^{T-t}}\mathbb{E}_{p^*}\left[\Phi_T|\mathcal{F}_t\right]. \tag{3.3.28}$$

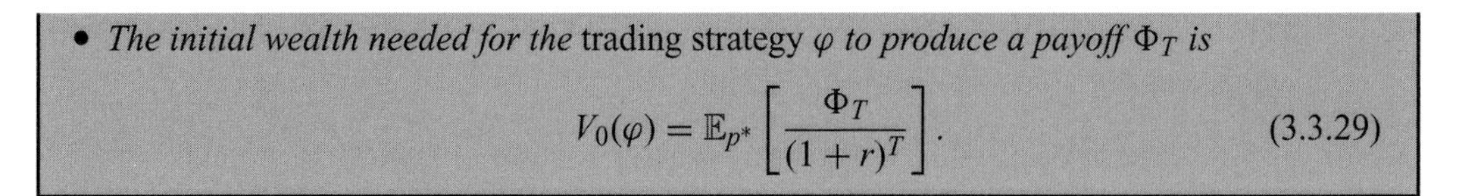

- *The initial wealth needed for the* trading strategy $\varphi$ *to produce a payoff* $\Phi_T$ *is*

$$V_0(\varphi) = \mathbb{E}_{p^*}\left[\frac{\Phi_T}{(1+r)^T}\right]. \tag{3.3.29}$$

Before we look at some examples, we would like to make some remarks about Theorem 3.3.12. The important parts of Theorem 3.3.12 are the existence of the replicating strategy $\varphi$ and the formulas for $V_t(\varphi)$. However, the explicit form of the replicating strategy in (3.3.27) and the computation leading to (3.3.27) are less important. We can use Theorem 3.3.12 to answer the following question:

*What is the arbitrage-free price of the contingent claim $\Phi_T$ at time $t \in \{0, \dots, T\}$?*

If we denote by $P_t$ the price of the contingent claim $\Phi_T$ at time $t$, then we show in Theorem 4.4.1 that the market is arbitrage-free if and only if $P_t = V_t(\varphi)$. It is therefore important that we are able to compute $\mathbb{E}_{p^*}[\Phi_T|\mathcal{F}_t]$ efficiently. In the binomial model, there is usually the direct way, and sometimes a more elegant way. Let us first give an example of the direct way.

**Example 3.3.13.** *Consider the case $T = 3$ and $S_0 = 20$, $B_0 = 1$, $u = 1.2$, $d = 0.9$, $r = 0.1$ and $p = \frac{1}{2}$. The corresponding probability tree for $S_t$ is given in Figure 3.5. Suppose that we have a European contingent claim $\Phi_3$ with maturity $T = 3$ and payoff illustrated in Figure 3.5.*

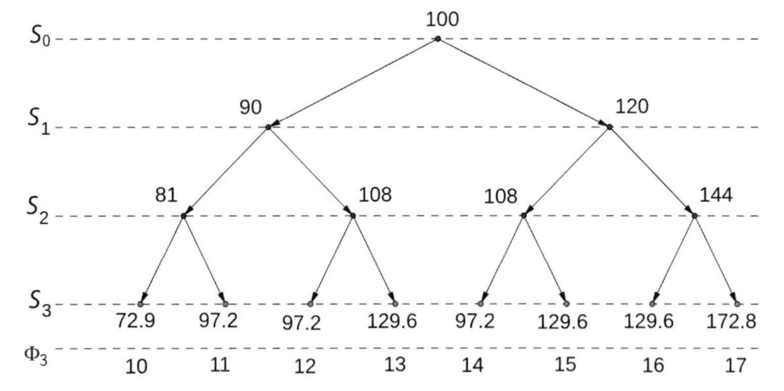


**Figure 3.5** Probability tree for the stock price and $\Phi_3$ in Example 3.3.13.

*We determine the arbitrage-free price $P_2$ of $\Phi_3$ at time $t = 2$. According to Theorem 3.3.12 and the comments afterwards,*

$$P_2 = V_2(\varphi) = \frac{1}{(1+r)}\mathbb{E}_{p^*}\left[\Phi_T \middle| \mathcal{F}_2\right] = \frac{1}{(1+r)}\mathbb{E}_{p^*}\left[\Phi_T \middle| S_0, S_1, S_2\right].$$

*The risk-neutral measure in this case is*

$$p^* = \frac{1.1 - 0.9}{1.2 - 0.9} = \frac{2}{3}.$$

*Since $S_0$ is a constant,* $\mathbb{E}_{p^*}\left[\Phi_T \,|S_0, S_1, S_2\right] = \mathbb{E}_{p^*}\left[\Phi_T \,|S_1, S_2\right]$. *We have to distinguish the four cases:*

(*i*) $S_1 = 90,\ S_2 = 81$; (*ii*) $S_1 = 90, S_2 = 108$; (*iii*) $S_1 = 120, S_2 = 108$; (*iv*) $S_1 = 120, S_2 = 144$.

*Now*

$$\mathbb{E}_{p^*}\left[\Phi_T \,|S_1 = 90, S_2 = 81\right] = 10 \cdot \frac{1}{3} + 11 \cdot \frac{2}{3} = \frac{32}{3},$$
$$\mathbb{E}_{p^*}\left[\Phi_T \,|S_1 = 90, S_2 = 108\right] = 12 \cdot \frac{1}{3} + 13 \cdot \frac{2}{3} = \frac{35}{3}.$$

*Similarly,*

$$\mathbb{E}_{p^*}\left[\Phi_T \,|S_1 = 120, S_2 = 108\right] = \frac{38}{3},\ \mathbb{E}_{p^*}\left[\Phi_T \,|S_1 = 120, S_2 = 144\right] = \frac{41}{3}.$$

*Combining these four cases, we obtain*

$$P_2 = \frac{10}{11}\mathbb{E}_{p^*}\left[\Phi_T \,|\mathcal{F}_2\right] = \begin{cases} \frac{320}{33}, & \textit{if } S_1 = 90,\ S_2 = 81, \\ \frac{350}{33}, & \textit{if } S_1 = 90,\ S_2 = 108, \\ \frac{380}{33}, & \textit{if } S_1 = 120,\ S_2 = 108, \\ \frac{410}{33}, & \textit{if } S_1 = 120,\ S_2 = 144. \end{cases}$$

*We therefore see that the arbitrage-free price $P_t$ for $t > 0$ is not in general a constant. The arbitrage-free price $P_1$ of $\Phi_3$ at time $t = 1$ can be determined similarly. We have*

$$\mathbb{E}_{p^*}\left[\Phi_T \,|S_1 = 90\right] = 10 \cdot \frac{1}{3} \cdot \frac{1}{3} + 11 \cdot \frac{1}{3} \cdot \frac{2}{3} + 12 \cdot \frac{2}{3} \cdot \frac{1}{3} + 13 \cdot \frac{2}{3} \cdot \frac{2}{3} = 12,$$
$$\mathbb{E}_{p^*}\left[\Phi_T \,|S_1 = 120\right] = 14 \cdot \frac{1}{3} \cdot \frac{1}{3} + 15 \cdot \frac{1}{3} \cdot \frac{2}{3} + 16 \cdot \frac{2}{3} \cdot \frac{1}{3} + 17 \cdot \frac{2}{3} \cdot \frac{2}{3} = 16.$$

*Therefore*

$$P_1 = \frac{1}{1.1^2}\mathbb{E}_{p^*}\left[\Phi_T \,|\mathcal{F}_1\right] = \frac{1}{1.1^2}\mathbb{E}_{p^*}\left[\Phi_T \,|S_1\right] = \begin{cases} \frac{12}{1.1^2} & \textit{if } S_1 = 90, \\ \frac{16}{1.1^2} & \textit{if } S_1 = 120. \end{cases}$$

It should not be particularly surprising that the arbitrage-free price $P_t$ is dependent on the stock prices up to time $t$. For instance, suppose three months ago we bought a call option with maturity tomorrow and a strike price £100. If the price of the stock today is £1 then the call option tomorrow will most likely give us no payoff and is almost worthless. If, on the other hand, the price of the stock today is £500 then in most cases the call option tomorrow will give us a payoff of around £400.

We saw in Example 3.3.13 how to compute $\mathbb{E}_{p^*}\left[\Phi_T \,|\mathcal{F}_t\right]$ directly. However, in many cases $\Phi_T$ is given in the form $f(S_T)$. Examples are call and put options. In this situation, we can use a more elegant argument to compute $\mathbb{E}_{p^*}\left[\Phi_T|\mathcal{F}_t\right]$. We use the definition of $S_t$ in (3.1.2) and reformulate $\Phi_T$ in the following way:

$$\Phi_T = f(S_T) = f(S_{T-1}\xi_T) = f(S_{T-1}u) \cdot \mathbb{1}_{\{\xi_T = u\}} + f(S_{T-1}d) \cdot \mathbb{1}_{\{\xi_T = d\}}. \tag{3.3.30}$$

Now $f(S_{T-1}u)$ and $f(S_{T-1}d)$ only depend on $S_{T-1}$, and $\xi_T$ is independent of $S_0,\dots,S_{T-1}$. Therefore, using Lemmas 2.2.21 and 2.2.23,

$$
\begin{aligned}
&\mathbb{E}_{p^*}[\Phi_T|\mathcal{F}_{T-1}] = \mathbb{E}_{p^*}[f(S_T)|S_0,\dots,S_{T-1}]\\
&= \mathbb{E}_{p^*}\left[f(S_{T-1}u)\cdot \mathbb{1}_{\{\xi_T=u\}}\,\middle|S_0,\dots,S_{T-1}\right] + \mathbb{E}_{p^*}\left[f(S_{T-1}d)\cdot \mathbb{1}_{\{\xi_T=d\}}\,\middle|S_0,\dots,S_{T-1}\right]\\
&= f(S_{T-1}u)\cdot \mathbb{E}_{p^*}\left[\mathbb{1}_{\{\xi_T=u\}}\,\middle|S_0,\dots,S_{T-1}\right] + f(S_{T-1}d)\cdot \mathbb{E}_{p^*}\left[\mathbb{1}_{\{\xi_T=d\}}\,\middle|S_0,\dots,S_{T-1}\right]\\
&= f(S_{T-1}u)\cdot \mathbb{E}_{p^*}\left[\mathbb{1}_{\{\xi_T=u\}}\right] + f(S_{T-1}d)\cdot \mathbb{E}_{p^*}\left[\mathbb{1}_{\{\xi_T=d\}}\right]\\
&= f(S_{T-1}u)p^* + f(S_{T-1}d)(1-p^*).
\end{aligned}
\tag{3.3.31}
$$

Compare the last equation with, for example, (3.3.24). Let us look at an example.

**Example 3.3.14.** *Suppose that*

$$T=2,\ B_0=100,\ r=0,\ S_0=100,\ \mathbb{P}[\xi_1=1.2]=0.3 \ \textit{and}\ \mathbb{P}[\xi_1=0.9]=0.7.$$

*Consider a* put option *with strike price $K=100$. The payoff of the* put option *is* $\Phi_2=\max\{100-S_2,0\}$. *Let $\varphi$ be a replicating strategy for the* put option. *We first determine the value of the portfolio $V_t(\varphi)$ at $t=0$, $t=1$ and $t=2$. The* risk-neutral measure *is given by*

$$p^* = \frac{(1+r)-d}{u-d} = \frac{0.1}{0.3} = \frac{1}{3}.$$

*By the definition of a replicating strategy, $V_2(\varphi)=\Phi_2$. Next, consider the case when $t=0$. Then*

$$
\begin{aligned}
V_0(\varphi) &= \mathbb{E}_{p^*}\left[\frac{\Phi_2}{(1+r)^2}\right] = \mathbb{E}_{p^*}[\max\{100-S_2\}]\\
&= \max\{100-100\cdot 1.2^2,0\}\left(\frac{1}{3}\right)^2 + \max\{100-100\cdot 1.2\cdot 0.9,0\}\cdot 2\cdot\frac{1}{3}\cdot\frac{2}{3}\\
&\quad + \max\{100-100\cdot 0.9^2,0\}\left(\frac{2}{3}\right)^2\\
&= \max\{100-100\cdot 0.9^2,0\}\left(\frac{2}{3}\right)^2 = 19\cdot\frac{4}{9}\approx 8.4.
\end{aligned}
$$

*It remains to look at the case $t=1$. Then*

$$
\begin{aligned}
V_1(\varphi) &= \mathbb{E}_{p^*}\left[\frac{\Phi_2}{(1+r)}\middle|S_1\right] = \mathbb{E}_{p^*}[\max\{100-S_2\}|S_1] = \mathbb{E}_{p^*}[\max\{100-S_1\xi_2\}|S_1]\\
&= \max\{100-S_1u\}\cdot p^* + \max\{100-S_1d\}\cdot(1-p^*).
\end{aligned}
$$

The computation in the case $\Phi_T = f(S_0,\dots,S_T)$ of $\mathbb{E}_{p^*}[\Phi_T|\mathcal{F}_t]$ can be done in a similar way to that in (3.3.30) and (3.3.31). The resulting expressions are just a bit more complex.

### Martingales and the Primary Market

We will use Theorem 3.3.12 in a similar way to how we used Theorem 3.3.5 in the one-period case. Explicitly, we will use Theorem 3.3.12 to show that the primary market is arbitrage-free and that the arbitrage-free price of a *European contingent claim* $\Phi_T$ is $V_0(\varphi) = \mathbb{E}_{p^*}\left[\frac{\Phi_T}{(1+r)^T}\right]$. However, before we do this, we would like to mention an important observation in the computations leading to Theorem 3.3.12. This will allow us to simplify our argument and to extend, in Chapter 4, the *risk-neutral measure* to more general models. If we define

$$V_t^*(\varphi) := V_t(\varphi)/(1+r)^t \tag{3.3.32}$$

then from (3.3.23) and (3.3.25),

$$\mathbb{E}_{p^*}\left[V_{t+1}^*(\varphi)|\mathcal{F}_t\right] = V_t^*(\varphi). \tag{3.3.33}$$

Sequences with this property have a special name.

**Definition 3.3.15.** Let $(\Omega, \mathbb{P})$ be a (discrete) probability space and $\mathbb{F} = \{\mathcal{F}_t\}_{0\leq t\leq T}$ with $T \in \mathbb{N} \cup \{\infty\}$ a filtration. A sequence $(X_t)_{t=0}^T$ of random variables is called a *martingale* if

- $X_t$ is $\mathcal{F}_t$-measurable for all $0 \leq t \leq T$, that is $(X_t)_{t=0}^T$ is $\mathbb{F}$-adapted,
- $\mathbb{E}[|X_t|] < \infty$ for all $0 \leq t \leq T$ and
- $\mathbb{E}[X_{t+1}|\mathcal{F}_t] = X_t$ for all $0 \leq t \leq T-1$.

As already mentioned at the beginning of this section, we work here with the filtration $\mathcal{F}_t = \sigma(S_0, \ldots, S_t)$ and therefore Lemma 2.3.29 implies

$$\mathbb{E}\left[X|(S_s)_{s=0}^t\right] = \mathbb{E}[X|\mathcal{F}_t].$$

From (3.3.33) we get that the sequence $V_t^*(\varphi)$ is indeed a martingale under the risk-neutral measure. One of the reasons why martingales are so important is that one can easily compute the conditional expectation.

Other martingales occur naturally in this setting.

**Lemma 3.3.16.** Consider the sequence $(S_t^*)_{t=0}^T$ with

$$S_t^* := \frac{S_t}{(1+r)^t}. \tag{3.3.34}$$

Then this sequence is a *martingale* with respect to $\mathbb{P}_{p^*}[\,.\,]$.

Before we prove this lemma, we would like to point out that the *martingale* property depends heavily on the probability measure $\mathbb{P}$. For instance, $(S_t^*)_{t=0}^T$ is not a martingale for $\mathbb{P}_p[\,.\,]$ if $p \neq p^*$. In order to be a martingale, we would need at least

$$S_0 \overset{!}{=} \mathbb{E}_p\left[S_1^*|S_0\right].$$

Inserting the definitions leads to the equation

$$(1+r)S_0 \overset{!}{=} \mathbb{E}_p[S_1|S_0] = \mathbb{E}_p[S_1] = S_0\,u\cdot p + S_0\,d\cdot(1-p). \tag{3.3.35}$$

It is straightforward to see that (3.3.35) is fulfilled if and only if $p = p^*$, so $(S_t^*)_{t=0}^T$ can only be a martingale for $p = p^*$.

*Proof of Lemma 3.3.16.* It is immediate that $S_t^*$ is $\mathcal{F}_t$-measurable since $r$ is a constant and hence the sequence $(S_t^*)_{t=0}^T$ is $\mathcal{F}_t$-adapted. Also $\mathbb{E}_{p^*}\left[|S_t^*|\right] < \infty$ since the sample space is finite. Therefore the first two properties of Definition 3.3.15 are fulfilled. Recall from (3.1.2) that $S_t = S_{t-1} \cdot \xi_t = S_0 \prod_{i=1}^t \xi_i$ with $(\xi_t)_{t=1}^T$ a sequence of i.i.d. random variables. Using Lemma 2.2.21 and 2.2.23,

$$\begin{aligned}
\mathbb{E}_{p^*}\left[S_{t+1}^* | \mathcal{F}_t\right] &= \frac{1}{(1+r)^{t+1}} \mathbb{E}_{p^*}\left[S_t\, \xi_{t+1} | (S_s)_{s=0}^t\right] = \frac{1}{(1+r)^{t+1}}\, S_t\, \mathbb{E}_{p^*}\left[\xi_{t+1} | (S_s)_{s=0}^t\right] \\
&= \frac{S_t}{(1+r)^{t+1}} \cdot \mathbb{E}_{p^*}\left[\xi_{t+1}\right] = \frac{S_t^*}{(1+r)} \cdot \mathbb{E}_{p^*}\left[\xi_{t+1}\right] = S_t^*\, \frac{up^* + d(1-p^*)}{1+r} \\
&= S_t^*.
\end{aligned}$$

In the last equality we used that $p^* = \frac{(1+r)-d}{u-d}$. □

The following lemma is along the same lines.

**Lemma 3.3.17.** Let $\varphi$ be a self-financing *trading strategy* and $V_t(\varphi)$ be the value of the portfolio at time $t$. Then the sequence $\left(V_t^*(\varphi)\right)_{t=0}^T$ with

$$V_t^*(\varphi) := \frac{V_t(\varphi)}{(1+r)^t} \quad \text{is a martingale under } p^*.$$

*Proof.* Recall, $V_t(\varphi) = \beta_t B_t + \alpha_t S_t$ and thus $V_t^*(\varphi) = \beta_t B_0 + \alpha_t S_t^*$. Since $\alpha_t$ and $\beta_t$ are $\mathcal{F}_t$-measurable, we immediately get that $\left(V_t^*(\varphi)\right)_{t=0}^T$ is $\mathcal{F}_t$-adapted. Also $\mathbb{E}_{p^*}\left[|V_t^*(\varphi)|\right] < \infty$ since the sample space is finite. By Lemma 2.2.23 and 3.3.16,

$$\begin{aligned}
\mathbb{E}_{p^*}\left[V_{t+1}^*(\varphi) | (S_s)_{s=0}^t\right] &= \mathbb{E}_{p^*}\left[\beta_t B_0 + \alpha_t S_{t+1}^* | (S_s)_{s=0}^t\right] \\
&= \beta_t B_0 + \alpha_t \mathbb{E}_{p^*}\left[S_{t+1}^* | (S_s)_{s=0}^t\right] \\
&= \beta_t B_0 + \alpha_t S_t^* \\
&= \frac{\beta_t B_t + \alpha_t S_t}{(1+r)^t} \\
&= V_t^*(\varphi).
\end{aligned}$$

In the first line we used the self-financing condition, see Definition 3.2.5. □

Our next aim is to prove that the primary market is arbitrage-free. For this we need the following property of martingales.

**Lemma 3.3.18.** Let $(\Omega, \mathbb{P})$ be a (discrete) probability space and $(X_t)_{t=0}^T$ be a *martingale* with respect to the filtration $\mathbb{F} = \{\mathcal{F}_t\}_{0 \le t \le T}$ with $T \in \mathbb{N} \cup \{\infty\}$. Then

$$\mathbb{E}[X_t] = \mathbb{E}[X_0] \;\; \forall 0 \leq t \leq T. \tag{3.3.36}$$

Furthermore, for $0 \leq s \leq t \leq T$,

$$\mathbb{E}[X_t|\mathcal{F}_s] = X_s. \tag{3.3.37}$$

*Proof.* We prove this lemma for the filtration $\mathcal{F}_t = \sigma(S_0, \ldots, S_t)$ only. We start with (3.3.36) and use induction over $t$. For $t = 0$, the statement is trivial. For $t \geq 1$, by the tower property, see Theorem 2.2.28, and then the martingale property,

$$\mathbb{E}[X_t] = \mathbb{E}\Big[\mathbb{E}[X_t|S_0, \ldots, S_{t-1}]\Big] = \mathbb{E}[X_{t-1}].$$

This completes the proof of (3.3.36). We prove (3.3.37) also by induction over $t$. For $t = s$, $X_s$ is $\sigma(S_0, \ldots, S_t)$-measurable by definition. Dynkin's lemma implies that $X_s = f(S_0, \ldots, S_s)$ and therefore

$$\mathbb{E}[X_s|S_0, \ldots, S_s] = \mathbb{E}[f(S_0, \ldots, S_s)|S_0, \ldots, S_s] = f(S_0, \ldots, S_s) = X_s.$$

Thus (3.3.36) is true for $t = s$. Suppose now the statement holds for some $t \geq s$. By the tower property and the definition of a martingale,

$$\mathbb{E}[X_{t+1}|S_0, \ldots, S_s] = \mathbb{E}\Big[\mathbb{E}[X_{t+1}|S_0, \ldots, S_t]\,|S_0, \ldots, S_s\Big] = \mathbb{E}[X_t|S_0, \ldots, S_s] = X_s.$$

□

We now show that the primary market is arbitrage-free. The argument is almost the same as in the one-period case, see page 83. To show this, assume there is an arbitrage opportunity in the primary market. This means there is a trading strategy $\varphi = (\varphi_t)_{t=0}^{T-1}$ with $\varphi_t = (\beta_t, \alpha_t)$ such that

$$V_0(\varphi) = 0, \; V_T(\varphi) \geq 0 \;\text{ and }\; \mathbb{E}_p[V_T(\varphi)] > 0. \tag{3.3.38}$$

We now compute the expectation of $V_T(\varphi)$ with respect to $p^*$. Since $\left(V_t^*(\varphi)\right)_{t=0}^{T}$ is a martingale, by Lemma 3.3.17,

$$\mathbb{E}_{p^*}[V_T(\varphi)] = (1+r)^T\,\mathbb{E}_{p^*}\left[V_T^*(\varphi)\right] = (1+r)^T\,\mathbb{E}_{p^*}\left[V_0^*(\varphi)\right] = (1+r)^T\,\mathbb{E}_{p^*}[V_0(\varphi)] = 0.$$

Therefore, $V_T(\varphi) \geq 0$ and $\mathbb{E}_{p^*}[V_T(\varphi)] = 0$. It follows that

$$\mathbb{P}_{p^*}[V_T(\varphi) > 0] = 0.$$

Note that $0 < p^* < 1$, so $\mathbb{P}_{p^*}[A] = 0 \iff \mathbb{P}_p[A] = 0$ for all $A \subset \Omega$. Therefore

$$\mathbb{P}_{p^*}[V_T(\varphi) > 0] = 0 \iff \mathbb{P}_p[V_T(\varphi) > 0] = 0. \tag{3.3.39}$$

This implies that $\mathbb{P}_p[V_T(\varphi) > 0] = 0$ and therefore $\mathbb{E}_p[V_T(\varphi)] = 0$. This is a contradiction and so there is no arbitrage opportunity in the primary market. The argument above used

- $\mathbb{P}_p[A] = 0 \iff \mathbb{P}_{p^*}[A] = 0$ for $A \subset \Omega$ and
- $\left(V_t^*(\varphi)\right)_{t=0}^{T}$ is a martingale (and therefore $\mathbb{E}_{p^*}[V_T(\varphi)]$ is easily computable).

Combining the computation above with Proposition 3.2.7 gives the following result.

**Theorem 3.3.19.** *The binomial model is arbitrage-free if and only if* $d < 1 + r < u$.

### The Arbitrage-Free Price of a Contingent Claim

We next show that the arbitrage-free price of a *European contingent claim* $\Phi_T$ at time $t = 0$ is $\mathbb{E}_{p^*}\left[\Phi_T/(1+r)^T\right]$. As in the one-period case, we need an extended market consisting of a stock $S_t$, a bond $B_t$ and a *contingent claim* $\Phi_T$. We assume here that

- the dynamics of the stock $S_t$ and the bond $B_t$ are given by the binomial model,
- the *contingent claim* has no influence on the stock $S_t$ and the bond $B_t$,
- the *maturity* of the *contingent claim* is $T$ and
- the *contingent claim* $\Phi_T$ is $\mathcal{F}_T$-measurable.

Furthermore, $\alpha_t$ and $\beta_t$ are as before (see Section 3.2) and therefore $\mathcal{F}_t$-measurable. Lastly, we denote by

$$\gamma_t := \text{the number of } \textit{contingent claims} \text{ we hold during the time period } [t, t+1).$$

Similar to $\alpha_t$ and $\beta_t$, we allow $\gamma_t \in \mathbb{R}$ and require that $\gamma_t$ is $\mathcal{F}_t$-measurable. A *trading strategy* $\psi$ in this extended market has the form

$$\psi = (\psi_t)_{t=0}^{T-1} \text{ with } \psi_t = (\varphi_t, \gamma_t) \text{ and } \varphi_t = (\beta_t, \alpha_t).$$

For simplicity, we allow trading of the *contingent claim* only at time $t = 0$ and thus

$$\gamma_0 = \gamma_1 = \cdots = \gamma_T.$$

We will see in Section 4.4 how one has to argue if one allows trading of the contingent claim at all $t \in \{0, 1, \ldots, T-1\}$. Denote by $P$ the price of the *contingent claim* at time $t = 0$. The value of our portfolio at time $t = 0$ and time $t = T$ is

$$\begin{aligned} V_0(\psi) &= \alpha_0 S_0 + \beta_0 B_0 + \gamma_0 P, \\ V_T(\psi) &= \alpha_{T-1} S_T + \beta_{T-1} B_T + \gamma_0 \Phi_T. \end{aligned}$$

We consider here only self-financing *trading strategies* $\psi$. Since we cannot change the number of contingent claims we hold, we call a *trading strategy* $\psi$ self-financing if

$$\beta_{t-1} B_t + \alpha_{t-1} S_t = \beta_t B_t + \alpha_t S_t \text{ for all } t \in \{1, 2, \ldots, T-1\}. \tag{3.3.40}$$

Similar to Definition 3.2.6 and (3.3.12), we call a *trading strategy* $\psi$ an arbitrage opportunity if

$$V_0(\psi) = 0, \; V_T(\psi) \geq 0 \text{ and } \mathbb{E}_p[V_T(\psi)] > 0.$$

We can now show that

**Theorem 3.3.20.** *Let $T < \infty$ be given. Consider the extended market described above and denote by $P$ the initial price of the* European contingent claim $\Phi_T$. *This market is arbitrage-free if and only if*

$$P = \mathbb{E}_{p^*}\left[\frac{\Phi_T}{(1+r)^T}\right].$$

*In other words, the unique arbitrage-free initial price of $\Phi_T$ is $\mathbb{E}_{p^*}\left[\frac{\Phi_T}{(1+r)^T}\right]$.*

*Proof.* The proof is very similar to the proof of Theorem 3.3.8. The *contingent claim* $\Phi_T$ is replicable by Theorem 3.3.12. Let $\varphi = (\varphi_t)_{t=0}^{T-1}$ with $\varphi_t = (\beta_t, \alpha_t)$ be the replicating strategy for the *contingent claim* $\Phi_T$ and denote by $V_0(\varphi) = \mathbb{E}_{p^*}\left[\frac{\Phi_T}{(1+r)^T}\right]$ the initial wealth needed for the strategy $\varphi$.

We first show that there is an arbitrage opportunity if $P \neq \mathbb{E}_{p^*}\left[\frac{\Phi_T}{(1+r)^T}\right]$. Suppose that $P > V_0(\varphi)$. We have two ways to produce the payoff $\Phi_T$: with the *contingent claim* itself and also with the replicating strategy $\varphi$. As before, we can construct an arbitrage opportunity $\psi$ by producing $\Phi_T$ the cheap way and selling it the expensive way. Since $P > V_0(\varphi)$, the arbitrage opportunity $\psi$ consists of the following three pieces.

- At time $t = 0$, sell one *contingent claim*, that is $\gamma_0 = -1$. This gives

$$\psi^{(1)} = \left(\psi_t^{(1)}\right)_{t=0}^{T-1} \quad \text{with} \quad \psi_t^{(1)} = (0, 0, -1).$$

- Invest the amount $V_0(\varphi)$ into the trading strategy $\varphi$. This gives

$$\psi^{(2)} = \left(\psi_t^{(2)}\right)_{t=0}^{T-1} \quad \text{with} \quad \psi_t^{(2)} = (\beta_t, \alpha_t, 0).$$

- For an arbitrage opportunity, we require $V_0(\psi) = 0$. However, buying $\psi^{(2)}$ costs less money than we earned by selling $\psi^{(1)}$. In order to obtain $V_0(\psi) = 0$, we invest the remaining money, that is $P - V_0(\varphi)$, into bonds. This gives

$$\psi^{(3)} = \left(\psi_t^{(3)}\right)_{t=0}^{T-1} \quad \text{with} \quad \psi_t^{(3)} = \left(\frac{P - V_0(\varphi)}{B_0}, 0, 0\right).$$

Combining these three pieces gives $\psi = (\psi_t)_{t=0}^{T-1}$ with

$$\psi_t = \psi_t^{(1)} + \psi_t^{(2)} + \psi_t^{(3)} = \left(\beta_t + \frac{P - V_0(\varphi)}{B_0}, \alpha_t, -1\right). \tag{3.3.41}$$

Note that $\psi$ is self-financing since $\psi_t^{(1)}$, $\psi_t^{(2)}$ and $\psi_t^{(3)}$ are self-financing. Now

$$V_0(\psi) = 0 \quad \text{and} \quad V_T(\psi) = \beta_{T-1}B_T + \alpha_{T-1}S_T + \frac{P - V_0(\varphi)}{B_0}B_T - \Phi_T.$$

However, we have chosen $\varphi$ such that $\Phi_T = \beta_{T-1}B_T + \alpha_{T-1}S_T$ so the value of our portfolio at time $T$ is

$$V_T(\psi) = \big(P - V_0(\varphi)\big)(1+r)^T > 0.$$

We therefore see that this strategy gives us an arbitrage opportunity. The argument for $P < V_0(\varphi)$ is similar. In this case, $\psi = (\psi_t)_{t=0}^{T-1}$ with

$$\psi_t = \left(-\beta_t + \frac{V_0(\varphi) - P}{B_0}, -\alpha_t, 1\right)$$

gives an arbitrage opportunity.

It remains to show that there is no arbitrage opportunity if $P = V_0(\varphi)$. For this assume that $\psi = (\psi_t)_{t=0}^{T-1}$ with $\psi_t = \big(\widehat{\varphi}_t, \widehat{\gamma}_t\big)$ and $\widehat{\varphi}_t = (\widehat{\beta}_t, \widehat{\alpha}_t)$ is an arbitrage opportunity. Now

$$V_0(\psi) = \widehat{\beta}_0 B_0 + \widehat{\alpha}_0 S_0 + \widehat{\gamma}_0 P = 0 \text{ and}$$
$$V_T(\psi) = \widehat{\beta}_{T-1} B_T + \widehat{\alpha}_{T-1} S_T + \widehat{\gamma}_{T-1}\Phi_T.$$

We compute the expectation of $V_T(\psi)$ with respect to $\mathbb{P}_{p^*}[\,.\,]$. It follows immediately from (3.3.40) that the trading strategy $\widehat{\varphi} = (\widehat{\varphi}_t)_{t=0}^{T-1}$ is self-financing. By Lemma 3.3.17, $\big(V_t^*(\widehat{\varphi})\big)_{t=0}^{T-1}$ with $V_t^*(\widehat{\varphi}) = (1+r)^{-t} V_t(\widehat{\varphi})$ is a martingale under $\mathbb{P}_{p^*}[\,.\,]$. Using $\widehat{\gamma}_0 = \cdots = \widehat{\gamma}_{T-1}$, Lemma 3.3.18 and $P = \mathbb{E}_{p^*}\left[\frac{\Phi_T}{(1+r)^T}\right]$,

$$\begin{aligned}\mathbb{E}_{p^*}[V_T(\psi)] &= (1+r)^T \mathbb{E}_{p^*}\left[V_T^*(\widehat{\varphi})\right] + \widehat{\gamma}_{T-1}\mathbb{E}_{p^*}[\Phi_T] = (1+r)^T \mathbb{E}_{p^*}\left[V_0^*(\widehat{\varphi})\right] + \widehat{\gamma}_0 \mathbb{E}_{p^*}[\Phi_T] \\ &= (1+r)^T V_0(\varphi) + \widehat{\gamma}_0 (1+r)^T P = (1+r)^T V_0(\psi) = 0.\end{aligned}$$

Therefore, $V_T(\psi)$ is a non-negative random variable with zero expectation. Arguing as on pages 75 and 90, we immediately get $V_T(\psi) \equiv 0$ so there cannot be any arbitrage opportunities. □

Let us consider two examples.

**Example 3.3.21.** *Let us determine the arbitrage-free price of a stock. Using Theorem 3.3.20 and Lemma 3.3.16, we immediately obtain*

$$P = \mathbb{E}_{p^*}\left[\frac{S_T}{(1+r)^T}\right] = \mathbb{E}_{p^*}\left[S_T^*\right] = S_0.$$

*This is not really surprising as the price of the stock at time $t = 0$ is $S_0$, so Theorem 3.3.20 clearly gives the right price of the stock.*

*Recall, in the introduction we tried to determine the price of a contingent claim with the help of the expectation. The idea behind this was to use a 'fair' price. Let us try this here too. Using the definition of $S_T$ in* (3.1.2) *gives*

$$\mathbb{E}_p[S_T] = \mathbb{E}\left[S_0 \prod_{t=1}^{T} \xi_t\right] = S_0 \prod_{t=1}^{T} \mathbb{E}_p[\xi_t] = S_0\big(u \cdot p + d \cdot (1-p)\big)^T.$$

*We therefore see that in general,* $\mathbb{E}_p[S_T] \neq S_0$. *This is clearly not the right price of the stock (unless* $p = p^*$*). If, for instance, the price of the stock is £100 today, why should I sell it for £90?*

**Example 3.3.22.** *Let us now use Theorem 3.3.12 to determine the arbitrage-free price $P$ of a* European call option *at time $t = 0$. Recall, the payoff of a* European call option *is* $\max\{S_T - K, 0\}$*, where $K$ is the strike price. The distribution of $S_T$ (see* (3.1.8)*) is*

$$\mathbb{P}\left[S_T = S_0\, u^n\, d^{T-n}\right] = \binom{T}{n} p^n (1-p)^{T-n} \quad \text{for } n \in \{0, 1, \ldots, T\}.$$

*Therefore*

$$P = \frac{\mathbb{E}_{p^*}[\max\{S_T - K, 0\}]}{(1+r)^T} = \frac{1}{(1+r)^T} \sum_{n=0}^{T} \binom{T}{n} (p^*)^n (1-p^*)^{T-n} \max\{S_0\, u^n\, d^{T-n} - K, 0\}.$$

*Using the comments after Theorem 3.3.12, we can also determine the arbitrage-free $P_t$ of the call option at time $t$ with $0 \le t \le T$. The definition of $S_T$ implies that we have*

$$\mathbb{P}\left[S_T = S_0\, u^n\, d^{T-t-n} \,\middle|\, S_t = s_t\right] = \binom{T-t}{n} p^n (1-p)^{T-t-n} \quad \text{for } n \in \{0, 1, \ldots, T-t\}$$

*for all* $s_t = S_0\, u^m d^{t-m}$ *with* $0 \le m \le t$. *We get*

$$\begin{aligned} P_t &= \mathbb{E}_{p^*}\left[\frac{\max\{S_T - K, 0\}}{(1+r)^{T-t}} \middle| \mathcal{F}_t\right] = \frac{\mathbb{E}_{p^*}\left[\max\{S_T - K, 0\} \middle| (S_s)_{s=0}^t\right]}{(1+r)^{T-t}} \\ &= \frac{1}{(1+r)^{T-t}} \sum_{n=0}^{T-t} \binom{T-t}{n} (p^*)^n (1-p^*)^{T-t-n} \max\{S_t\, u^n\, d^{T-n} - K, 0\}. \end{aligned}$$

*At this point, we see that the price of a call option is not a nice or short expression. In particular, we get that in general the price of a call option at time $t$ is not equal to* $\max\{S_t - K, 0\}$*, see also Example 1.2.4.*

## 3.4 American Contingent Claims

In this section, we look briefly at *American contingent claims*. Recall that the difference between an *American contingent claim* and a *European contingent claim* is that it can be exercised at any time $\tau \in \{0, 1, 2, \ldots, T\}$. Therefore the payoff of an *American contingent claim* is modelled by a sequence $(\Phi_t)_{0 \le t \le T}$, where $\Phi_t$ is the payoff at time $t$. Since $\Phi_t$ can only depend on the information available up to time $t$, each $\Phi_t$ is $\mathcal{F}_t$-measurable. Therefore the sequence $(\Phi_t)_{0 \le t \le T}$ is $\mathcal{F}_t$-adapted, see Definition 3.2.2.

**Example 3.4.1.**

- Stocks*: The payoff of one stock at time $t$ is* $\Phi_t = S_t$.

- Bonds*: The payoff of one bond at time t is* $\Phi_t = B_t$.
- American call option*: When the* strike price *is* $K \in (0, \infty)$*, the payoff is*

$$\Phi_t = \max\{S_t - K, 0\}. \tag{3.4.1}$$

*The explanation of this payoff is similar to that for* European contingent claims*, see Example 1.2.3.*

- American put option*: When the* strike price *is* $K \in (0, \infty)$*, the payoff is*

$$\Phi_t = \max\{K - S_t, 0\}. \tag{3.4.2}$$

If we have an *American contingent claim*, we need to decide when to exercise the *contingent claim*. This is done mathematically with stopping times. Also, the pricing of *American contingent claims* is more difficult since *American contingent claims* are in general not replicable. This requires the co-called *Snell envelope* and the *optional stopping theorem* and goes beyond the level of this book. However, for completeness, we would like to mention a couple of results.

**Theorem 3.4.2.** *The arbitrage-free price of an* American call option *is the same as for a* European call option *with the same strike price and maturity, see Example 3.3.10.*

Note that Theorem 3.4.2 cannot be generalised to all *European contingent claims* and in particular does not hold for *American put options*. We do not prove Theorem 3.4.2 here, but the next proposition gives an explanation of why *American* and *European call options* have the same price.

**Proposition 3.4.3.** It is never optimal to exercise an *American call option* before maturity.

*Proof.* For simplicity, we assume that $r = 0$. Suppose we are at time $t$ with $t < T$. There are two cases. First, suppose that $S_t \leq K$. In this case, we do not exercise the call option as we would lose money. Therefore, it is not optimal to exercise in this case. Suppose $S_t > K$. We consider two possible strategies and compare the payoffs.

- We exercise the call option. The payoff in this case is $\max\{S_t - K, 0\} = S_t - K$.
- We keep the call option. In addition, we borrow one stock and exchange it for bonds. Then we do nothing until time $T$. At time $T$, we own a call option, bonds with the value $S_t$ and owe somebody a stock. We will now return the borrowed stock. Since we do not own a stock, we have to buy one.
  - If $S_T \geq K$ we exercise the call option. This allows us to buy one stock for price $K$, which we do by selling bonds up to a value of $K$. After this, we still own bonds with the value $S_t - K$. Our payoff in this case is therefore $S_t - K$.

– If $S_T < K$ then we do not exercise the call option. Instead, we buy one stock on the market for price $S_T$ by selling bonds. After this, we still own bonds with the value $S_t - S_T$. Thus our payoff in this case is $S_t - S_T > S_t - K$.

The payoff of the second strategy is larger than the payoff of the first strategy, therefore it is never optimal to exercise the option before maturity. This completes the proof. □

## 3.5 Exercises

**Exercise 3.1.** *Write down the formulas for the self-financing trading strategies $\varphi$ in each of the following cases:*

*(a) We start at time $t = 0$ with one stock and one bond and we hold this portfolio for the rest of the time.*
*(b) We start at time $t = 0$ with three stocks and one bond. We hold this portfolio until the price of the stock is at least £20 for the first time. Then we sell all stocks and invest the money into bonds and keep this portfolio for the rest of the time.*
*(c) We start with seven stocks and no bonds. Whenever the price of the stock increases, we add three stocks to our stock holdings by borrowing bonds. If the price of the stock decreases, we sell three units, unless we are holding a negative number of stocks, in which case we increase our position to $+1$. If the price remains constant, we do not change the portfolio.*

**Exercise 3.2.** *Consider the binomial model. A trading strategy $\varphi = (\varphi_t)_{t=0}^T$ with $\varphi_t = (\beta_t, \alpha_t)$ is called affordable if*

$$\beta_{t-1}B_t + \alpha_{t-1}S_t \leq \beta_t B_t + \alpha_t S_t \text{ for all } t \in \{1, \ldots, T\}. \tag{3.5.1}$$

*(a) Give an economic interpretation of an affordable trading strategy.*
*(b) Show that an affordable trading strategy $\varphi$ can be turned into a self-financing trading strategy $\varphi'$ by buying additional bonds. Give an explicit formula for $\varphi'$.*
*(c) Is buying additional bonds the only way to turn the affordable trading strategy $\varphi$ into a self-financing trading strategy $\varphi'$?*

**Exercise 3.3.** *Consider the one-period binomial model. We assume that the bond $B_t$ has an interest rate $r = 0$ and $B_0 = 1$. Furthermore, we assume $S_0 = 100$ and that $S_1$ can take the two values 90 and 120, each with probability 1/2. Finally, consider a* call option *written on the stock with a strike price $K = 100$. Find a replicating strategy $\varphi = (\beta_0, \alpha_0)$ for the* call option *and compute the initial wealth $V_0(\varphi)$ needed for this strategy.*

**Exercise 3.4.** *A strap is a combined option consisting of two European call options and a European put option. More precisely, a strap consists of a long position (we are the buyer of the option) in two* European call options *with strike price $K > 0$ and a long position in a* European put option *with the same strike price $K$. We assume that both options are written on the same stock and have the same* maturity.

(a) *Determine the payoff of the strap when the price of the stock at* maturity *is $S$, and draw a payoff diagram of the strap.*

(b) *Consider the two-period binomial model. We assume that the bond $B_t$ has an interest rate $r = 0.1$. Furthermore, we assume that*

$$T = 2,\ S_0 = B_0 = 58,\ u = 1.3,\ d = 1,\ \mathbb{P}[S_1 = uS_0] = 0.79 \text{ and } \mathbb{P}[S_1 = dS_0] = 0.21.$$

*Now consider a strap with strike price $K = 75$ written on the stock $S_t$ with* maturity $T = 2$.

(i) *Determine the* risk-neutral probability $p^*$.

(ii) *Determine the arbitrage-free price of the strap.*

(iii) *Let $\varphi$ be a replicating strategy for the strap. Determine the value of the portfolio $V_t(\varphi)$ for all $0 \leq t \leq 2$.*

**Exercise 3.5.** *Consider the two-period binomial model (in which each period is 1 year) consisting of a bond $B_t$, paying an annual risk-free rate of $r = 0.2$ and a risky asset $S_t$, with current spot price $S_0 = 100$. In the next year the price of the risky asset can increase, with increase rate $u = 1.5$, with probability $p = 0.3$, and decrease, with decrease rate $d = 0.6$, with probability $1 - p = 0.7$. Now consider a European call option $C$ written on $S_t$, with strike price* 108 *and maturity of* 2 *years.*

(a) *Explain why the call option is replicable.*

(b) *Determine the arbitrage-free price of the* call option.

(c) *Would you buy this option for a price $P = 55$ at time $t = 0$?*

(d) *Would you buy this option for a price $P = 31$ at time $t = 0$?*

(e) *Suppose now that the dynamics of $S_t$ are given by $p = 0.99$ and $1-p = 0.01$. Does this affect the decision to buy the option in (c) and (d)?*

# 4 Finite Market Model

In this section we study a more general discrete model, of which the binomial model is a special case. This model is called the *finite market model*. As for the binomial model, we are interested in the arbitrage-free price of a *contingent claim*. To study this question, we introduced two important tools in Chapter 3. These were

- The *risk-neutral probability measure*, see Definition 3.3.4.
- Replicating strategies for *contingent claims*, see page 72.

A large part of this section will be used to determine the properties the model must have in order to be able to use these two tools, see Section 4.2 and Section 4.3. For instance, in the case of the binomial model, $0 < p^* < 1$ if and only if $d < r < u$, so the *risk-neutral probability measure* is not always well-defined. Combining all the results in Chapter 3, we see that the binomial model is arbitrage-free if and only if the *risk-neutral probability measure* exists. We show in Section 4.2 that this is also true for the *finite market model*.

## 4.1 Model Specification and Notation

### 4.1.1 Model Specification

As with the binomial model, the finite market model is defined over a finite set of times $t \in \{0, 1, 2, \ldots, T\}$ with $T < \infty$. This time, however, there are $d + 1$ assets: a bond $B_t$ and $d$ stocks $S_t^1, \ldots, S_t^d$ with $d \geq 1$. We write the prices at time $t$ as a vector

$$\vec{S}_t := (B_t, S_t^1, \ldots, S_t^d) \in \mathbb{R}^{d+1}.$$

We make the following assumptions.

- The *bond* is assumed to have a constant interest rate $r > -1$. This means that

$$B_t = B_0 (1 + r)^t, \tag{4.1.1}$$

  where $B_0$ is the price of the bond at time $t = 0$. We normally assume that $B_0 = 1$.
- In the binomial model, see page 69, we made specific assumptions about the distribution of the stock price. This will no longer be the case. We assume only that each $S_t^j$ can take at most finitely many values.

- To model time in this financial model, we use the filtration $\mathbb{F} = (\mathcal{F}_t)_{t=0}^T$ with

$$\mathcal{F}_t = \sigma\left(\vec{S}_0, \vec{S}_1, \ldots, \vec{S}_t\right) = \sigma\left(S_s^j,\ 1 \le j \le d,\ 0 \le s \le t\right). \tag{4.1.2}$$

- We assume that the sample space $\Omega$ (which is the basis for this experiment) is finite. In formulas: $|\Omega| < \infty$. Furthermore, we assume that $\mathbb{P}[\omega] > 0$ for all $\omega \in \Omega$ and that $\mathcal{F}_T = \mathcal{P}\Omega$.

The last assumption is not required for the dynamics of the model, but will be needed in some proofs. For example, in the binomial model, $\Omega = \{u, d\}^T$ and $|\Omega| = 2^T$.

### 4.1.2 Trading Strategies

We would like once again to buy/sell *stocks* and *bonds*. Denote by

$\alpha_t^j :=$ the number of shares of the $j$th stock we hold during the time period $[t, t+1)$,

$\beta_t :=$ the number of bonds we hold during the same time period.

We write this as a vector

$$\varphi_t := (\beta_t, \alpha_t^1, \ldots, \alpha_t^d). \tag{4.1.3}$$

Here we allow $\beta_t \in \mathbb{R}$ and $\alpha_t^j \in \mathbb{R}$ for all $j$ and thus $\varphi_t \in \mathbb{R}^{d+1}$. As for the binomial model, $\beta_t$ and all $\alpha_t^j$ are determined at time $t$ and so can only take into account the information available at time $t$. Therefore $\beta_t$ and all $\alpha_t^j$ have to be $\mathcal{F}_t$-measurable. Since our model runs over the time interval $[0, T]$, we stop trading at time $T$. For simplicity, we assume that $\varphi_T = \varphi_{T-1}$. This assumption is mainly to simplify the notation, see for instance (4.1.4). This leads to the following definition.

**Definition 4.1.1.** A *trading strategy* in the *primary market* is an *adapted* process $\varphi = (\varphi_t)_{t=0}^{T-1}$ (see Definition 3.2.2) with $\varphi_t$ as in (4.1.3) with respect to the filtration $\mathbb{F} = (\mathcal{F}_t)_{t=0}^{T-1}$ with $\mathcal{F}_t$ as in (4.1.2).

If a trading strategy $\varphi$ is given, the portfolio at time $t$ has the following value:

$$V_t(\varphi) = \beta_t B_t + \sum_{j=1}^{d} \alpha_t^j S_t^j = \varphi_t \cdot \vec{S}_t, \quad \text{for all } t \ge 0, \tag{4.1.4}$$

where $\cdot$ stands for the standard scalar product on $\mathbb{R}^{d+1}$. As before, we are interested in *self-financing trading strategies* only. Using an argument similar to that on page 68, we get the following definition.

**Definition 4.1.2** (Self-financing condition). A *self-financing trading strategy* is a *trading strategy* $\varphi$ such that

$$\varphi_{t-1} \cdot \vec{S}_t = \varphi_t \cdot \vec{S}_t \quad \text{for all } t \in \{1, 2, \ldots, T\}.$$

We can now define an arbitrage opportunity in the primary market.

**Definition 4.1.3** (Arbitrage opportunity). An *arbitrage opportunity* (in the primary market) is a self-financing trading strategy $\varphi = (\varphi_t)_{t=0}^{T-1}$ with $\varphi_t$ as in (4.1.3) which satisfies the following properties.

- No initial cost: $V_0(\varphi) = 0$.
- Always non-negative final value: $V_T(\varphi) \geq 0$.
- The possibility of a positive final gain:

$$\mathbb{E}[V_T(\varphi)] > 0.$$

As in the binomial model, under the assumption $V_T(\varphi) \geq 0$, we have the equivalence

$$\mathbb{E}[V_T(\varphi)] > 0 \iff \mathbb{P}[V_T(\varphi) > 0] > 0. \tag{4.1.5}$$

We will use whichever statement in (4.1.5) is the most convenient for the particular situation. Note that, under the model assumption $\mathbb{P}[\omega] > 0$, $\mathbb{P}[V_T(\varphi) > 0] > 0$ is equivalent to $V_T(\varphi)(\omega) > 0$ for at least one $\omega \in \Omega$.

### 4.1.3 Discounted Asset Prices

As for the binomial model (see (3.3.33) and Lemma 3.3.16), define $\vec{S}_t^*$ and $V_t^*(\varphi)$ by

$$\vec{S}_t^* := \frac{\vec{S}_t}{(1+r)^t} \quad \text{and} \quad V_t^*(\varphi) := \frac{V_t(\varphi)}{(1+r)^t} = \varphi_t \cdot \vec{S}_t^*. \tag{4.1.6}$$

These are called *discounted asset prices*. We can write

$$\vec{S}_t^* := (B_0, S_t^{1,*}, \dots, S_t^{d,*}) \text{ with } S_t^{j,*} := \frac{S_t^j}{(1+r)^t}.$$

We give another expression for $V_t^*(\varphi)$, which we need later.

**Lemma 4.1.4.** For all $t$ and all self-financing trading strategies $\varphi$,

$$V_t^*(\varphi) = V_0(\varphi) + \sum_{s=0}^{t-1} \varphi_s \cdot \Delta\vec{S}_s^*, \text{ with } \Delta\vec{S}_s^* := \vec{S}_{s+1}^* - \vec{S}_s^*. \tag{4.1.7}$$

*Remark* 4.1.5. This expression for $V_t^*(\varphi)$ can be viewed as a discrete version of a stochastic integral. The framework of discrete time has the advantage of avoiding all the technicalities associated with the construction of this integral.

*Proof.* We use a telescoping sum followed by the self-financing condition:

$$V_t^*(\varphi) = V_0^*(\varphi) + \sum_{s=0}^{t-1} \left[V_{s+1}^*(\varphi) - V_s^*(\varphi)\right] = V_0^*(\varphi) + \sum_{s=0}^{t-1} \left[\varphi_s \cdot \vec{S}_{s+1}^* - \varphi_s \cdot \vec{S}_s^*\right]$$

$$= V_0^*(\varphi) + \sum_{s=0}^{t-1} \varphi_s \cdot \left[\vec{S}_{s+1}^* - \vec{S}_s^*\right].$$

□

Note that

$$\Delta \vec{S}_t^* = (B_0, S_{t+1}^{1,*}, \ldots, S_{t+1}^{d,*}) - (B_0, S_t^{1,*}, \ldots, S_t^{d,*}) = (0, S_{t+1}^{1,*} - S_t^{1,*}, \ldots, S_{t+1}^{d,*} - S_t^{d,*}).$$

It follows that

$$\varphi_t \cdot \Delta \vec{S}_t^* = \sum_{j=1}^{d} \alpha_t^j (S_{t+1}^{j,*} - S_t^{j,*}),$$

so $\varphi_t \cdot \Delta \vec{S}_t^*$ only involves the number of stocks $\alpha_t = (\alpha_t^1, \ldots, \alpha_t^d)$, but not the number of bonds $\beta_t$. We therefore see that the expression for $V_t^*(\varphi)$ in (4.1.7) only involves the initial value of the portfolio $V_0(\varphi)$ and the number of stocks $\alpha_t$ held. This suggests that the number of bonds $\beta_t$ is not needed for $V_t^*(\varphi)$ and that the self-financing condition does not impose any restrictions on the allowed values of the $\alpha_t^j$. The following lemma shows that this is indeed the case.

**Lemma 4.1.6.** Suppose that the initial value of the portfolio $V_0$ and an arbitrary trading strategy involving only stocks $(\alpha_t)_{t=0}^{T-1}$ with $\alpha_t = (\alpha_t^1, \ldots, \alpha_t^d)$ is given. Then there exists a unique sequence $(\beta_t)_{t=0}^{T-1}$ such that each $\beta_t$ is $\mathcal{F}_t$-measurable and

$$\varphi = (\varphi_t)_{t=0}^{T-1} \text{ with } \varphi_t = (\beta_t, \alpha_t^1, \ldots, \alpha_t^d)$$

is a self-financing trading strategy.

*Proof.* As the initial value of the portfolio is $V_0$, by (4.1.4),

$$V_0 = \beta_0 B_0 + \sum_{j=1}^{d} \alpha_0^j S_0^j \implies \beta_0 = \frac{V_0 - \sum_{j=1}^{d} \alpha_0^j S_0^j}{B_0}.$$

Since each $\alpha_0^j$ is $\mathcal{F}_0$-measurable, $\beta_0$ is also $\mathcal{F}_0$-measurable. The argument for the remaining $\beta_t$ is by induction. Suppose that we have determined $\beta_0, \ldots, \beta_{t-1}$ for some $t \geq 1$. Since the trading strategy is self-financing, by (4.1.4),

$$\beta_t B_t + \sum_{j=1}^{d} \alpha_t^j S_t^j = \beta_{t-1} B_t + \sum_{j=1}^{d} \alpha_{t-1}^j S_t^j \implies \beta_t = \beta_{t-1} + \frac{\sum_{j=1}^{d} (\alpha_{t-1}^j - \alpha_t^j) S_t^j}{B_t}.$$

Clearly, $\beta_t$ is $\mathcal{F}_t$-measurable since $\alpha_t^j$, $\alpha_{t-1}^j$, $S_t^j$ and $\beta_{t-1}$ are all $\mathcal{F}_t$-measurable. □

## 4.2 First Fundamental Theorem of Asset Pricing

In this section we extend the *risk-neutral probability measure* to the finite market model. Recall that in the binomial model the risk-neutral probability is given by

$$\mathbb{P}_{p^*}[\xi_t = u] = p^* = \frac{(1+r) - d}{u - d} \text{ and } \mathbb{P}_{p^*}[\xi_t = d] = 1 - p^* = \frac{u - (1+r)}{u - d}. \quad (4.2.1)$$

A naive approach to extending the *risk-neutral probability measure* to the finite market model could be to make an educated guess based on (4.2.1). Consider, for instance, a one-period model

($T = 1$) with one stock such that $S_1^1$ takes three values $d$, $m$ and $u$ with $d < m < u$. What would be the *risk-neutral probability measure* in this case? At this point it becomes apparent that it is not straightforward to extend (4.2.1) even to this simple case. We try another approach: What properties should the *risk-neutral probability measure* have? Checking the computations in the binomial model, see for instance pages 83 and 98, we see that we need the following two properties:

- $S_t^* = \frac{S_t}{(1+r)^t}$ is a martingale with respect to $\mathbb{P}_{p^*}[\,.\,]$, see Lemma 3.3.16 and
- $\mathbb{P}_{p^*}[A] = 0 \iff \mathbb{P}_p[A] = 0$, where $A \subset \Omega$.

Let us now introduce some notation.

**Definition 4.2.1.** Consider two probability measures $\mathbb{P}$ and $\mathbb{P}_*$ on the same sample space $\Omega$. Then $\mathbb{P}$ and $\mathbb{P}_*$ are called *equivalent* (written $\mathbb{P} \approx \mathbb{P}_*$) if and only if

$$\mathbb{P}[A] = 0 \iff \mathbb{P}_*[A] = 0, \quad \text{where } A \subset \Omega \text{ is arbitrary.}$$

**Definition 4.2.2 (Risk-neutral probability measures).** Suppose that the finite market model is realised on the discrete probability space $(\Omega, \mathbb{P})$. A *risk-neutral probability measure* (or *equivalent martingale measure*) is a probability measure $\mathbb{P}_*$ such that

- $\mathbb{P}_* \approx \mathbb{P}$.
- $\vec{S}_t^*$ is a $\mathbb{P}_*$-martingale. In formulas this means

$$\mathbb{E}_*\left[\vec{S}_{t+1}^* \middle| \mathcal{F}_t\right] = \vec{S}_t^*, \tag{4.2.2}$$

where $\mathbb{E}_*[\,.\,]$ is the expectation with respect to $\mathbb{P}_*[\,.\,]$.

Equation (4.2.2) should be understood component-wise. Explicitly, this means

$$\mathbb{E}_*\left[S_{t+1}^{j,*} | \mathcal{F}_t\right] = S_t^{j,*} \text{ for all } 1 \leq j \leq d. \tag{4.2.3}$$

Note that the first component of the vector $\vec{S}_t^*$ is $B_0$ for all $t$ and is therefore a constant. Trivially, $\mathbb{E}[B_0|\mathcal{F}_t] = B_0$ for all measures and $\sigma$-algebras. This implies that the first component of $\vec{S}_t^*$ is always a martingale, regardless of which measures or $\sigma$-algebras we are considering. We therefore only need to worry about the stocks $S_t^{j,*}$ in (4.2.3). Consider an example.

**Example 4.2.3.** *Let $T = 1$ and $d = 1$. This means that we have one stock and one time period. Furthermore, let $B_0 = 1$, $r = 0$, $\Omega = \{\omega_1, \omega_2, \omega_3\}$ and $\mathbb{P}[\omega_1] = \mathbb{P}[\omega_2] = \mathbb{P}[\omega_3] = 1/3$. Since there is only one stock, write $S_t$ instead of $S_t^1$. Suppose*

$$S_0(\omega_1) = 1 \ \text{ and } \ S_1(\omega_1) = 0.99,$$
$$S_0(\omega_2) = 1 \ \text{ and } \ S_1(\omega_2) = 1.03,$$

$$S_0(\omega_3) = 1 \text{ and } S_1(\omega_3) = 1.05.$$

*Let us now determine a* risk-neutral probability measure *(if this is possible). We need a measure* $\mathbb{P}_*$ *such that*

$$\mathbb{E}_*\left[S_1^*|S_0\right] \stackrel{!}{=} S_0^*.$$

*Note that* $S_0^* = S_0 = 1$. *Write* $p_i = \mathbb{P}_*[\omega_i]$ *for* $1 \le i \le 3$. *Then*

$$\begin{aligned}\mathbb{E}_*\left[S_1^*|S_0\right] = \mathbb{E}_*\left[S_1^*\right] &= S_1^*(\omega_1)\cdot p_1 + S_1^*(\omega_2)\cdot p_2 + S_1^*(\omega_3)\cdot p_3 \\ &= 0.99\cdot p_1 + 1.03\cdot p_2 + 1.05\cdot p_3.\end{aligned}$$

*We therefore have to find* $p_1$, $p_2$ *and* $p_3$ *with*

$$0.99\cdot p_1 + 1.03\cdot p_2 + 1.05\cdot p_3 = 1,\ p_1+p_2+p_3 = 1 \text{ and } 0 < p_i < 1. \tag{4.2.4}$$

*Using the substitution* $p_3 = 1 - p_1 - p_2$, (4.2.4) *is equivalent to*

$$5 = 6\cdot p_1 + 2\cdot p_2,\ p_1+p_2 < 1 \text{ and } 0 < p_i < 1. \tag{4.2.5}$$

*Equation* (4.2.5) *can be illustrated graphically, see Figure 4.1. Here* $p_1 + p_2 < 1$ *and* $0 < p_i < 1$ *together give the triangle and* $5 = 6\cdot p_1 + 2\cdot p_2$ *gives the line.*

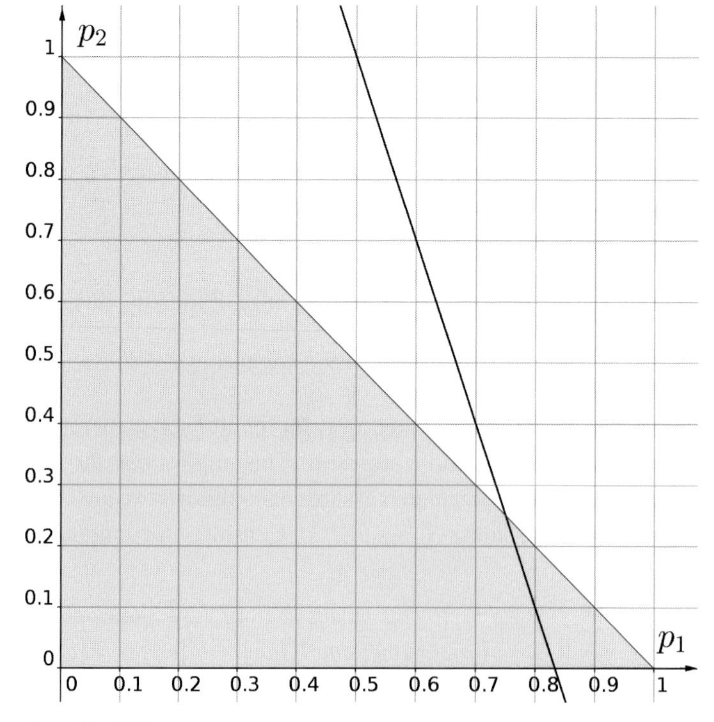


**Figure 4.1** Illustration of (4.2.5).

*We see in Figure 4.1 that there are infinitely many solutions of* (4.2.5) *and thus there are infinitely many* risk-neutral probability measures. *An example of a* risk-neutral probability measure *is, for instance,* $p_1 = 0.8$, $p_2 = 0.1$ *and* $p_3 = 0.1$.

In the situation of Example 4.2.3, we see that we have a *risk-neutral probability measure*. In fact we see that there are infinitely many *risk-neutral probability measures*. Using the argument on pages 75 and 89, one can show that the model in Example 4.2.3 is arbitrage-free. We omit this argument as we will give it soon in full generality. The question at this point is: Do we always have a *risk-neutral probability measure*? Looking again at Figure 4.1, one can see that a *risk-neutral probability measure* exists in Example 4.2.3 because the line $5 = 6 \cdot p_1 + 2 \cdot p_2$ intersects the triangle spanned by $p_1 + p_2 < 1$ and $0 < p_i < 1$. Is this also true for all other choices of $S_1$? The answer is no. For instance, replacing $S_1(\omega_1) = 0.99$ by $S_1(\omega_1) = 1.01$ in Example 4.2.3, it is straightforward to see that the line does not intersect the triangle and therefore no *risk-neutral probability measure* exists. Note that in this case $B_0 = S_0$ and $B_1 \leq S_1$ (since $r = 0$) and therefore intuition suggests that one can easily make money. Indeed, the same argument as in the proof of Proposition 3.2.7 shows that there is an arbitrage opportunity in this model.

The observation above suggests that the existence of a *risk-neutral probability measure* is closely related to the existence of arbitrage opportunities. This is indeed true. Before we can state and prove this, we need some preliminary results.

**Lemma 4.2.4.** Let the finite market model be realised on the discrete probability space $(\Omega, \mathbb{P})$ and let $\mathbb{P}_*$ be an arbitrary probability measure on $\Omega$. Then the following are equivalent:

(a) $\vec{S}_t^*$ is a martingale with respect to $\mathbb{P}_*$.
(b) $V_t^*(\varphi)$ is, under $\mathbb{P}_*$, a martingale for all self-financing trading strategies $\varphi$.
(c) $\mathbb{E}_*\left[V_T^*(\varphi)\right] = V_0^*(\varphi)$ for all self-financing trading strategies $\varphi$.
(d) $\mathbb{E}_*\left[V_T^*(\varphi)\right] = 0$ for all self-financing trading strategies $\varphi$ with $V_0^*(\varphi) = 0$.

*Proof.*
(a)$\Longrightarrow$ (b): The proof is almost identical to that of Lemma 3.3.17 and is therefore omitted.
(b)$\Longrightarrow$ (c): This follows immediately from Lemma 3.3.18 and that $V_0^*(\varphi)$ is constant.
(c)$\Longrightarrow$ (d): This is immediate.
(d)$\Longrightarrow$ (a): In order to show that $\vec{S}_t^*$ is a martingale, we need to verify that

$$\mathbb{E}_*\left[\vec{S}_{t+1}^*\middle|\mathcal{F}_t\right] = \vec{S}_t^* \quad \text{for all } t \in \{0, 1, \ldots, T-1\},$$

which, by (4.2.3), is equivalent to

$$\mathbb{E}_*\left[\vec{S}_{t+1}^{j,*}\middle|\mathcal{F}_t\right] = \vec{S}_t^{j,*} \quad \text{for all } 1 \leq j \leq d \ \text{ and } t \in \{0, 1, \ldots, T-1\}.$$

By the definition of conditional expectation, see Definition 2.3.28, we have to show for each event

$$A := \{\vec{S}_1 = \vec{s}_1, \vec{S}_2 = \vec{s}_2, \ldots, \vec{S}_t = \vec{s}_t\}$$

for which $\mathbb{P}[A] > 0$, where $\vec{s}_s = (s_s^1, s_s^2, \ldots, s_s^d) \in \mathbb{R}^d$, that

$$\mathbb{E}_*\left[S_{t+1}^{j,*}\mathbb{1}_A(\omega)\right] = \mathbb{E}_*\left[S_t^{j,*}\mathbb{1}_A(\omega)\right]. \tag{4.2.6}$$

Note that $S_t^{j,*}$ is constant on the event $A$ and the value of $S_t^{j,*}$ on $A$ is $\frac{s_t^j}{(1+r)^t}$. Then (4.2.6) can be reformulated as

$$\mathbb{E}_*\left[\mathbb{1}_A(\omega)\cdot(S_{t+1}^{j,*} - S_t^{j,*})\right] = 0. \tag{4.2.7}$$

We now determine a suitable self-financing trading strategy $\varphi$ to show that (4.2.7) is fulfilled. Using (4.1.7), for any self-financing trading strategy $\varphi$,

$$\begin{aligned}\mathbb{E}_*\left[V_T^*(\varphi)\right] &= V_0(\varphi) + \mathbb{E}_*\left[\sum_{s=0}^{T-1}\varphi_s\cdot\Delta S_s^*\right] = V_0(\varphi) + \mathbb{E}_*\left[\sum_{s=0}^{T-1}\varphi_s\cdot(S_{s+1}^* - S_s^*)\right] \\ &= V_0(\varphi) + \mathbb{E}_*\left[\sum_{s=0}^{T-1}\sum_{i=1}^{d}\alpha_s^i\cdot(S_{s+1}^{i,*} - S_s^{i,*})\right].\end{aligned} \tag{4.2.8}$$

To obtain (4.2.7) from (4.2.8), choose

$$\alpha_t^j = \mathbb{1}_A(\omega) \text{ and } \alpha_s^i = 0 \text{ for } (i,s) \neq (j,t).$$

Since $A \in \mathcal{F}_t$, $\alpha_t^j$ is $\mathcal{F}_t$-measurable. Furthermore, $\alpha_s^i$ is trivially $\mathcal{F}_t$-measurable for $(i,s) \neq (j,t)$. Using Lemma 4.1.6 one can construct a self-financing trading strategy $\varphi = (\varphi_t)_{t=1}^T$ with $\varphi_t = (\beta_t, \alpha_t^1, \ldots, \alpha_t^d)$ and $V_0(\varphi) = 0$. By property (d), (4.2.7) immediately follows from (4.2.8) for this trading strategy, as required. □

Next, we need a version of the separating hyperplane theorem.

**Theorem 4.2.5** (Separating hyperplane theorem). *Let $V$ be a linear subspace of $\mathbb{R}^n$ and let $C$ be a compact, convex, non-empty subset of $\mathbb{R}^n$ such that $V \cap C = \emptyset$. Then there exists a vector $p_* \in \mathbb{R}^n$ such that*

$$p_*\cdot v = 0 \text{ for all } v \in V \text{ and } p_*\cdot c > 0 \text{ for all } c \in C.$$

We omit the proof of Theorem 4.2.5, but it can be found, for instance, in [15, section 3.6].

We now come to the main result in this section.

**Theorem 4.2.6** (First fundamental theorem of asset pricing). *The finite market model is arbitrage-free if and only if there exists at least one* risk-neutral probability measure *$\mathbb{P}_*$ (see Definition 4.2.2).*

*Proof.* '$\Leftarrow$' Assume that there exists a *risk-neutral probability measure* $\mathbb{P}_*$. We have to show that there are no arbitrage opportunities. The argument used is almost identical to the argument on page 89. Suppose that there exists an arbitrage opportunity $\varphi$. Then $\varphi$ is a self-financing trading strategy for which

$$V_0(\varphi) = 0, \; V_T(\varphi) \geq 0 \text{ and } \mathbb{P}[V_T(\varphi) > 0] > 0. \tag{4.2.9}$$

By Lemma 4.2.4, the sequence $V_t^*(\varphi)$ is a martingale under $\mathbb{P}_*$. Furthermore, by Lemma 4.2.4(d),

$$\mathbb{E}_*\left[V_T^*(\varphi)\right] = \mathbb{E}_*\left[V_0^*(\varphi)\right] = 0.$$

Therefore, $\mathbb{P}_*[V_T(\varphi) > 0] = 0$. Since $\mathbb{P}_*$ and $\mathbb{P}$ are equivalent, $\mathbb{P}[V_T(\varphi) > 0] = 0$. This is a contradiction and therefore the model is arbitrage-free.

'$\Rightarrow$' This implication is harder to prove and the proof is based on convex analysis on the space of random variables on $\Omega$. Assume that the model is arbitrage-free. So we have to show the existence of a *risk-neutral probability measure*. For this we will use the *separating hyperplane theorem*.

First, make the observation that $\Omega$ is finite so we can write $\Omega = \{\omega_1, \ldots, \omega_n\}$ for some $n \in \mathbb{N}$. The important trick in this proof is that we identify the space of $\mathbb{R}$-valued random variables on $\Omega$ with the vector space $\mathbb{R}^n$. This can be done as follows: If a random variable $Y : \Omega \to \mathbb{R}$ is given, then identify it with the vector

$$\big(Y(\omega_1), \ldots, Y(\omega_n)\big) \in \mathbb{R}^n.$$

This is clearly a bijection. Using Lemma 4.2.4, we see that we have to find a probability measure $\mathbb{P}_*$ such that $\mathbb{E}_*\left[V_T^*(\varphi)\right] = 0$ for all self-financing trading strategies $\varphi$ with $V_0^*(\varphi) = 0$. Let us reformulate this using the above identification. For this, let $\mathbb{P}_*$ be an arbitrary probability measure and define $p_* := (p_1, \ldots, p_n)$ with $\mathbb{P}_*\left[\omega_j\right] := p_j$. We must have

$$0 \stackrel{!}{=} \mathbb{E}_*\left[V_T^*(\varphi)\right] = \sum_{i=1}^{n} V_T^*(\varphi)(\omega_i) \cdot \mathbb{P}[\omega_i] = \sum_{i=1}^{n} V_T^*(\varphi)(\omega_i) \cdot p_i = V_T^*(\varphi) \bullet p_*$$

where, for the last equality, we used the vector interpretation

$$V_T^*(\varphi) = \big(V_T^*(\varphi)(\omega_1), \ldots, V_T^*(\varphi)(\omega_n)\big).$$

In other words, we have to find a vector $p_*$, which is orthogonal to the space

$$V := \left\{V_T^*(\varphi) \mid \varphi \text{ is self-financing with } V_0^*(\varphi) = 0\right\}.$$

Clearly, $V$ is a linear subspace of $\mathbb{R}^n$ ($0 \in V$ and the dependence in $\varphi$ is linear). If it was the case that $V = \mathbb{R}^n$, then $V$ would contain the vector $(1, 1, \ldots, 1)$, which corresponds to a trading strategy with strictly positive payoff. This contradicts the assumption that the model is arbitrage-free. Therefore, $V \neq \mathbb{R}^n$ and there exist non-zero vectors orthogonal to $V$. However, $p_*$ cannot be arbitrary since it has to give a *risk-neutral probability measure*. In particular, we need $p_i > 0$ for all $i$. This can be reformulated as

$$p_i = p_* \cdot e_i > 0 \quad \text{with} \quad e_i = (0, \ldots, 0, \underbrace{1}_{i\text{th position}}, 0, \ldots, 0).$$

We would like to apply Theorem 4.2.5 to the vector space $V$ and the set $\{e_1, \ldots, e_n\}$. However, $\{e_1, \ldots, e_n\}$ is not convex so we introduce

$$C := \left\{ v = (v_1, \ldots, v_n) \in \mathbb{R}^n; \sum_{i=1}^{n} v_i = 1 \text{ and } v_i \geq 0 \right\}.$$

Clearly, $e_i \in C$ for all $i$. If there exists some $p_*$ such that $p_* \cdot c > 0$ for all $c \in C$, then also $p_* \cdot e_i > 0$ and hence $p_i > 0$. Now $C$ is compact, convex and non-empty. To apply Theorem 4.2.5, we have to verify that $V \cap C = \emptyset$. For this, we require the assumption that the model is arbitrage-free. Using (4.2.9), one sees immediately that an arbitrage opportunity $\varphi$ gives a vector

$$V_T^*(\varphi) = \big(v_1, \ldots, v_n\big) \in V \text{ with all } v_j \geq 0 \text{ and at least one } v_j > 0. \tag{4.2.10}$$

In other words, if there is an arbitrage opportunity $\varphi$ then there is a non-zero vector $V_T^*(\varphi) \in V$, which is also contained in the set

$$D := \{(v_1, \ldots, v_n) \in \mathbb{R}^n; \ v_i \geq 0 \text{ for all } 1 \leq i \leq n\}.$$

Therefore the assumption that the model is arbitrage-free implies that $V \cap D = \{0\}$. Since $C \subset D$, we immediately get $V \cap C = \emptyset$. This enables us to apply Theorem 4.2.5 to get a vector $p_* = (p_1, \ldots, p_n)$ such that

$$p_* \cdot v = 0 \text{ for all } v \in V \quad \text{and} \quad p_* \cdot c > 0 \text{ for all } c \in C. \tag{4.2.11}$$

Note that (4.2.11) is true for all vectors of the form $\lambda p_*$ with $\lambda > 0$. We therefore can assume that $\sum_{i=1}^{n} p_i = 1$ and so $p_*$ gives us the required *risk-neutral probability measure*. □

A careful inspection of the proof above shows that we don't require the condition $S_t^j \geq 0$. We can therefore also apply this theorem to markets in which some securities can have negative values. This includes the extended markets we consider in Sections 4.4 and 4.5. Let us now apply Theorem 4.2.6 to an example.

**Example 4.2.7.** *Let $T = 2$ and $d = 1$. This means that we have one stock and two time periods. Furthermore, let $B_0 = 1$, $r = 0$, $\Omega = \{\omega_1, \omega_2, \omega_3, \omega_4\}$ and $\mathbb{P}[\omega_i] > 0$ for all $i$. Since there is only one stock, write $S_t$ instead of $S_t^1$. Suppose that*

$$\begin{aligned} S_0(\omega_1) &= 5 \ \textit{and} \ S_1(\omega_1) = 7 \ \textit{and} \ S_2(\omega_1) = 8, \\ S_0(\omega_2) &= 5 \ \textit{and} \ S_1(\omega_2) = 7 \ \textit{and} \ S_2(\omega_2) = 4, \\ S_0(\omega_3) &= 5 \ \textit{and} \ S_1(\omega_3) = 6 \ \textit{and} \ S_2(\omega_3) = 9, \\ S_0(\omega_4) &= 5 \ \textit{and} \ S_1(\omega_4) = 6 \ \textit{and} \ S_2(\omega_4) = 3. \end{aligned}$$

*Is this model arbitrage-free? Using Theorem 4.2.6, we see that we have to check whether a* risk-neutral measure $\mathbb{P}_*$ *exists for this model. Write* $p_i := \mathbb{P}_*[\omega_i]$ *for* $1 \le i \le 4$ *to simplify the notation. We determine the conditions which these* $p_i$ *must fulfil.*

- $\mathbb{P}_*$ *is an equivalent probability measure, so*

$$p_1 + p_2 + p_3 + p_4 = 1 \text{ and } p_i > 0.$$

- *By the martingale property,* $\mathbb{E}_*\left[S_1^*|S_0\right] \stackrel{!}{=} S_0^*$. *Since* $r = 0$, $S_t^* = S_t$ *for all t. This gives*

$$5 = S_0 \stackrel{!}{=} \mathbb{E}_*\left[S_1^*|S_0\right] = \mathbb{E}_*\left[S_1\right] = 7p_1 + 7p_2 + 6p_3 + 6p_4.$$

- *Similarly,* $\mathbb{E}_*\left[S_2^*|S_1, S_0\right] \stackrel{!}{=} S_1^*$. *Note that* $S_0$ *is a constant so* $\mathbb{E}_*\left[S_2^*|S_1, S_0\right] = \mathbb{E}_*\left[S_2^*|S_1\right]$. *By Definition 2.2.14,*

$$S_1 \stackrel{!}{=} \mathbb{E}_*\left[S_2^*|S_1\right] = \mathbb{E}_*\left[S_2|S_1\right] = \mathbb{E}_*\left[S_2|S_1 = 7\right] \cdot \mathbb{1}_{\{S_1=7\}} + \mathbb{E}_*\left[S_2|S_1 = 6\right] \cdot \mathbb{1}_{\{S_1=6\}}.$$

*Therefore*

$$7 \stackrel{!}{=} \mathbb{E}_*\left[S_2|S_1 = 7\right] = 8 \cdot \frac{p_1}{p_1 + p_2} + 4 \cdot \frac{p_2}{p_1 + p_2},$$
$$6 \stackrel{!}{=} \mathbb{E}_*\left[S_2|S_1 = 6\right] = 9 \cdot \frac{p_3}{p_3 + p_4} + 3 \cdot \frac{p_4}{p_3 + p_4}.$$

*Combining these gives the system of equations*

$$p_1 + p_2 + p_3 + p_4 = 1, \quad 7p_1 + 7p_2 + 6p_3 + 6p_4 = 5,$$
$$p_1 - 3p_2 = 0, \quad 3p_3 - 3p_4 = 0.$$

*Using the substitution* $p_1 = 3p_2$ *and* $p_3 = p_4$,

$$4p_2 + 2p_4 = 1 \text{ and } 28p_2 + 12p_4 = 5.$$

*These two equations imply* $p_4 = 1$ *and* $p_2 = -1/4$, *contradicting our requirement that* $p_i > 0$. *Therefore the model is not arbitrage-free. This can also be seen directly. Note that the stock price goes up from* $t = 0$ *to* $t = 1$ *while the bond stays constant. Therefore, at time* $t = 0$, *we can buy one stock and borrow five bonds. This gives* $V_0(\varphi) = 0$ *and* $V_1(\varphi) \ge 1$. *To complete the trading strategy, at time* $t = 1$ *sell the stock and invest all the money in bonds. In formulas,*

$$(\beta_0, \alpha_0) = (-5, 1) \text{ and } (\beta_1, \alpha_1) = (S_1 - 5, 0).$$

*This trading strategy gives an arbitrage opportunity.*

Theorem 4.2.6 gives us a way to determine whether a market is arbitrage-free. In other words, if a market has arbitrage opportunities then Theorem 4.2.6 enables us to show the existence of arbitrage opportunities. However, neither Theorem 4.2.6 nor its proof give any information about what the actual arbitrage opportunities look like. We have to use practical experience and educated guesses to find them. We summarise the two main ways to construct arbitrage opportunities that we have seen in this course.

- There are two securities and the interest of one security is always larger than the interest of the other security. Examples are the proof of Proposition 3.2.7 and Example 4.2.7.
- There are two ways to produce the same payoff, but the initial prices of the two ways are different. See for instance page 78.

## 4.3 Second Fundamental Theorem of Asset Pricing

We saw in Section 3.3 that a *European contingent claim* is represented by an $\mathcal{F}_T$-measurable random variable $\Phi_T$, where $\Phi_T$ is the payoff of the *contingent claim*. An example is a *European call option* with *maturity* $T$ and strike price $K$ based on the first stock. In this case

$$\Phi_T = \max\{S_T^1 - K, 0\}.$$

The main question we are interested in is: What is the arbitrage-free price of a *contingent claim* $\Phi_T$? In Section 3.3 we used replicating strategies for *contingent claims* to answer this question. We try to use the same argument here.

**Definition 4.3.1** (Replicable contingent claims). Let $\Phi_T$ be an $\mathcal{F}_T$-measurable random variable. A contingent claim with payoff $\Phi_T$ is called *replicable* if there exists a self-financing trading strategy $\varphi$ satisfying

$$\Phi_T = V_T(\varphi) \quad \text{with} \quad V_T(\varphi) \text{ as in (4.1.4).}$$

We saw in the binomial model that all $\mathcal{F}_T$-measurable random variables $\Phi_T$ are replicable. Let us consider a simple example to see if this is still the case here.

**Example 4.3.2.** *Let $T = 1$ and $d = 1$. This means that we have one stock and one time period. Furthermore, let $B_0 = 1$, $r = 0$, $\Omega = \{\omega_1, \omega_2, \omega_3\}$ and $\mathbb{P}[\omega_i] > 0$ for all $i$. Since there is only one stock, write $S_t$ instead of $S_t^1$. Suppose that*

$$S_0 = 1,\ S_1(\omega_1) = 0.8,\ S_1(\omega_2) = 1 \ \text{ and } \ S_1(\omega_3) = 1.2.$$

*Is a* call option *with strike price $K = 1$ replicable?*

*Since $T = 1$ and $d = 1$, a trading strategy $\varphi$ has the form $\varphi = (\beta_0, \alpha_0)$ with $\alpha_0, \beta_0 \in \mathbb{R}$. We look for $\alpha_0, \beta_0 \in \mathbb{R}$ with*

$$\max\{S_1 - K, 0\} \stackrel{!}{=} V_1(\varphi) = \beta_0 B_1 + \alpha_0 S_1.$$

*Inserting $\omega_1$, $\omega_2$ and $\omega_3$ into this equation, we obtain*

$$0 = \beta_0 + \alpha_0 \cdot 0.8, \quad 0 = \beta_0 + \alpha_0 \cdot 1, \quad 0.2 = \beta_0 + \alpha_0 \cdot 1.2.$$

*The unique solution of the first two equations is $\alpha_0 = \beta_0 = 0$, but this does not solve the third equation. Therefore the* call option *is not replicable.*

Example 4.3.2 shows that a contingent claim is not always replicable, even when the market is arbitrage-free. In view of the above observations, we give the following definition.

**Definition 4.3.3** (Complete markets). A finite market model is called *complete* if all $\mathcal{F}_T$-measurable random variables $\Phi_T$ are replicable.

We now try to establish the conditions needed to ensure that a market is *complete*. Note that we do not require that the market is arbitrage-free in Definition 4.3.3. However, arbitrage-free markets are the most important case and we will only study this question in this situation.

Recall that, in the proof of Theorem 4.2.6, we identified the random variables on $\Omega$ with the vector space $\mathbb{R}^n$. Furthermore, we introduced

$$V := \left\{ V_T^*(\varphi) \mid \varphi \text{ is self-financing with } V_0^*(\varphi) = 0 \right\}.$$

Clearly, $V$ can be interpreted as the space of all random variables which can be replicated with a self-financing trading strategy $\varphi$ with $V_0(\varphi) = 0$. Introduce

$$V' := \left\{ V_T^*(\varphi) \mid \varphi \text{ is self-financing} \right\}. \tag{4.3.1}$$

Then $V'$ consists of all random variables which can be replicated with an arbitrary self-financing trading strategy $\varphi$. In other words, $V'$ is the space of all replicable random variables. Therefore, the market is *complete* if and only if $V' = \mathbb{R}^n$. Also, $V \subset V' \subset \mathbb{R}^n$. As $V'$ has one more degree of freedom than $V$, one expects that $\dim V = \dim V' - 1$. We can therefore guess that the market is complete if and only if $\dim V = n - 1$. Furthermore, the vector $p_*$ used for construction of the *risk-neutral measure* in the proof of Theorem 4.2.6 has to be orthogonal to $V$. If $\dim V = n - 1$ then the vector $p_*$ is unique (up to scaling) and therefore so is the *risk-neutral measure* $\mathbb{P}_*$. This consideration leads to the following theorem.

**Theorem 4.3.4** (Second fundamental theorem of asset pricing). *Consider a finite market model with no arbitrage. Then the market is* complete *if and only if the* risk-neutral measure $\mathbb{P}_*$ *is unique.*

There are some details we need to justify in the heuristic argument above. For this we need the following theorem.

**Theorem 4.3.5.** *Suppose that the finite market model is arbitrage-free and $\Phi_T$ is a* replicable *random variable.*

- *Let $\varphi$ and $\varphi'$ be two replicating strategies for $\Phi_T$. Then for all t,*

$$V_t(\varphi) = V_t(\varphi').$$

- *Moreover, let $\varphi$ be a replicating strategy for $\Phi_T$ and $\mathbb{P}_*$ and $\mathbb{P}'_*$ be two* risk-neutral measures. *Then for all t,*

$$V_t^*(\varphi) = \mathbb{E}_*\left[\frac{\Phi_T}{(1+r)^T}\middle|\mathcal{F}_t\right] = \mathbb{E}'_*\left[\frac{\Phi_T}{(1+r)^T}\middle|\mathcal{F}_t\right].$$

Before proving Theorem 4.3.5, we would like to point out that replicating strategies in a finite market model do not have to be unique. In fact, one can easily construct examples where replicating strategies are not unique. Consider, for instance, a market consisting of a bond and two stocks for which $B_1 + S_1^1 = S_1^2$. Then there exist at least two distinct replicating strategies for $S_1^2$, namely $\varphi = (1, 1, 0)$ and $\varphi' = (0, 0, 1)$.

*Proof of Theorem 4.3.5.* Let $\varphi$ be a replicating strategy for $\Phi_T$. By Theorem 4.2.6 there exists at least one *risk-neutral measure* $\mathbb{P}_*$. Lemma 4.2.4 shows that $V_t^*(\varphi)$ is a martingale under $\mathbb{P}_*$. Therefore

$$V_t(\varphi) = (1+r)^t V_t^*(\varphi) = (1+r)^t \mathbb{E}_*\left[V_T^*(\varphi)|\mathcal{F}_t\right] = (1+r)^{-T+t}\mathbb{E}_*\left[\Phi_T|\mathcal{F}_t\right]. \quad (4.3.2)$$

The RHS of (4.3.2) depends only on $\Phi_T$ and $\mathbb{P}_*$ and does not involve $\varphi$. So $V_t(\varphi)$ must be the same for all trading strategies $\varphi$. On the other hand, the LHS of (4.3.2) depends on $\varphi$ only and does not involve $\mathbb{P}_*$. So $\mathbb{E}_*\left[\frac{\Phi_T}{(1+r)^T}|\mathcal{F}_t\right]$ has to be the same for all *risk-neutral measures*. □

We are now able to prove Theorem 4.3.4.

*Proof of Theorem 4.3.4.* '$\Rightarrow$' Assume that the market is *complete*. We have to show that the *risk-neutral measure* $\mathbb{P}_*$ is unique. For this, let $\mathbb{P}_*$ be a *risk-neutral measure* and $\mathbb{P}'_*$ be another arbitrary *risk-neutral measure*. We will now prove that $\mathbb{P}_*$ and $\mathbb{P}'_*$ are equal. Let $A \in \mathcal{F}_T$ be arbitrary and define $\Phi_T := \mathbb{1}_A$. Then $\Phi_T$ is an $\mathcal{F}_T$-measurable random variable. By hypothesis, the market is complete and so $\Phi_T$ is replicable. Therefore there exists a trading strategy $\varphi$ such that $\Phi_T = V_T(\varphi)$. Using Theorem 4.3.5, we obtain

$$V_0(\varphi) = \mathbb{E}_*\left[\frac{\Phi_T}{(1+r)^T}\right] = \mathbb{E}'_*\left[\frac{\Phi_T}{(1+r)^T}\right],$$

where $\mathbb{E}'_*$ is the expectation with respect to $\mathbb{P}'_*$. Inserting the definition of $\Phi_T$, we obtain

$$\mathbb{P}_*[A] = \mathbb{E}_*[\mathbb{1}_A] = \mathbb{E}'_*[\mathbb{1}_A] = \mathbb{P}'_*[A].$$

Since $A$ was arbitrary, we immediately get $\mathbb{P}_* = \mathbb{P}'_*$.

'$\Leftarrow$' We will proceed by contraposition. We assume that the market is not complete and show that there exist at least two *risk-neutral measures* $\mathbb{P}_*$ and $\mathbb{P}'_*$ such that $\mathbb{P}_* \neq \mathbb{P}'_*$. We will use a similar argument to that in the proof of Theorem 4.2.6, that is we identify the space of $\mathbb{R}$-valued random variables on $\Omega$ with the vector space $\mathbb{R}^n$. In the proof of Theorem 4.2.6 we defined

$$V := \left\{V_T^*(\varphi) \mid \varphi \text{ is self-financing with } V_0^*(\varphi) = 0\right\},$$

$$C := \left\{v = (v_1, \ldots, v_n) \in \mathbb{R}^n; \sum_{i=1}^{n} v_i = 1 \text{ and } v_i \geq 0\right\}$$

and showed that there exists a vector $p_* \in \mathbb{R}^n$ such that

$$p_* \cdot v = 0 \text{ for all } v \in V \text{ and } p_* \cdot c > 0 \text{ for all } c \in C. \tag{4.3.3}$$

We now show that there is another vector $p'_*$ fulfilling (4.3.3), which is not a scalar multiple of $p_*$. For $V'$ defined in (4.3.1), one immediately gets that $V \subseteq V' \subseteq \mathbb{R}^n$. We claim that

$$V \subsetneq V' \subsetneq \mathbb{R}^n. \tag{4.3.4}$$

Since $V'$ is the vector space of all replicable random variables and, by assumption, our model is not complete, there is a random variable $\Phi_T \notin V'$, so $V' \subsetneq \mathbb{R}^n$. Suppose that $V = V'$ and let $\varphi$ be an arbitrary trading strategy with $V_0(\varphi) = 1$. Define $\Phi_T := V_T(\varphi)$. This $\Phi_T$ is clearly replicable and $\Phi_T \in V'$. Since $V = V'$, $\Phi_T \in V$ and so there exists a trading strategy $\varphi'$ with $V_0(\varphi') = 0$ and $\Phi_T = V_T(\varphi')$. But by Theorem 4.3.5, since $\Phi_T$ is replicable, $V_0(\varphi') = V_0(\varphi)$, giving a contradiction.

Equation (4.3.4) clearly implies that $\dim V \leq n - 2$ and so there exists a non-zero vector $v$, which is orthogonal to $V$ and $p_*$. Define

$$p'_* = p_* + \lambda v,$$

where $\lambda > 0$ is so small that $p'_* \cdot c > 0$ for all $c \in C$. This is possible since $C$ is compact and therefore $\min_{c \in C}\{p_* \cdot c\} > 0$. By construction.

$$p'_* \cdot v = 0 \text{ for all } v \in V \text{ and } p'_* \cdot c > 0 \text{ for all } c \in C,$$

so $p'_*$ gives a *risk-neutral measure* $\mathbb{P}'_*$ (after rescaling). The vectors $p'_*$ and $p'_*$ are linearly independent and therefore $\mathbb{P}'_* \neq \mathbb{P}_*$. □

Let us now look at an example.

**Example 4.3.6.** *Consider a finite market model with two time periods ($T = 2$) and two assets, a bond $B_t$ and a risky asset $S_t$. The price of the risky asset $S_t$ follows the probability tree in Figure 4.2 and the bond $B_t$ has the interest rate $r = 0$ and $B_0 = 1$.*

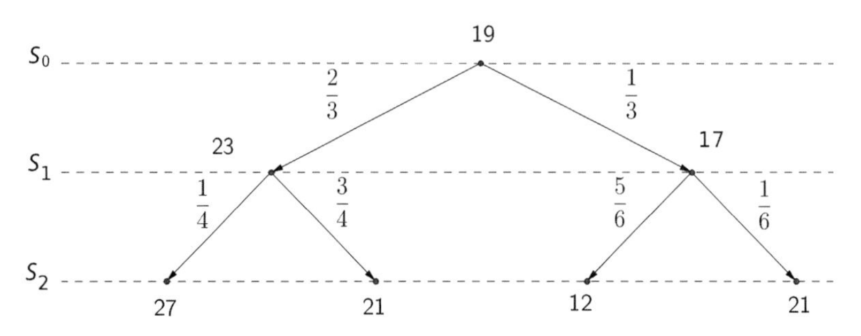


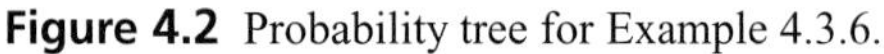

**Figure 4.2** Probability tree for Example 4.3.6.

*Is this market arbitrage-free and complete? To answer this question, we will look for a* risk-neutral measure $\mathbb{P}_*$. *Since* $r = 0$, *we need*

$$\mathbb{E}_* [S_1] = S_0 \text{ and } \mathbb{E}_* [S_2 | S_1] = S_1.$$

*Since* $S_1$ *can have only the two values* 17 *and* 23, *we need*

$$\mathbb{E}_* [S_1] = S_0,\ \mathbb{E}_* [S_2 | S_1 = 17] = 17 \text{ and } \mathbb{E}_* [S_2 | S_1 = 23] = 23. \tag{4.3.5}$$

*Let* $p_1$, $p_2$ *and* $p_3$ *be as in Figure 4.3.*

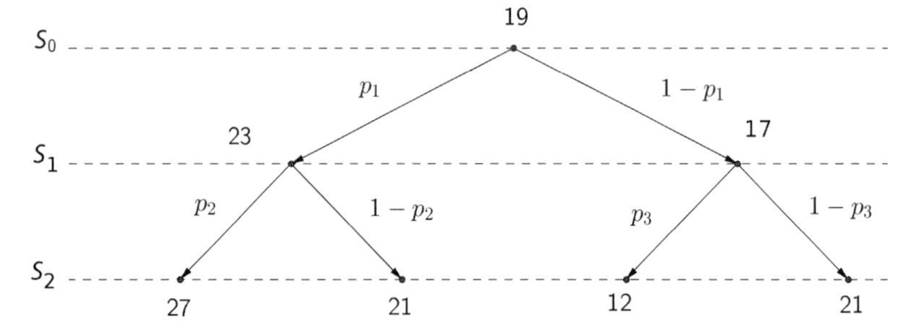


**Figure 4.3** Probability tree for *risk-neutral measure.*

*Inserting* $p_1$, $p_2$ *and* $p_3$ *into* (4.3.5) *gives*

$$19 = 23 \cdot p_1 + 17 \cdot (1 - p_1),\ 23 = 27 \cdot p_2 + 21 \cdot (1 - p_2),\ 17 = 12 \cdot p_1 + 21 \cdot (1 - p_3).$$

*Solving gives* $p_1 = 1/3$, $p_2 = 2/3$ *and* $p_3 = 4/9$. *Therefore the* risk-neutral measure $\mathbb{P}_*$ *exists and is unique. Thus the first and second fundamental theorems of asset pricing show that the market is arbitrage-free and complete.*

**Example 4.3.7.** *Consider the same market model as in Example 4.3.6, but this time with* $r = 1$. *For the* risk-neutral measure $\mathbb{P}_*$, *we need*

$$\mathbb{E}_* \left[ \frac{S_1}{1+r} \right] = S_0 \text{ and } \mathbb{E}_* \left[ \frac{S_2}{(1+r)^2} \middle| S_1 \right] = \frac{S_1}{1+r}.$$

*Because $S_1$ can only have the two values 17 and 23, we are looking for a measure with*

$$\mathbb{E}_* [S_1] = 2 \cdot S_0, \ \mathbb{E}_* [S_2|S_1 = 17] = 2 \cdot 17 \ \text{and} \ \mathbb{E}_* [S_2|S_1 = 23] = 2 \cdot 23. \tag{4.3.6}$$

*This leads to the equations*

$$38 = 23 \cdot p_1 + 17 \cdot (1 - p_1), \ 46 = 27 \cdot p_2 + 21 \cdot (1 - p_2), \ 34 = 12 \cdot p_1 + 21 \cdot (1 - p_3).$$

*Solving gives $p_1 = 7/2$, $p_2 = 25/6$ and $p_3 = 13/9$. Since these numbers are not in the interval $(0, 1)$, there does not exist any* risk-neutral measure. *The market is therefore not arbitrage-free.*

Note that we cannot apply Theorem 4.3.4 in Example 4.3.7 as the market is not arbitrage-free. Therefore the computations in Example 4.3.7 do not show us whether the market is complete or incomplete. However, one can show that all *contingent claims* are replicable in the setting of Example 4.3.7 and so the market is complete.

## 4.4 Pricing of Replicable European Contingent Claims

We assume in this section that the finite market model is arbitrage-free and determine the arbitrage-free price of a replicable contingent claim $\Phi_T$. The argument will be similar to the argument in Section 3.3.2.

To do this, we need to extend the finite market model by adding a *contingent claim* $\Phi_T$. We assume that

- the dynamics of the stocks $S_t^j$ and the bond $B_t$ are given by the finite market,
- the *contingent claim* has no influence on the stocks $S_t^j$ and the bond $B_t$,
- the *maturity* of the *contingent claim* is $T$ and
- the *contingent claim* $\Phi_T$ is $\mathcal{F}_T$-measurable.

In contrast to Section 3.3.2, we will allow the *contingent claim* to be traded at all times $t = 0, \ldots, T-1$. The question is: What is the right price of the contingent claim at time $t$? One possibility could be to use the initial price, but this turns out to be a bad idea. To see this, suppose that one month ago we bought a call option with maturity tomorrow with a strike price £100. If the price of the stock today is £250 then, in most cases, the option would give us tomorrow a gain of around £150. If, on the other hand, the price of the stock today is £1 then the option will most likely give us nothing tomorrow. We therefore see that the price of the option should depend on the time and the stock prices. Denote the price of the contingent claim at time $t$ by $P_t$. Clearly, we must have $P_T = \Phi_T$. Also, the price $P_t$ at time $t$ can only depend on the information available at time $t$. Thus $P_t$ has to be $\mathcal{F}_t$-measurable. In general, $P_t$ is not a constant, see for instance Example 3.3.13. Set

$$\begin{aligned} \alpha_t^j &:= \text{the number of shares of the } j\text{th stock we hold during the time period } [t, t+1), \\ \beta_t &:= \text{the number of bonds we hold during the same time period,} \\ \gamma_t &:= \text{the number of } \textit{contingent claims} \text{ we hold during the same time period.} \end{aligned}$$

Thus a *trading strategy* $\psi$ in this extended market has the form

$$\psi = (\psi_t)_{t=0}^{T-1} \text{ with } \psi_t = (\varphi_t, \gamma_t) \text{ and } \varphi_t = (\beta_t, \alpha_t^1, \dots, \alpha_t^d).$$

As usual, we have to assume that $\alpha_t$, $\beta_t$ and $\gamma_t$ are $\mathcal{F}_t$-measurable and that we do not trade at time $T$ so $\psi_T = \psi_{T-1}$. The value of our portfolio at time $t$ is

$$V_t(\psi) = \varphi_t \cdot \vec{S}_t + \gamma_t P_t \text{ for all } t \geq 0. \tag{4.4.1}$$

A trading strategy $\psi$ is self-financing if

$$\varphi_{t-1} \cdot \vec{S}_t + \gamma_{t-1} P_t = \varphi_t \cdot \vec{S}_t + \gamma_t P_t \tag{4.4.2}$$

for all $t \geq 1$, and an arbitrage opportunity is a trading strategy $\psi$ such that

$$V_0(\psi) = 0,\ V_T(\psi) \geq 0 \text{ and } \mathbb{P}[V_T(\psi) > 0] > 0.$$

We now prove the following result.

**Theorem 4.4.1.** *Consider the extended market described above. Assume that the underlying finite market model is arbitrage-free and that $\Phi_T$ is replicable. Let $\mathbb{P}_*$ be a* risk-neutral measure *and $\varphi$ be a replicating strategy for $\Phi_T$.*

*Then this extended market is arbitrage-free if and only if*

$$P_t = V_t(\varphi) = \frac{1}{(1+r)^{T-t}} \mathbb{E}_* \left[\Phi_T | \mathcal{F}_t\right] \text{ for all } 0 \leq t \leq T.$$

*In particular, the unique arbitrage-free initial price of $\Phi_T$ is $\frac{1}{(1+r)^T}\mathbb{E}_*[\Phi_T]$.*

Note that in Theorem 4.4.1 we do not assume that the market is *complete*. Therefore there can be more than one *risk-neutral measure* $\mathbb{P}_*$. However, Theorem 4.3.5 ensures that the expression $\mathbb{E}_*[\Phi_T|\mathcal{F}_t]$ is the same for all *risk-neutral measures* $\mathbb{P}_*$.

*Proof.* By assumption, $\Phi_T$ is replicable and $\varphi = (\varphi_t)_{t=0}^T$, with $\varphi_t = (\beta_t, \alpha_t^1, \dots, \alpha_t^d)$, is a replicating strategy for $\Phi_T$. Then

$$V_T(\varphi) = \Phi_T.$$

By Theorem 4.3.5, the value of the portfolio for the replicating strategy $\varphi$ at time $t$ is given by

$$V_t(\varphi) = \frac{1}{(1+r)^{T-t}} \mathbb{E}_* \left[\Phi_T | \mathcal{F}_t\right].$$

We now use the same argument as in the proof of Theorem 3.3.20. Note that, at each time $t \in \{0, \dots, T\}$, there are two ways to produce a payoff $\Phi_T$ at time $T$:

- Buy the contingent claim at time $t$.
- Use the trading strategy $\varphi$, starting at time $t$ with $\varphi_t$.

If $P_t \neq V_t(\varphi)$ for some $t \in \{0, \dots, T\}$ then we have two ways to produce the same payoff, but with different prices. This creates an arbitrage opportunity. Let us construct this arbitrage opportunity explicitly. Suppose that $\mathbb{P}[P_t > V_t(\varphi)] > 0$. Clearly $t < T$ since $V_T(\varphi) = \Phi_T$. Then our trading strategy $\psi$ is

- Do nothing until time $t$.
- If $P_t \leq V_t(\varphi)$ then continue to do nothing.
- If $P_t > V_t(\varphi)$ then
  - sell one *contingent claim* and keep it for the rest of the time, that is $\gamma_s = -1$ for $t \leq s \leq T-1$,
  - invest $V_t(\varphi)$ into the trading strategy $\varphi$ (starting with $\varphi_t$ at time $t$) and
  - invest the remaining money, that is $P_t - V_t(\varphi)$, into bonds.

The *trading strategy* $\psi$ has the form

$$\psi_s = (\varphi_s, \gamma_s) = \begin{cases} 0, & \text{for } s < t, \\ 0, & \text{for } s \geq t \text{ if } P_t \leq V_t(\varphi), \\ \left(\beta_s + \frac{P_t - V_t(\varphi)}{B_t}, \alpha_s^1, \dots, \alpha_s^d, -1\right), & \text{for } s \geq t \text{ if } P_t > V_t(\varphi). \end{cases}$$

By construction, $\psi_s$ is $\mathcal{F}_s$-measurable for all $s \in \{0, \dots, T-1\}$. It remains to check that $\psi$ is a self-financing trading strategy. We have to check that (4.4.2) is fulfilled. If $P_t \leq V_t(\varphi)$ then $\psi_s = 0$ for all $s$ and so the strategy is trivially self-financing. If $P_t > V_t(\varphi)$ then we have to be more careful. For $s < t$, this is still obvious. For $s = t$ we have

$$\begin{aligned}\left(\left(\beta_t + \frac{P_t - V_t(\varphi)}{B_t}\right) B_t + \sum_{j=1}^{d} \alpha_t^j S_t^j\right) - P_t &= \left(\beta_t B_t + \sum_{j=1}^{d} \alpha_t^j S_t^j\right) - V_t(\varphi) \\ &= \varphi_t \cdot \vec{S}_t - V_t(\varphi) = 0.\end{aligned}$$

In the last line we used the definition of $V_t(\varphi)$. Therefore, $\psi$ is self-financing at $t = s$ since $\psi_s = 0$ for $s < t$. The computation for $s > t$ is similar and is omitted.

We now show that $\psi$ is an arbitrage opportunity. If $P_t \leq V_t(\varphi)$ then $V_T(\psi) = 0$. If, on the other hand, $P_t > V_t(\varphi)$ then we use that $P_T = \Phi_T = V_T(\varphi)$ to get

$$\begin{aligned}V_T(\psi) &= \left(\left(\beta_T + \frac{P_t - V_t(\varphi)}{B_t}\right) B_T + \sum_{j=1}^{d} \alpha_T^j S_T^j\right) - P_T \\ &= \left(\beta_T B_T + \sum_{j=1}^{d} \alpha_T^j S_T^j\right) + \big(P_t - V_t(\varphi)\big)(1+r)^{T-t} - \Phi_T \\ &= \varphi_T \cdot \vec{S}_T + \big(P_t - V_t(\varphi)\big)(1+r)^{T-t} - V_T(\varphi) \\ &= \big(P_t - V_t(\varphi)\big)(1+r)^{T-t}.\end{aligned}$$

Combining the computations in the cases $P_t > V_t(\varphi)$ and $P_t \leq V_t(\varphi)$, we get

$$\begin{aligned} V_T(\psi) &= 0 \cdot \mathbb{1}_{\{P_t \leq V_t(\varphi)\}} + \big(P_t - V_t(\varphi)\big)(1+r)^{T-t} \cdot \mathbb{1}_{\{P_t > V_t(\varphi)\}} \\ &= \big(P_t - V_t(\varphi)\big)(1+r)^{T-t} \mathbb{1}_{\{P_t > V_t(\varphi)\}}. \end{aligned}$$

Therefore $V_0(\psi) = 0$, $V_T(\psi) \geq 0$ and $\mathbb{P}[V_T(\psi) > 0] = \mathbb{P}[P_t > V_t(\varphi)] > 0$, so $\psi$ is an arbitrage opportunity.

Let us now show that there is no arbitrage opportunity if $P_t = V_t(\varphi)$ for all $t$. For this, suppose that $\widehat{\psi} = (\widehat{\psi}_t)_{t=0}^{T-1}$ with $\widehat{\psi}_t = (\widehat{\varphi}_t, \widehat{\gamma}_t)$ is an arbitrage opportunity. We claim that $V_t^*(\widehat{\psi}) := V_t(\widehat{\psi})/(1+r)^t$ is a martingale with respect to $\mathbb{P}_*$. We have

$$\begin{aligned} \mathbb{E}_*\left[V_{t+1}^*(\widehat{\psi})\middle|\mathcal{F}_t\right] &= \mathbb{E}_*\left[\widehat{\varphi}_t \cdot \vec{S}_{t+1}^* + \widehat{\gamma}_t P_{t+1}/(1+r)^{t+1}\middle|\mathcal{F}_t\right] \\ &= \widehat{\varphi}_t \cdot \mathbb{E}_*\left[\vec{S}_{t+1}^*\middle|\mathcal{F}_t\right] + \widehat{\gamma}_t \mathbb{E}_*\left[V_{t+1}^*(\varphi)\middle|\mathcal{F}_t\right] \\ &= \widehat{\varphi}_t \cdot \vec{S}_t^* + \widehat{\gamma}_t V_t^*(\varphi) \\ &= V_t^*(\widehat{\psi}). \end{aligned}$$

In this computation we used that the trading strategy $\widehat{\psi}$ is self-financing and that $S_t^*$ and $V_t^*(\varphi)$ are martingales under $\mathbb{P}_*$. Therefore, $V_t^*(\widehat{\psi})$ is a martingale. By Lemma 3.3.18,

$$\mathbb{E}_*\left[V_T^*(\widehat{\psi})\right] = \mathbb{E}_*\left[V_0^*(\widehat{\psi})\right] = 0.$$

For the last equality we used that $V_0^*(\widehat{\psi}) = 0$, since $\widehat{\psi}$ is an arbitrage opportunity. Furthermore, by assumption $V_T(\widehat{\psi}) \geq 0$ and so

$$\mathbb{P}_*\left[V_T^*(\widehat{\psi}) > 0\right] = \mathbb{P}_*\left[V_T(\widehat{\psi}) > 0\right] = 0.$$

Since $\mathbb{P}_*$ and $\mathbb{P}$ are equivalent, $\mathbb{P}\left[V_T(\widehat{\psi}) > 0\right] = 0$. This is a contradiction and so there exist no arbitrage opportunities in this model. □

## 4.5 Incomplete Markets

If the market is *arbitrage-free* and *complete* then Theorem 4.4.1 gives the arbitrage-free price of each *contingent claim*. However, there are *arbitrage-free* markets, which are not *complete*. For instance, there are infinitely many *risk-neutral measures* in the model in Example 4.2.3. Theorem 4.3.4 therefore implies that this market is not *complete*. The main question in this section is: What is the arbitrage-free price of a *contingent claim* $\Phi_T$, which is not replicable? We consider the same extended market as described at the beginning of Section 4.4 and use the same assumptions and notations.

A possible approach to determine the arbitrage-free price of a non-replicable contingent claim is to choose a risk-neutral measure $\mathbb{P}_*$ and try to argue as for Theorem 4.4.1. This leads to the following theorem.

**Theorem 4.5.1.** *Consider the extended market as described in Section 4.4. Assume that the underlying finite market model is arbitrage-free and let* $\mathbb{P}_*$ *be a* risk-neutral measure. *Let* $\Phi_T$ *be a contingent claim and denote by* $P_t$ *the price of* $\Phi_T$ *at time t. Then this extended market is arbitrage-free if*

$$P_t = \frac{1}{(1+r)^{T-t}} \mathbb{E}_* \left[ \Phi_T \middle| \mathcal{F}_t \right] \text{ for all } 0 \le t \le T. \tag{4.5.1}$$

Note that an important difference between Theorem 4.5.1 and Theorem 4.4.1 is that in Theorem 4.5.1 there is just an '*if*' and not an '*if and only if*' statement. Before we discuss this further, we prove Theorem 4.5.1.

*Proof of Theorem 4.5.1.* We use a similar argument to that in the second part of the proof of Theorem 4.4.1.

Suppose that $\widehat{\psi} = (\widehat{\psi}_t)_{t=0}^{T-1}$ with $\widehat{\psi}_t = (\widehat{\varphi}_t, \widehat{\gamma}_t)$ is an arbitrage opportunity and define $V_t(\widehat{\psi})$ as in (4.4.1). We claim that $V_t^*(\widehat{\psi}) := V_t(\widehat{\psi})/(1+r)^t$ is a martingale with respect to $\mathbb{P}_*$. Indeed, using the self-financing property,

$$\begin{aligned}
\mathbb{E}_* \left[ V_{t+1}^*(\widehat{\psi}) \middle| \mathcal{F}_t \right] &= \mathbb{E}_* \left[ \widehat{\varphi}_t \cdot \vec{S}_{t+1}^* + \widehat{\gamma}_t P_{t+1}/(1+r)^{t+1} \middle| \mathcal{F}_t \right] \\
&= \widehat{\varphi}_t \cdot \mathbb{E}_* \left[ \vec{S}_{t+1}^* \middle| \mathcal{F}_t \right] + \widehat{\gamma}_t \frac{1}{(1+r)^{t+1}} \mathbb{E}_* \left[ P_{t+1} \middle| \mathcal{F}_t \right].
\end{aligned}$$

We now use the assumption (4.5.1) and the tower property to obtain

$$\begin{aligned}
\mathbb{E}_* \left[ P_{t+1} \middle| \mathcal{F}_t \right] &= \frac{1}{(1+r)^{T-(t+1)}} \mathbb{E}_* \left[ \mathbb{E}_* \left[ \Phi_T \middle| \mathcal{F}_{t+1} \right] \middle| \mathcal{F}_t \right] \\
&= \frac{1}{(1+r)^{T-t-1}} \mathbb{E}_* \left[ \Phi_T \middle| \mathcal{F}_t \right] = (1+r) P_t.
\end{aligned}$$

Inserting this into the above equation and using that $\vec{S}_t^*$ is a martingale,

$$\begin{aligned}
\mathbb{E}_* \left[ V_{t+1}^*(\widehat{\psi}) \middle| \mathcal{F}_t \right] &= \widehat{\varphi}_t \cdot \vec{S}_t^* + \widehat{\gamma}_t \frac{P_t}{(1+r)^t} = \frac{\widehat{\varphi}_t \cdot \vec{S}_t + \widehat{\gamma}_t P_t}{(1+r)^t} \\
&= \widehat{\varphi}_t \cdot \vec{S}_t^* + \widehat{\gamma}_t \frac{P_t}{(1+r)^t} = V_t^*(\widehat{\psi}).
\end{aligned}$$

Therefore $V_t^*(\widehat{\psi})$ is a martingale and by Lemma 3.3.18,

$$\mathbb{E}_* \left[ V_T^*(\widehat{\psi}) \right] = \mathbb{E}_* \left[ V_0^*(\widehat{\psi}) \right] = 0.$$

In the last equality we used that $\widehat{\psi}$ is an arbitrage opportunity. This implies that $\mathbb{P}_* \left[ V_T(\widehat{\psi}) > 0 \right] = 0$ and therefore $\mathbb{P} \left[ V_T(\widehat{\psi}) > 0 \right] = 0$. This is a contradiction and so there exist no arbitrage opportunities in this model. □

Theorem 4.5.1 shows that the extended market is arbitrage-free if we choose the prices $P_t$ as in (4.5.1). However, in Theorem 4.5.1 we first have to choose a *risk-neutral measure*.

A natural and important question at this point is: Are the prices $P_t$ independent of this choice? Theorem 4.3.5 shows that this is the case if $\Phi_T$ is replicable. We now consider an example to see whether this is also the case for a non-replicable contingent claim $\Phi_T$.

**Example 4.5.2.** *Consider again the situation in Example 4.3.2. Recall that $T = 1$, $d = 1$, $B_0 = 1$, $r = 0$, $\Omega = \{\omega_1, \omega_2, \omega_3\}$ with $\mathbb{P}[\omega_i] > 0$ for all i and*

$$S_0 = 1,\ S_1(\omega_1) = 0.8,\ S_1(\omega_2) = 1 \ \text{and}\ S_1(\omega_3) = 1.2.$$

*We have shown that a* call option *with strike price $K = 1$ is not replicable in this model. We begin by determining the risk-neutral measures. Let $p_i := \mathbb{P}_*[\omega_i]$ for $1 \le i \le 3$. To have a risk-neutral measure $\mathbb{P}_*$, we need*

$$\mathbb{E}_*\left[S_1 \middle| S_0\right] = S_0.$$

*Since $S_0$ is a constant, $\mathbb{E}_*\left[S_1 \middle| S_0\right] = \mathbb{E}_*[S_1]$. So $\mathbb{P}_*$ has to fulfil*

$$0.8 \cdot p_1 + 1 \cdot p_2 + 1.2 \cdot p_3 = 1, \qquad p_1 + p_2 + p_3 = 1.$$

*This is equivalent to*

$$p_1 = p_3, \qquad p_2 = 1 - 2p_1. \tag{4.5.2}$$

*It follows immediately that the solution of* (4.5.2) *gives a risk-neutral measure $\mathbb{P}_*$ if and only if $p_1 \in (0, 1/2)$. We next compute the initial price for a* call option *with strike price $K = 1$ using* (4.5.1). *We get*

$$P_0 = \mathbb{E}_*[\max\{S_1 - 1, 0\}] = (1.2 - 1) \cdot p_1 + (1 - 1) \cdot p_2 + 0 \cdot p_3 = \frac{p_1}{5}.$$

*Therefore $\mathbb{E}_*[\max\{S_1 - 1, 0\}]$ can take all values in the open interval* $(0, 1/10)$.

The results in Example 4.5.2 look quite strange and surprising and lead to the following question: Can a contingent claim in an arbitrage-free market have more than one price? The answer surprisingly is yes and no. Let us explain this.

- **Yes**. Suppose we have an arbitrage-free market and we would like to add a non-replicable contingent claim $\Phi_T$. Then the price of $\Phi_T$ is not uniquely determined by the 'no arbitrage' condition.
- **No**. Suppose we have an arbitrage-free market with two European contingent claims $\Phi_T^{(1)}$ and $\Phi_T^{(2)}$ with the same payoff at time $T$, that is $\Phi_T^{(1)} = \Phi_T^{(2)}$. Then $\Phi_T^{(1)}$ and $\Phi_T^{(2)}$ must have the same price at all $t \le T$. Otherwise we could use the usual '*buy cheap and sell expensive*' strategy to construct an arbitrage opportunity. The argument in this case is essentially the same as in the first part of the proof of Theorem 4.4.1.

The next lemma shows that the observation in Example 4.5.2 is usual for non-replicable contingent claims $\Phi_T$.

**Lemma 4.5.3.** Consider an arbitrage-free finite market model, realised on the probability space $(\Omega, \mathbb{P})$ and let $\Phi_T$ be an $\mathcal{F}_T$-measurable random variable, which is not replicable. Then there exist *risk-neutral measures* $\mathbb{P}_*$ and $\mathbb{P}'_*$ such that

$$\mathbb{E}_*[\Phi_T] \neq \mathbb{E}'_*[\Phi_T].$$

*Proof.* We use the same argument as in the proofs of Theorems 4.2.6 and 4.3.4. Recall the notation

$$\begin{aligned} V &= \left\{V_T^*(\varphi) \mid \varphi \text{ is self-financing with } V_0^*(\varphi) = 0\right\}, \\ V' &= \left\{V_T^*(\varphi) \mid \varphi \text{ is self-financing}\right\}, \\ C &= \left\{v = (v_1, \dots, v_n) \in \mathbb{R}^n; \sum_{i=1}^n v_i = 1 \text{ and } v_i \geq 0\right\}. \end{aligned}$$

In the proof of Theorem 4.2.6 we constructed a measure $\mathbb{P}_*$ and a vector $p_* = (p_1, \dots, p_n)$ such that $\mathbb{P}_*[\omega_i] = p_i$ and

$$p_* \cdot v = 0 \text{ for all } v \in V \text{ and } p_* \cdot c > 0 \text{ for all } c \in C. \tag{4.5.3}$$

Note that if one identifies random variables with vectors in $\mathbb{R}^n$, for all random variables $Y$,

$$\mathbb{E}_*[Y] = \sum_{i=1}^n Y(\omega_i) \cdot \mathbb{P}_*[\omega_i] = \sum_{i=1}^n Y(\omega_i) \cdot p_i = Y \cdot p_*.$$

Since the random variable $\Phi_T$ is not replicable, it follows that $\Phi_T \notin V'$ so we can write

$$\Phi_T = \Phi_{TV} + v,$$

where $\Phi_{TV} \in V'$ and $v = (v_1, \dots, v_n) \neq 0$ is orthogonal to $V'$. We can assume at this point that $v \neq \lambda p_*$ for all $\lambda \in \mathbb{R}$, otherwise replace $p_*$ by $p'_*$ in the proof of Theorem 4.3.4. We now use the same argument as in the proof of Theorem 4.3.4 together with the vector $v$ to construct another measure $\mathbb{P}'_*$, with $\mathbb{E}_*[\Phi_T] \neq \mathbb{E}'_*[\Phi_T]$. Explicitly, define

$$p'_* = p_* + \lambda v,$$

where $\lambda \neq 0$ is so small that $p'_* \cdot c > 0$ for all $c \in C$. Clearly, $p'_*$ is also orthogonal to $V$. Therefore $p'_* = (p'_1, \dots, p'_n)$ induces a *risk-neutral measure* after a possible rescaling. We now claim that no rescaling is necessary, that is we have to show that

$$\begin{aligned} 1 &\stackrel{!}{=} p'_1 + \dots + p'_n = p_1 + \dots + p_n + \lambda v_1 + \lambda v_2 + \dots + \lambda v_n \\ &= 1 + v \cdot \lambda e^* \text{ with } e^* = (1, \dots, 1). \end{aligned}$$

Now $e^* \in V'$ since the trading strategy corresponding to $e^*$ is to invest some money into bonds at time $t = 0$ and to hold these bonds for the rest of the time. Since $v$ is orthogonal to $V'$, it follows that $v \cdot e^* = 0$. Therefore $p'_*$ does not need rescaling to induce a probability measure. Then

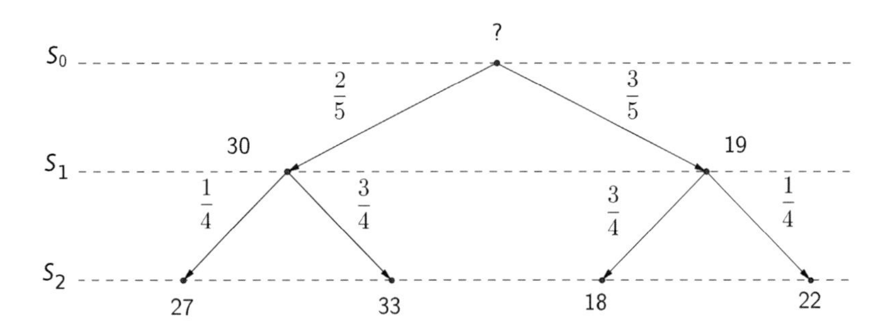


**Figure 4.4** Probability tree for Exercise 4.1.

$$\begin{aligned}\mathbb{E}'_*[\Phi_T] &= \Phi_T \cdot p'_* = \Phi_T \cdot (p_* + \lambda v) = \Phi_T \cdot p_* + \lambda\, \Phi_T \cdot \lambda v = \mathbb{E}_*\,[\Phi_T] + \Phi_T \cdot \lambda v \\ &= \mathbb{E}_*\,[\Phi_T] + \lambda(\Phi_{TV} + v) \cdot v = \mathbb{E}_*\,[\Phi_T] + \lambda v \cdot v \neq \mathbb{E}_*\,[\Phi_T]\end{aligned}$$

since $v \neq 0$ and $\lambda \neq 0$. This completes the proof. □

We get an immediate corollary.

**Corollary 4.5.4.** Consider an arbitrage-free finite market model and let $\Phi_T$ be an $\mathcal{F}_T$-measurable random variable. The random variable $\Phi_T$ is replicable if and only if $\mathbb{E}_*\,[\Phi_T]$ has the same value for all *risk-neutral measures* $\mathbb{P}_*$.

Let us summarise the observations in this section. A non-replicable contingent claim $\Phi_T$ has more than one arbitrage-free price. One way to determine these prices is to argue as in Example 4.5.2: first determine all *risk-neutral measures* and then evaluate the expression in (4.5.1). Another way is to extend the market as in Section 4.4 with the contingent claim $\Phi_T$ with prices $P_t$ at time $t$ and then to determine all $P_t$ such that a risk-neutral measure exists in this extended market.

## 4.6 Exercises

**Exercise 4.1.** *Consider an asset, and denote by $S_0$ the price of this asset today, by $S_1$ the price in one year and by $S_2$ the price in two years. Suppose that the price in the next two years follows the probability tree in Figure 4.4. Suppose further that there is a bond $B_t$ with interest rate $r = 0$.*

(*a*) *Do the prices in this market follow the binomial model?*
(*b*) *Determine $\mathbb{E}\,[S_2]$ and $\mathbb{E}\,[S_2|S_1]$.*
(*c*) *Show that the model is not arbitrage-free if $S_0 = 12$.*
(*d*) *Show that the model is arbitrage-free if $S_0 = 25$ and determine the* risk-neutral probability measure.
(*e*) *Let $C$ be a call option written on the asset $S_t$ with maturity 2 years and strike price $K = 25$. Determine the arbitrage-free initial price $P$ of the call option, under the assumption $S_0 = 25$.*

**Exercise 4.2.** *Let $T > 1$ and $\mathcal{S} = \{1, 2, 3, \dots, 10\}$. Let $(Y_t)_{t=0}^T$ be a sequence of discrete random variables taking values only in $\mathcal{S}$. Consider the filtration $\mathbb{F} = (\mathcal{F}_t)_{t=0}^T$ with $\mathcal{F}_t = \sigma(Y_0, \dots Y_t)$.*

*Suppose that for each $t \in \{0, \dots, T-1\}$ there exists a function $f_t : \mathcal{S} \to \mathcal{S}$ such that $Y_{t+1} = f_t(Y_t)$. Find a necessary and sufficient condition on the sequence $(f_t)_{t=0}^{T-1}$ so that $Y_t$ is a martingale. (You can assume that $\mathbb{P}[Y_0 = n] > 0$ for all $n \in \mathcal{S}$.)*

**Exercise 4.3.** *Consider a one-period financial market model on $\Omega = \{\omega_1, \omega_2, \omega_3\}$ consisting of a bond $B_t$ with interest rate $r = 0$ and $B_0 = 1$ and two risky assets $S_t^1$, $S_t^2$ with*

| Time | $t = 0$ | $t = 1$ |
|---|---|---|
| Risky asset $S_t^1$ | $S_0^1 = 9$ | $S_1^1 = \begin{cases} 13, & \text{if } \omega_1 \text{ occurs,} \\ 6, & \text{if } \omega_2 \text{ occurs,} \\ 9, & \text{if } \omega_3 \text{ occurs,} \end{cases}$ |
| Risky asset $S_t^2$ | $S_0^2 = 12$ | $S_1^2 = \begin{cases} 16, & \text{if } \omega_1 \text{ occurs,} \\ 10, & \text{if } \omega_2 \text{ occurs,} \\ 12, & \text{if } \omega_3 \text{ occurs.} \end{cases}$ |

(*a*) *Determine a risk-neutral measure $\mathbb{P}_*$ with $\mathbb{P}_*[\omega_1] = 1/3$ and another risk-neutral measure $\widetilde{\mathbb{P}}_*$ with $\widetilde{\mathbb{P}}_*[\omega_1] = 1/4$.*
(*b*) *Show that this model is arbitrage-free, but not complete.*
(*c*) *Consider a European call option $C$ with strike price $K = 6$ and maturity $T = 1$ written on $S_t^1$.*
  (i) *Show that $C$ is replicable and determine a replicating strategy.*
  (ii) *Is this replicating strategy unique? Justify your answer.*
  (iii) *Show that $\mathbb{E}_*[C] = \widetilde{\mathbb{E}}_*[C]$.*
  (iv) *Determine all arbitrage-free prices $Q$ of the call option $C$.*
(*d*) *Consider a European put option $P$ with strike price $K = 6$ and maturity $T = 1$ written on $S_t^1$.*
  (i) *Compute $\mathbb{E}_*[P]$ and $\widetilde{\mathbb{E}}_*[P]$.*
  (ii) *Show that $P$ is not replicable.*
  (iii) *Determine all arbitrage-free prices $Q$ of the put option $P$.*
  Hint: Apply the first fundamental theorem of asset pricing to the market consisting of the bond $B_t$ and the three risky assets $S_t^1$, $S_t^2$ and $P$.

**Exercise 4.4** (Betting on horse racing). *In this question we consider betting on horse racing. If a bookmaker offers you bet of $a$ to $b$ on a horse this means:*

- *If you bet £$b$ on that horse and it wins then you get your invested £$b$ back plus a gain of £$a$.*
- *If the horse loses then the bookmaker will keep your money.*

*Suppose that the horses Alydar and Man o' War are starting in the same race today and a bookmaker offers the bets 6 to 2 on Alydar and 10 to 2 on Man o' War.*

(*a*) *Suppose that only Alydar or Man o' War can win the race. Use the first fundamental theorem of asset pricing to show that there is an arbitrage opportunity.*

(*b*) *Construct an arbitrage opportunity.*
Hint: Assume you have £100 and bet £$x$ on Alydar and £$(100 - x)$ on Man o' War.

(*c*) *Suppose now that it is possible that neither Alydar nor Man o' War win the race (and we always bet on Alydar or Man o' War). Show that there is no arbitrage opportunity.*

Hint: For (a) and (c) consider a market consisting of a bond and two stocks. The bond corresponds to your wallet and the two stocks to the value of the horse bets. Ignore here that stocks cannot normally be 0.

# 5 Discrete Black–Scholes Model

The main aim in this chapter is to generalise the binomial model so that $S_t$ better reflects actual stock prices in real markets. The model considered in this section still has finitely many time steps, but we replace the discrete random variables $\xi_t$ in the definition of the stock price $S_t$ in the binomial model by suitably chosen continuous random variables. We will study the pricing of European *contingent claims* within this model through determining the existence of a *risk-neutral probability measure* and replicating strategies. The results and computations in this chapter serve as motivation and preparation for the Black–Scholes model.

In this chapter, we require some familiarity with the normal distribution. For convenience, we have summarised key properties of the normal distribution in Section A.1.2.

## 5.1 Heuristic Considerations on the Stock Price

Suppose we would like to model the prices in the time interval $[0, T]$. To avoid technical difficulties, we would like to have (as in the binomial model) only finitely many trading times. Therefore choose a large $n \in \mathbb{N}$ and assume that we can only trade at the times $t \in \{0, \frac{T}{n}, \frac{2T}{n}, \ldots, T\}$. We wish to keep $n$ fixed, but ideally choose $n$ large enough so that the steps, $h := \frac{T}{n}$, are small. The steps $h$ should be sufficiently small that we can reasonably trade at 'any time'. In order to simplify the notation, define

$$[0, T; n] := \left\{0, \frac{T}{n}, \frac{2T}{n}, \ldots, T\right\}. \tag{5.1.1}$$

The question at this point is: *What are realistic assumptions to make on the stock price $S_t$?* First observe that it is more natural to model the proportional change of the stock price, rather than the absolute change. Indeed, if the stock price increases by £1 then this is a substantial change if the stock price was £2, but only a minor change if the stock price was £375. Therefore, it is more natural to consider the distribution of $\frac{S_{t+h}}{S_t}$ instead of $S_t$. The following assumptions turn out to be reasonable:

- $S_t$ is always larger than 0 and can take all values $\mathbb{R}_+$.
- The relative change in stock price is caused by many forces.
- These forces, considered at different times, are independent of each other.

The forces mentioned are investors trading in the market. A starting point is therefore that the price $S_t$ of the *stock* at time $t$ should have the form

$$S_t = S_{t-h} \cdot \xi_k = S_0 \prod_{j=1}^{k} \xi_i \quad \text{for } t = \frac{kT}{n}, \tag{5.1.2}$$

where $S_0 > 0$ is the price of the stock at $t = 0$ and $(\xi_j)_{j=1}^n$ is an i.i.d. sequence of non-negative real-valued random variables, corresponding to the forces mentioned above. Taking log of (5.1.2), gives

$$\log(S_t) = \log(S_0) + \sum_{j=1}^{k} \log(\xi_j). \tag{5.1.3}$$

By the central limit theorem (see Theorem A.1.5), $\log(S_T)$ is approximately normal when $n$ is large. This approximation becomes exact if we set

$$\xi_j = \exp(I_j),$$

where $(I_j)_{j=1}^n$ is an i.i.d. sequence of normally distributed random variables. This assumption is reasonable as each price change $\xi_j$ is the result of many people trading in the market. From (5.1.3),

$$\mathbb{E}\left[\log(S_T)\right] = \mathbb{E}\left[\log(S_0)\right] + n\mathbb{E}\left[I_1\right],$$
$$\mathrm{Var}(\log(S_T)) = n\,\mathrm{Var}(I_1),$$

so in order for these to be finite, the mean and variance of $I_j$ should both be proportional to $h$. Therefore, we choose to set

$$I_j := \left(\mu - \frac{\sigma^2}{2}\right) h + \sigma N_j, \tag{5.1.4}$$

with $\mu, \sigma \in \mathbb{R}$, $\sigma > 0$ and $(N_j)_{j=1}^n$ an i.i.d. sequence of normally distributed random variables with expectation 0 and variance $h$. Note that, with this choice,

$$\mathbb{E}\left[\xi_j\right] = e^{\mu h}.$$

Combining everything, we obtain

$$S_t = S_0\, e^{\left(\mu - \frac{\sigma^2}{2}\right)t} \cdot \prod_{j=1}^{k} \exp(\sigma N_j) \quad \text{for } t = \frac{kT}{n}. \tag{5.1.5}$$

The constant term in (5.1.4) is chosen so that

$$\mathbb{E}\left[S_t\right] = S_0 e^{\mu t} \quad \text{for all } t \in [0, T; n], \tag{5.1.6}$$

so $\mu$ can be interpreted as the average growth rate of the stock price. The parameter $\mu$ is called the *drift rate* and $\sigma$ the *volatility* of $S_t$. Note that in real financial markets $\mu$ and $\sigma$ depend also on $t$, but we assume for simplicity that they are constant. An alternative formulation for (5.1.5) is

$$S_t = S_0\, e^{\left(\mu - \frac{\sigma^2}{2}\right)t} \cdot \exp(\sigma W_t^{(n)}) \quad \text{with} \quad W_t^{(n)} = \sum_{j \le nt} N_j. \tag{5.1.7}$$

This reformulation is more useful than (5.1.5). In particular, we use it in Chapter 7 as motivation for defining Brownian motion, which leads to the Black–Scholes model in Chapter 9.

## 5.2 Model Specification

In this section we specify our model for the time interval $[0, T]$. Assume that we can trade at all $t \in [0, T; n]$, with $[0, T; n]$ as in (5.1.1) and $n \in \mathbb{N}$. As before, assume that the market is liquid. The primary market consists of two assets, a *bond* $B_t$ and a *stock* $S_t$, where

- The *bond* $B_t$ is assumed to have a constant compound interest rate $r \in \mathbb{R}$, that is
$$B_t = B_0\, e^{rt}, \tag{5.2.1}$$
where $B_0$ is the price of the bond at time $t = 0$.
- The *stock* $S_t$ has the form
$$S_t = S_0 \exp\left(\left(\mu - \frac{\sigma^2}{2}\right) t + \sigma W_t^{(n)}\right) \qquad \text{for } t \in [0, T; n], \tag{5.2.2}$$
with $W_t^{(n)}$ as in (5.1.7), $\mu \in \mathbb{R}$ and $\sigma > 0$.
- To model the time, we use the filtration $\mathbb{F} = (\mathcal{F}_s)_{s \in [0,T;n]}$ with
$$\mathcal{F}_t = \sigma\left(S_s, s \leq t \text{ and } s \in [0, T; n]\right) \qquad \text{for } t \in [0, T; n]. \tag{5.2.3}$$

The *drift rate* $\mu$ can be equal to the interest $r$, but does not have to be. Indeed, if we consider the stock of an innovative startup company on the road to success, then typically $\mu > r$. On the other hand, the stock of a company which has serious problems would typically have $\mu < r$.

The process $\{W_t^{(n)}, t \in [0, T; n]\}$, defined as in (5.1.7), is a discretisation of Brownian motion, which will be encountered in Chapter 7. It has the property that, for any $t_1 < \cdots < t_k \in [0, T; n]$, the distribution of the random vector $(W_{t_1}^{(n)}, \ldots, W_{t_k}^{(n)})$ is multivariate Gaussian.

**Definition 5.2.1** (Multivariate Gaussian distribution). A random vector $X = (X_1, \ldots, X_n)$ on probability space $(\Omega, \mathcal{F}, \mathbb{P})$ is said to have an $n$-dimensional *multivariate Gaussian* or *multivariate normal* distribution if
$$a_1 X_1 + \cdots + a_n X_n$$
has a normal distribution for any choice of $a_1, a_2, \ldots, a_n \in \mathbb{R}$. We denote by $\mu \in \mathbb{R}^n$ the vector of means with $i$th element $\mu_i = \mathbb{E}[X_i]$ and by $\Sigma$ the $n \times n$ covariance matrix with $(i,j)$th element $\Sigma_{ij} = \mathrm{Cov}(X_i, X_j)$.

We often denote the distribution of $X$ by $X \sim \mathrm{MVN}(\mu, \Sigma)$.

A multivariate Gaussian, $X \sim \mathrm{MVN}(\mu, \Sigma)$, has the following properties.

(i) For each $i$, $X_i \sim \mathcal{N}(\mu_i, \Sigma_{ii})$.
(ii) The random variables $X_1, \ldots, X_n$ are independent if and only if $\Sigma$ is a diagonal matrix.
(iii) If $A$ is any real-valued $m \times n$ matrix, then $AX^t \sim MVN(A\mu^t, A\Sigma A^t)$, where $(.)^t$ indicates the transpose.

(iv) Provided $\det \Sigma \neq 0$, the probability density function of $X$ is given by

$$f_X(x) = \frac{1}{(2\pi)^{n/2}\sqrt{\det \Sigma}} \exp\left\{-\frac{1}{2}(x-\mu)\Sigma^{-1}(x-\mu)'\right\}, \quad x \in \mathbb{R}^n.$$

The next lemma summarises some important properties of $W_t^{(n)}$, which we will need later.

**Lemma 5.2.2.** Let $W_t^{(n)}$ be as in (5.1.7). For any $0 = t_0 < t_1 < \cdots < t_k \in [0, T; n]$, the distribution of the random vector $(W_{t_1}^{(n)}, \ldots, W_{t_k}^{(n)})$ is multivariate Gaussian, determined by the following properties.

(a) $W_0^{(n)} = 0$.
(b) $W_{t_i}^{(n)} - W_{t_{i-1}}^{(n)} \sim \mathcal{N}(0, t_i - t_{i-1})$ for all $i = 1, \ldots, k$.
(c) $W_{t_i}^{(n)} - W_{t_{i-1}}^{(n)}$ and $W_{t_j}^{(n)} - W_{t_{j-1}}^{(n)}$ are independent for all $i \neq j$.

We omit the proof of Lemma 5.2.2 as it follows immediately from elementary properties of independent normal distributions. Furthermore, if a process $W_t^{(n)}$ with $t \in [0, T; n]$ satisfies the properties in Lemma 5.2.2, then $W_t^{(n)}$ can be written as in (5.1.7).

## 5.3 Trading Strategies and Discounted Asset Prices

Trading strategies in this model look very similar to trading strategies in the binomial model, but there are two small differences. The first is that our time step is now $h = T/n$ instead of 1, and we thus have to adjust the notation a little bit. The second is that the sample space $\Omega$ (forming the basis of this model) is not finite anymore. This implies that expressions like $\mathbb{E}[V_t(\varphi)]$ are not automatically well-defined. For this reason, we have to impose some additional assumptions on our trading strategies.

As already mentioned, trading is only allowed at times $t \in [0, T; n]$. Therefore, the number of bonds and stocks we hold is constant in each of the intervals $[t, t+h)$ with $t \in [0, T; n]$. For $t \in [0, T; n]$, with $h = T/n$, denote

$$\alpha_t := \text{the number of shares of stock we hold in the time period } [t, t+h) \tag{5.3.1}$$

$$\beta_t := \text{the number of bonds we hold during that same period.} \tag{5.3.2}$$

As before, we allow $\beta_t \in \mathbb{R}$ and $\alpha_t \in \mathbb{R}$ for all $t \in [0, T; n]$. Set $\varphi_t = (\beta_t, \alpha_t)$. We assume that we do not trade at time $T$, so $\varphi_T = \varphi_{T-h}$. Furthermore, $\beta_t$ and all $\alpha_t$ are determined at time $t$, so $\varphi_t$ has to be $\mathcal{F}_t$-measurable. Additionally, we would like to be able to work with expressions like $\mathbb{E}[V_t(\varphi)]$. As our model only runs over a finite time interval $[0, T]$, a natural assumption is to request that the variances of all expressions occurring in $\varphi$ are finite. This leads to the following definition.

**Definition 5.3.1.** A *trading strategy* in the *primary market* is a process $\varphi = (\varphi_t)_{t\in[0,T;n]}$ with $\varphi_t = (\beta_t, \alpha_t)$ such that

- $\varphi$ is *measurable* with respect to the filtration $\mathbb{F}$ in (5.2.3) and
- $\mathbb{E}\left[(\alpha_t)^2\right] < \infty$ and $\mathbb{E}\left[(\beta_t)^2\right] < \infty$ for all $t \in [0, T; n]$.

If a trading strategy $\varphi$ is given, then at time $t$ the portfolio has the following value:

$$V_t(\varphi) = \beta_t B_t + \alpha_t S_t \quad \text{for } t \in [0, T; n]. \tag{5.3.3}$$

As in previous models, we are interested in *self-financing trading strategies* only. Using an argument similar to that on page 68 leads to the following definition.

**Definition 5.3.2** (Self-financing condition). A *self-financing trading strategy* is a *trading strategy* $\varphi$ such that

$$\beta_t B_t + \alpha_t S_t = \beta_{t-h} B_t + \alpha_{t-h} S_t \quad \text{for all } t \in [0, T; n] \setminus \{0, T\}.$$

We can now define an arbitrage opportunity in the primary market.

**Definition 5.3.3** (Arbitrage opportunity). An *arbitrage opportunity* (in the primary market) is a self-financing trading strategy $\varphi = (\varphi_t))_{t\in[0,T;n]}$ with $\varphi_t$ satisfying the following conditions.

- No initial cost: $V_0(\varphi) = 0$.
- Always non-negative final value: $V_T(\varphi) \geq 0$.
- The possibility of a positive final gain:

$$\mathbb{E}\left[V_T(\varphi)\right] > 0.$$

As in the binomial model, we have the equivalence

$$\mathbb{E}\left[V_T(\varphi)\right] > 0 \iff \mathbb{P}\left[V_T(\varphi) > 0\right] > 0 \tag{5.3.4}$$

since $V_T(\varphi) \geq 0$. We also need to introduce the discounted asset prices. These are defined as

$$S_t^* := \frac{S_t}{e^{rt}} \quad \text{and} \quad V_t^*(\varphi) := \frac{V_t(\varphi)}{e^{rt}}. \tag{5.3.5}$$

Note that the normalisation in (5.3.5) is different from that in the binomial and finite market models, see (3.3.34) and (4.1.6). The origin of this difference is that the bond has a different form in this model. The normalisation in (5.3.5) is chosen so that we can use a similar argument to that in the previous models. Specifically, in the binomial model we used that $V_t^*(\varphi)$ is a martingale if $S_t^*$ is a martingale, see Lemma 3.3.17. This result was then used to show that there is no arbitrage opportunity in the model. Consider the trading strategy $\varphi$ with $\beta_t = 1$ and $\alpha_t = 0$ for all $t$. This trading strategy is self-financing with $V_t(\varphi) = B_t$. Furthermore, $\mathbb{E}[B_{t+h}|\mathcal{F}_t] = B_{t+h}$ as all $B_t$ are constants. It immediately follows that the normalisation in (5.3.5) is required to turn $V_t(\varphi) = B_t$ into a martingale.

## 5.4 Risk-Neutral Measure

The topic of this section is risk-neutral measures in this model. We use (almost) the same definition of a risk-neutral measure as in the finite market model in Definition 3.3.4.

**Definition 5.4.1** (Risk-neutral probability measures). A *risk-neutral probability measure* (or *equivalent martingale measure*) is a probability measure $\mathbb{P}_*$ such that

- $\mathbb{P}_* \approx \mathbb{P}$, that is both measures have the same null sets.
- The discounted stock price $S_t^*$ in (5.3.5) is a $\mathbb{P}_*$-martingale. In formulas this means

$$\mathbb{E}_*\left[S_{t+h}^*\middle|\mathcal{F}_t\right] = S_t^* \quad \text{for all } t \in [0, T; n] \setminus \{T\}. \tag{5.4.1}$$

Note that $\mathcal{F}_t$ is generated in this model by continuous random variables, see (5.2.3). Therefore, we cannot use Definition 2.2.14 for the expression $\mathbb{E}\left[S_{t+h}^*\middle|\mathcal{F}_t\right]$. Instead, we have to use Definition 2.3.28. In Section 6.3 we will look at general conditional expectations and the properties needed in this book. For now, the reader can just assume that $\mathbb{E}\left[S_{t+h}^*\middle|\mathcal{F}_t\right]$ is a well-defined expression and that general conditional expectations have similar properties to those in the discrete case.

Our first aim is to show that there is a risk-neutral measure. We begin by taking a look at the underlying probability space of this model. In the construction of $S_t$ and $W_t^{(n)}$ in (5.1.7), we use a sequence of i.i.d. random variables $(N_i)_{i=1}^n$ with $N_i \sim \mathcal{N}(0, h)$. As the model only depends on the distribution of the sequence $(N_i)_{i=1}^n$ and not on the sample space itself, we can choose a sample space which suits us best. We will assume that $\Omega = \mathbb{R}^n$ and the probability measure $\mathbb{P}$ is given by

$$\mathbb{P}[A] = \frac{1}{(2\pi h)^{n/2}} \int_A \exp\left(-\frac{1}{2h}\sum_{j=1}^n x_j^2\right) dx_1 \dots dx_n \quad \text{for } A \in \mathcal{B}^n.^1 \tag{5.4.2}$$

Further, we assume that the random variables $N_j : \mathbb{R}^n \to \mathbb{R}$ are defined as

$$N_j(x_1, \dots, x_n) = x_j \quad \text{for } 1 \le j \le n \text{ and } (x_1, \dots, x_n) \in \mathbb{R}^n. \tag{5.4.3}$$

[1] Here $\mathcal{B}^n$ denotes the Borel $\sigma$-algebra. The definition of $\mathcal{B}^n$ and the reason why we have to restrict to events $A \in \mathcal{B}^n$ is explained in Section 6.1.

We now look for a risk-neutral measure. A naive approach is to hope that $S_t^*$ is already a martingale with respect to the measure $\mathbb{P}$. To check whether this is true, we have to compute $\mathbb{E}\left[S_{t+h}^* \middle| \mathcal{F}_t\right]$. Recall that if a normal-distributed random variable $Z \sim \mathcal{N}(m, \lambda)$ is given then, for all $s \in \mathbb{R}$,

$$\mathbb{E}\left[e^{sZ}\right] = e^{ms+\frac{1}{2}\lambda s^2}. \tag{5.4.4}$$

By (5.3.5), (5.2.2) and Lemma 5.2.2,

$$\begin{aligned}
\mathbb{E}\left[S_{t+h}^* \middle| \mathcal{F}_t\right] &= \mathbb{E}\left[e^{-r(t+h)} S_0\, e^{\left(\mu-\frac{\sigma^2}{2}\right)(t+h)+\sigma W_{t+h}} \middle| \mathcal{F}_t\right] \\
&= e^{\left(\mu-\frac{\sigma^2}{2}-r\right)h} \mathbb{E}\left[e^{-rt} S_0\, e^{\left(\mu-\frac{\sigma^2}{2}\right)t+\sigma W_t} e^{\sigma(W_{t+h}-W_t)} \middle| \mathcal{F}_t\right] \\
&= e^{\left(\mu-\frac{\sigma^2}{2}-r\right)h} \mathbb{E}\left[S_t^*\, e^{\sigma(W_{t+h}-W_t)} \middle| \mathcal{F}_t\right] \\
&= S_t^* e^{\left(\mu-\frac{\sigma^2}{2}-r\right)h} \mathbb{E}\left[e^{\sigma(W_{t+h}-W_t)}\right] = S_t^* e^{h(\mu-r)}.
\end{aligned} \tag{5.4.5}$$

In this computation we used that $W_{t+h} - W_t$ is independent of all random variables which generate $\mathcal{F}_t$. Therefore, $S_t^*$ is a martingale under $\mathbb{P}$ if and only if $\mu = r$. However, it is not necessarily the case in this model that $\mu = r$ and so $S_t^*$ is typically not a martingale under $\mathbb{P}$. The trick to obtaining a risk-neutral measure is to shift the sequence $(N_i)_{i=1}^n$ and to compare the induced probability measure of the shifted sequence with the original one. Explicitly, let $\nu \in \mathbb{R}$ be given and define, for $A \in \mathcal{B}^n$,

$$\mathbb{P}_\nu[A] := \mathbb{P}[A - \nu h] \quad \text{with } A - \nu h := \{(x_1 - \nu h, \ldots, x_n - \nu h);\ (x_1, \ldots, x_n) \in A\}. \tag{5.4.6}$$

We now determine the distribution of $(N_i)_{i=1}^n$ under $\mathbb{P}_\nu$. The definition of $N_j$ in (5.4.3) and the definition of $\mathbb{P}_\nu$ give

$$\mathbb{P}_\nu[(N_1, \ldots, N_n) \in A] = \mathbb{P}_\nu[A] = \mathbb{P}[A - \nu h] = \mathbb{P}[(N_1, \ldots, N_n) \in A - \nu h] \tag{5.4.7}$$

for all $A \in \mathcal{B}^n$. We immediately get that, under $\mathbb{P}_\nu$, $(N_j)_{j=1}^n$ is still an i.i.d. normally distributed sequence, but now with $N_j \sim \mathcal{N}(\nu h, h)$. Inserting the definition of $W_t^{(n)}$ in (5.1.7), we get that $W_t^{(n)} \sim \mathcal{N}(\nu t, t)$. Repeat the computation in (5.4.5), but this time with respect to $\mathbb{P}_\nu$. Writing $\mathbb{E}_\nu$ for the expectation with respect to $\mathbb{P}_\nu$, we get as in (5.4.5),

$$\mathbb{E}_\nu\left[S_{t+h}^* \middle| \mathcal{F}_t\right] = S_t^* e^{\left(\mu-\frac{\sigma^2}{2}-r\right)h} \mathbb{E}_\nu\left[e^{\sigma(W_{t+h}-W_t)}\right] = S_t^* e^{h(\mu+\sigma\nu-r)}. \tag{5.4.8}$$

We immediately get the following result.

**Theorem 5.4.2.** • *The measure $\mathbb{P}_\nu$ in (5.4.6) is a risk-neutral measure if and only if $\nu = \frac{r-\mu}{\sigma}$.*

• *The primary market in this model is arbitrage-free for all $\mu \in \mathbb{R}$ and $\sigma > 0$.*

*Proof.* We begin with the first point. By (5.4.8), condition (5.4.1) is fulfilled if and only if $\nu = \frac{r-\mu}{\sigma}$. It remains to show that $\mathbb{P}_\nu \approx \mathbb{P}$. Note that

$$\mathbb{P}[A] = \int_A f(x_1, \ldots, x_n)\, dx_1 \ldots dx_n \text{ and } \mathbb{P}_\nu[A] = \int_A g(x_1, \ldots, x_n)\, dx_1 \ldots dx_n, \qquad (5.4.9)$$

with $f$ and $g$ strictly positive functions. It is straightforward to see that $\mathbb{P}_\nu$ and $\mathbb{P}$ are both equivalent to the Lebesgue measure, and therefore equivalent to each other. This completes the proof of the first point.

The proof of the second point is (almost) the same as in the binomial and finite market models, see Section 3.3.2 or the proof of Theorem 4.2.6. The only point that we have to justify is $\mathbb{E}_\nu[|V_T(\varphi)|] < \infty$, where $\varphi$ is an arbitrage opportunity. For this we use Lemma 5.4.3 below to given an explicit expression for $\mathbb{E}_\nu[|V_T(\varphi)|]$. This then gives

$$\mathbb{E}_\nu[|V_T(\varphi)|] = \mathbb{E}\left[e^{\nu W_T^{(n)} - \frac{\nu^2}{2}T} |V_T(\varphi)|\right]. \qquad (5.4.10)$$

The Cauchy–Schwartz inequality together with $\mathbb{E}\left[(\alpha_T)^2\right] < \infty$, $\mathbb{E}\left[(\beta_T)^2\right] < \infty$ and (5.4.5) then implies that the expression in (5.4.10) is finite. The remaining arguments of this proof are the same as in the binomial and finite market models, so we omit them. □

We will see that the construction above of a risk-neutral measure in principle also works in the continuous Black–Scholes model introduced in Chapter 9. However, in the argument above we required an explicit choice of the sample space. In this model we are able to choose $\Omega = \mathbb{R}^n$. Unfortunately, in the continuous Black–Scholes model one cannot choose as nice a sample space as $\mathbb{R}^n$. We therefore give an alternative characterisation of the risk-neutral measure defined above, which does not depend on the choice of an explicit sample space.

**Lemma 5.4.3.** For all $\nu \in \mathbb{R}$ and all events $A \in \mathcal{F}_T$ with $\mathcal{F}_T$ as in (5.2.3),

$$\mathbb{P}_\nu[A] = \mathbb{E}\left[e^{\nu W_T^{(n)} - \frac{\nu^2}{2}T} \mathbb{1}_A(\omega)\right], \qquad (5.4.11)$$

with $W_t^{(n)}$ as in (5.1.7) and $\mathbb{1}_A(\omega)$ as in (2.2.6). Furthermore, for all random variables $X$ with $\mathbb{E}_\nu[|X|] < \infty$,

$$\mathbb{E}_\nu[X] = \mathbb{E}\left[e^{\nu W_T^{(n)} - \frac{\nu^2}{2}T} X\right]. \qquad (5.4.12)$$

*Proof.* By the definition of $\mathbb{P}_\nu$,

$$\mathbb{P}_\nu[A] = \frac{1}{(2\pi h)^{n/2}} \int_{A-\nu h} \exp\left(-\frac{1}{2h}\sum_{j=1}^n x_j^2\right) dx_1 \ldots dx_n.$$

Using the variable substitution $y_j = x_j - \nu h$ and that $h = 1/n$, we obtain

$$\begin{aligned}\mathbb{P}_\nu[A] &= \frac{1}{(2\pi h)^{n/2}} \int_A \exp\left(-\frac{1}{2h}\sum_{j=1}^n (y_j - \nu h)^2\right) dy_1 \dots dy_n \\ &= \frac{1}{(2\pi h)^{n/2}} \int_A e^{-\frac{n}{2h}(\nu h)^2 + \nu \sum_{j=1}^n y_j} \exp\left(-\frac{1}{2h}\sum_{j=1}^n y_j^2\right) dy_1 \dots dy_n \\ &= \mathbb{E}\left[e^{-\frac{\nu^2}{2} + \nu \sum_{j=1}^n N_j} \mathbb{1}_A(\omega)\right].\end{aligned}$$

This completes the proof of (5.4.11). The proof of (5.4.12) uses some ideas from later chapters so is deferred to Lemma 9.3.7. □

## 5.5 Black–Scholes Formula

In this section we use the results from Section 5.4 for the pricing of a given European contingent claim $\Phi_T$. In particular, we deduce the famous Black–Scholes formula for the pricing of European call and put options. The argument will be similar to the argument in Section 4.5.

As before, we need to extend the primary market and add the *contingent claim* $\Phi_T$. Therefore our market consists of the three assets $B_t$, $S_t$ and $\Phi_T$. We assume here that

- the dynamics of $S_t$ and $B_t$ are given by the model assumptions in Section 5.2,
- the *contingent claim* $\Phi_T$ has no influence on the stocks $S_t$ and the bond $B_t$,
- the *maturity* of the *contingent claim* is $T$ and
- the *contingent claim* $\Phi_T$ is $\mathcal{F}_T$-measurable.

We immediately get the following theorem.

**Theorem 5.5.1.** *Let $\mathbb{P}_*$ be a risk-neutral measure. Denote by $P_t$ the price of $\Phi_T$ at time $t$ for $t \in [0, T; n]$. This extended market is arbitrage-free if*

$$P_t = e^{-r(T-t)}\mathbb{E}_*\left[\Phi_T \middle| \mathcal{F}_t\right] \quad \textit{for all } t \in [0, T; n]. \tag{5.5.1}$$

The proof of this theorem is (almost) the same as the proof of Theorem 4.5.1 so we omit it. Applying this theorem to put and call options for the risk–neutral measure in Section 5.4, we immediately get the famous Black–Scholes formula.

**Theorem 5.5.2** (Black–Scholes formula). *Consider the primary market extended with a call and a put option. Assume that both options have strike price $K$ and maturity $T$. For $t \in [0, T; n]$ and $x \in \mathbb{R}$ set*

$$d_1(t,x) := \frac{\log(x/K) + (r + \sigma^2/2)(T-t)}{\sigma\sqrt{T-t}}, \tag{5.5.2}$$

$$d_2(t,x) := d_1(t,x) - \sigma\sqrt{T-t} \quad \textit{and} \tag{5.5.3}$$

$$C(t,x) := x\Phi(d_1(t,x)) - Ke^{-r(T-t)}\Phi(d_2(x,t)), \tag{5.5.4}$$

*where $\Phi$ is the cumulative distribution function of the standard $\mathcal{N}(0,1)$ random variable. This extended market is arbitrage-free if*

- *the price of the call option at time $t \in [0,T;n]$ is $C(t,S_t)$ and*
- *the price of the put option at time $t \in [0,T;n]$ is $C(t,S_t) - S_t + Ke^{-r(T-t)}$.*

*Proof.* We apply Theorem 5.5.1 with the risk-neutral measure in Section 5.4. First consider the call option. For this we have to compute for all $t \in [0,T;n]$,

$$e^{-r(T-t)}\mathbb{E}_\nu[\max(S_T - K, 0) \mid \mathcal{F}_t] \tag{5.5.5}$$

with $\nu = \frac{r-\mu}{\sigma}$. By (5.2.2),

$$\max(S_T - K, 0) = \max\left(S_t e^{\left(\mu - \frac{\sigma^2}{2}\right)(T-t) + \sigma(W_T^{(n)} - W_t^{(n)})} - K, 0\right).$$

Observe that $W_T^{(n)} - W_t^{(n)}$ is normal $\mathcal{N}(\nu(T-t), T-t)$-distributed under $\mathbb{P}_\nu$. Furthermore, by Lemma 5.2.2, $W_T^{(n)} - W_t^{(n)}$ is independent of $W_s^{(n)}$ for all $s \leq t$. Therefore $W_T^{(n)} - W_t^{(n)}$ is independent of $\mathcal{F}_t$. Using these observations,

$$\mathbb{E}_\nu[\max(S_T - K, 0) \mid \mathcal{F}_t] = \frac{1}{\sqrt{2\pi(T-t)}} \int_{-\infty}^{\infty} \max\left(S_t e^{\left(\mu - \frac{\sigma^2}{2}\right)(T-t) + \sigma x} - K, 0\right) e^{\frac{-\left(x - \nu(T-t)\right)^2}{2(T-t)}} \, dx.$$

Applying the variable substitution $y = \frac{x - \nu(T-t)}{\sqrt{2(T-t)}}$ and inserting $\nu = \frac{r-\mu}{\sigma}$, we get

$$\frac{1}{\sqrt{2\pi}} \int_{-\infty}^{\infty} \max\left(S_t e^{\left(r - \frac{\sigma^2}{2}\right)(T-t) + \sigma y\sqrt{2(T-t)}} - K, 0\right) e^{-y^2} \, dy. \tag{5.5.6}$$

The integrand in (5.5.6) is strictly positive for all $y$ with

$$y > \frac{\log(K/S_t) - \left(r - \frac{\sigma^2}{2}\right)(T-t)}{\sigma\sqrt{2(T-t)}}. \tag{5.5.7}$$

Comparing this expression with $d_1(x,t)$ in (5.5.2), we can write (5.5.6) as

$$\begin{aligned} &= \frac{1}{\sqrt{2\pi}} \int_{-d(t,S_t)}^{\infty} \left(S_t e^{\left(r - \frac{\sigma^2}{2}\right)(T-t) + \sigma y\sqrt{2(T-t)}} - K\right) e^{-y^2} \, dy \\ &= S_t e^{\left(r - \frac{\sigma^2}{2}\right)(T-t)} \frac{1}{\sqrt{2\pi}} \int_{-d(t,S_t)}^{\infty} e^{\sigma y\sqrt{2(T-t)} - y^2} \, dy - K \int_{-d(t,S_t)}^{\infty} e^{-y^2} \, dy. \end{aligned}$$

Completing the square in the first integrand and inserting the definition of $\Phi$, we get

$$\begin{aligned}\mathbb{E}_\nu[\max(S_T - K, 0) \mid \mathcal{F}_t] &= S_t e^{r(T-t)} \frac{1}{\sqrt{2\pi}} \int_{-d(t,S_t)+\sigma\sqrt{T-t}}^{\infty} e^{-y^2}\, dy - K\Phi\big(d(t,S_t)\big) \\ &= S_t e^{r(T-t)} \Phi\big(d(t,S_t) - \sigma\sqrt{T-t}\big) e^{-y^2}\, dy - K\Phi\big(d(t,S_t)\big).\end{aligned}$$

Combining this with (5.5.5) completes the computation for the call option. To compute the price of the put option, use that

$$S_T - K = \max(S_T - K, 0) - \max(K - S_T, 0).$$

Since $S_t$ is a martingale with respect to $\mathbb{P}_\nu$,

$$\mathbb{E}_\nu[\max(K - S_T, 0) \mid \mathcal{F}_t] = \mathbb{E}_\nu[\max(S_T - K, 0) \mid \mathcal{F}_t] + K - S_t.$$

Inserting the computation for the call option completes the proof. □

## 5.6 Replicating Strategies

We have seen in the finite market model in Section 4.5 that the arbitrage-free price of a contingent claim $\Phi_T$ is unique if and only $\Phi_T$ is replicable, see Corollary 4.5.4. Further, we have seen in Theorem 4.3.4 that all contingent claims in an arbitrage-free market are replicable if and only if there exists exactly one risk-neutral measure. We cannot immediately apply the results from Chapter 4 to the discrete Black–Scholes model as we have proven them only for models with finite sample spaces. However, it is natural to expect that these results can be extended to more general models, including the one in this section. We will show that the model in this section follows the expected pattern. Consider an example.

**Example 5.6.1.** *Let $n = 1$, $r = \mu = 0$ and $S_0 = B_0 = 1$.*

*(a) Let $\Phi_T := (S_T)^2$. Is $\Phi_T$ replicable? Since $n = 1$, we have only one time step and thus a trading strategy in this model has the form $\varphi = (\beta_T, \alpha_T)$ with $\alpha_T$, $\beta_T \in \mathbb{R}$. We are therefore looking for $\alpha_T$, $\beta_T \in \mathbb{R}$ with*

$$\Phi_T \stackrel{!}{=} V_T(\varphi) = \beta_T B_T + \alpha_T S_T.$$

*This equation has to hold for all possible outcomes. Since $S_T$ can take all values in $\mathbb{R}_+$ and $B_T = 1$, we are looking for $\alpha_T$, $\beta_T \in \mathbb{R}$ such that*

$$x^2 = \beta_T + \alpha_T x \qquad \text{for all } x > 0.$$

*This is not possible, so $\Phi_T$ is not replicable.*

*(b) Is the risk-neutral measure unique? The* stock *$S_T$ has the form*

$$S_T = \exp\left(\left(-\frac{\sigma^2}{2}\right) T + \sigma W_T\right),$$

*with $W_T \sim \mathcal{N}(0, T)$. We can assume that $\Omega = \mathbb{R}$, the probability measure $\mathbb{P}$ has the density $\frac{1}{\sqrt{2\pi}} e^{-x^2/2T}$ and $W_T : \mathbb{R} \to \mathbb{R}$ is defined as $W_T(x) = x$.*

*Let $b > 0$ be arbitrary and $v \in \mathbb{R}$. We now endow $\Omega$ with the probability measure $\mathbb{P}_{v,b}$ with the density $\frac{1}{\sqrt{2\pi b}} e^{-(x-Tv)^2/2Tb}$. Then $W_T \sim \mathcal{N}(Tv, Tb)$ under $\mathbb{P}_{v,b}$. Since we have only one time step and $r = 0$, we get that $\mathbb{P}_{v,b}$ is a risk-neutral measure if and only if*

$$\mathbb{E}_{v,b}[S_T] = S_0.$$

*Inserting the definition of $S_T$, by (5.4.4),*

$$\mathbb{E}_{v,b}[S_T] = e^{-\frac{\sigma^2 T}{2}} \mathbb{E}_{v,b}[e^{\sigma W_T}] = e^{-\frac{\sigma^2 T}{2} + \sigma v T + \sigma^2 T b/2} \overset{!}{=} 1.$$

*Therefore, $\mathbb{P}_{v,b}$ is a risk-neutral measure if and only if $v = \sigma \frac{1-b}{2}$. Since we need only $b > 0$ and $v \in \mathbb{R}$, there are infinitely many risk-neutral measures.*

We did the computations in Example 5.6.1 only for the case $n = 1$, but one can do the same computations for $n > 1$. This shows that the discrete Black–Scholes model is arbitrage-free, but not complete, for all $n$. A possible question at this point is: *Can we approximate a non-replicable contingent claim $\Phi_T$ with the help of a self-financing trading strategy $\varphi$?* In other words, is there a trading strategy $\varphi$ such that $\Phi_T$ and $V_T(\varphi)$ have a similar value? Also, does this approximation improve as $n \to \infty$?

We do not give a rigorous answer to this question here. However, we give a heuristic construction for such an approximation.

Suppose that $\Phi_T = f(S_T)$, with $f : \mathbb{R} \to \mathbb{R}$ a twice-differential function with $|f''| < M$ for some $M > 0$. Suppose further that $n$ is very large, so $h = T/n$ is very small. If we are looking for a replicating strategy $\varphi$ for $\Phi_T$, then the first step is to find $\mathcal{F}_{T-h}$-measurable $\alpha_{T-h}$ and $\beta_{T-h}$ such that

$$\Phi_T = V_T(\varphi) = \beta_{T-h} B_T + \alpha_{T-h} S_T.$$

Since $h$ is very small, one can expect that $S_T$ and $S_{T-h}$ are almost equal. Using the Taylor expansion of $f$, write

$$f(S_T) = f(S_{T-h}) + f'(S_{T-h})(S_T - S_{T-h}) + R_T$$

with $|R_T| \leq M(S_T - S_{T-h})^2$. Determine $\alpha_{T-h}$ and $\beta_{T-h}$ so that the first two terms in this Taylor expansion are cancelled. Explicitly, choose

$$\alpha_{T-h} = f'(S_{T-h}) \qquad \text{and} \qquad \beta_{T-h} = \frac{f(S_{T-h}) - f'(S_{T-h}) S_{T-h}}{B_T}.$$

If we choose $\alpha_{T-h}$ and $\beta_{T-h}$ in this way then they are $\mathcal{F}_{T-h}$-measurable and

$$|\Phi_T - V_T(\varphi)| = |R_T| \leq M(S_T - S_{T-h})^2.$$

Since the stock price is random, the question at this point is: How large is $R_T$? By (5.2.2),

$$\mathbb{E}\left[M(S_T - S_{T-h})^2\right] = M\mathbb{E}\left[(S_{T-h})^2 \left(\frac{S_T}{S_{T-h}} - 1\right)^2\right] = M\mathbb{E}\left[(S_{T-h})^2\right]\mathbb{E}\left[\left(\frac{S_T}{S_{T-h}} - 1\right)^2\right].$$

In the last equation, we used that $\frac{S_T}{S_{T-h}}$ and $S_{T-h}$ are independent, which follows immediately from the independence of $W_T - W_{T-h}$ and $W_{T-h}$ and the definition of $S_t$. Using (5.4.4) and the definition of $S_t$,

$$\mathbb{E}[S_t] = e^{\mu t} \text{ and } \mathbb{E}\left[(S_t)^2\right] = e^{2\mu t + 2\sigma^2 t}.$$

Furthermore, $\frac{S_T}{S_{T-h}}$ has the same distribution as $S_h$ and thus

$$\mathbb{E}\left[\left(\frac{S_T}{S_{T-h}} - 1\right)^2\right] = \mathbb{E}\left[(S_h - 1)^2\right] = \mathbb{E}\left[(S_h)^2\right] - 2\mathbb{E}[S_h] + 1 = e^{2\mu h + \sigma^2 h} - 2e^{\mu h} + 1.$$

Note that $|e^x - 1 - x| \leq 2x^2$ for $|x| \leq 2$. Combining this with the fact that $|t| \leq T$, and that $\mu$ and $\sigma$ are constants,

$$\mathbb{E}\left[M(S_T - S_{T-h})^2\right] \leq M'h^2$$

for some $M' = M'(M, \mu, \sigma, T)$. Therefore, $\mathbb{E}[|R_T|] \leq M'h^2$. Repeat this argument for $f(S_{T-h})$ and choose $\alpha_{T-2h}$ and $\beta_{T-2h}$ with

$$f(S_{T-h}) = \beta_{T-2h}B_{T-h} + \alpha_{T-2h}S_{T-h} + R_{T-h},$$

with $|R_{T-h}| \leq M'h^2$. Proceeding in this way, we obtain a trading strategy $\varphi = (\varphi_t)_{t\in[0,T;n]}$ with $\varphi_t = (\beta_t, \alpha_t)$. This trading strategy $\varphi$ is in general not self-financing. However, we can estimate how much money we have to add or remove from the market with this strategy. We have

$$\left|\mathbb{E}\left[\sum_{t\in[0,T;n]} R_t\right]\right| \leq \sum_{t\in[0,T;n]} \mathbb{E}[|R_t|] \leq \sum_{t\in[0,T;n]} M'h^2 = M'h = \frac{M'}{n},$$

which tends to 0 as $n$ tends to infinity. This computation suggests that there is a complete model such that the discrete Black–Scholes model converges to this model as $n \to \infty$. Such a model indeed exists and is called the Black–Scholes model. Establishing this will be the topic of the remainder of the book.

## 5.7 Exercises

**Exercise 5.1.** *Consider the discrete Black–Scholes model with $T > 0$, $n = 1$, $\mu = r = 0$, $\sigma > 0$ and $B_0 = S_0 = 1$. Determine all arbitrage-free prices of the contingent claim $\Phi_T = (S_T)^2$.*

# Part II

# Continuous-Time Models for Finance

# 6 Continuous Probability

In this chapter we will review the basic knowledge of continuous probability theory and stochastic processes. We will take a look at general probability spaces and random variables as well as convergence of random variables in Section 6.1. In Section 6.2, we will take a look at stochastic processes and in Section 6.3, we will look at filtrations and conditional expected values in the general situation. A reader having a good understanding of those concepts can skip this chapter.

## 6.1 Basics

### 6.1.1 General Probability Spaces

If one is working with a finite or countable sample space $\Omega$, then it is easy to construct a probability measure $\mathbb{P}$ which assigns a probability to every subset $A \subset \Omega$. If one has a countable sample space $\Omega = \{\omega_1, \omega_2, \ldots, \}$, then for any sequence of non-negative numbers $p_i$, $i \in \mathbb{N}$, with $\sum_{i=1}^{\infty} p_i = 1$, define the function $\mathbb{P} : \mathcal{P}\Omega \to [0, 1]$ by

$$\mathbb{P}[A] = \sum_{i:\omega_i \in A} p_i. \tag{6.1.1}$$

It is straightforward to check that $\mathbb{P}$ satisfies the axioms of probability (details are given in Section A.2). However, if $\Omega$ is uncountable then this is no longer possible in general.

We illustrate the problem with a classical example. It is natural to want to be able to construct a probability measure which is 'uniform' on an interval $[a, b]$; such a probability measure should necessarily be translation-invariant. Recall that a probability measure $\mathbb{P}$ on a set $\Omega \subset \mathbb{R}$ is called translation-invariant if $\mathbb{P}[A] = \mathbb{P}[x + A]$ for all events $A$ and $x \in \mathbb{R}$, with $A \subset \Omega$ and $x + A \subset \Omega$. Unfortunately, the following result from 1905, due to Vitali, says that if we would like our probability measure to assign a value to *every* subset $A \subset [a, b]$, this is impossible.

**Proposition 6.1.1.** There exists no translation-invariant probability measure on the interval $[a, b]$ with $-\infty < a < b < \infty$ which assigns a value to every subset $A \subset [a, b]$ so that the axioms of probability are fulfilled.

The proof of this proposition is based on so-called *Vitali sets*. The construction of these sets and the proof of Proposition 6.1.1 are given in Section A.4.1.

We can resolve the problem exposed in Proposition 6.1.1 by restricting the collection of sets to which we can assign probabilities to a family $\mathcal{F}$ of well-behaved events. Of course, we would like a general probability measure $\mathbb{P}$ to have similar properties to those that held in the discrete case. For instance, we would like to have $\mathbb{P}[A^c] = 1 - \mathbb{P}[A]$. In other words, if we can assign a probability to a set $A$ then we should also be able to assign a probability to $A^c$. In formulas, if $A \in \mathcal{F}$ then we must also have $A^c \in \mathcal{F}$. Similar considerations can be made with unions and intersections. Therefore, it is natural to require that the family $\mathcal{F}$ is a $\sigma$-algebra. This leads to the definition of a probability space.

**Definition 6.1.2.** A *(general) probability space* is a triple $(\Omega, \mathcal{F}, \mathbb{P})$ consisting of a set $\Omega$, a $\sigma$-algebra $\mathcal{F} \subset \mathcal{P}\Omega$ and a function $\mathbb{P} : \mathcal{F} \to [0, 1]$ satisfying

- $\mathbb{P}[A] \geq 0$ for all $A \in \mathcal{F}$,
- $\mathbb{P}[\Omega] = 1$ and
- for any sequence of disjoint sets $(A_j)_{j=1}^{\infty}$ with $A_j \in \mathcal{F}$,

$$\mathbb{P}\left[\bigcup_{i=1}^{\infty} A_i\right] = \sum_{i=1}^{\infty} \mathbb{P}[A_i]. \tag{6.1.2}$$

We call the sets $A \in \mathcal{F}$ events and the value $\mathbb{P}[A]$ the probability of the event $A$ occurring.

We mentioned above that we cannot, in a reasonable way, assign a probability to each subset of a sample space $\Omega$ in the general case. Fortunately, in almost all situations, the $\sigma$-algebra $\mathcal{F}$ can be chosen to be so rich that it contains all of the events in which one might be interested. Thus the problem we mention above almost never plays a role in practical applications, including those cases of interest to us.

### 6.1.2 Measures on an Arbitrary Probability Space

For the sake of completeness, we will take a brief look at how to show the existence of a probability measure on a general sample space. However, we will not discuss it in detail and will omit most proofs.

A concept closely related to a probability measure $\mathbb{P}$ is that of a *measure* $\mu$. The main difference is that one replaces the condition $\mathbb{P}[\Omega] = 1$ in Definition 6.1.2 by $\mu(\Omega) > 0$. Most of the measures occurring in this book are probability measures. The main exception is the Lebesgue measure $\mathcal{L}$, which assigns to an interval $[a, b] \subset \mathbb{R}$ the length of the interval. In formulas, we have

$$\mathcal{L}\big[[a, b]\big] = b - a \quad \text{for all } a \leq b. \tag{6.1.3}$$

We have seen in Proposition 6.1.1 that it is in general not possible to define a probability measure on the whole power set $\mathcal{P}\Omega$. Thus it is natural to request that $\mathcal{F}$ contains the events one is interested in. For instance, if $\Omega = \mathbb{R}$ then one typically requires that $\mathcal{F}$ contains all intervals. This leads to the following definition.

**Definition 6.1.3** (Borel $\sigma$-algebra).

- The *Borel $\sigma$-algebra* $\mathcal{B}$ on $\mathbb{R}$ is the smallest $\sigma$-algebra containing all intervals $[a,b]$ with $a,b \in \mathbb{R}$ and $a \leq b$. In formulas,
$$\mathcal{B} := \sigma\big([a,b]\,;\ a \leq b \in \mathbb{R}\big).$$
- The Borel $\sigma$-algebra $\mathcal{B}^n$ on $\mathbb{R}^n$ with $n \in \mathbb{N}$ is the smallest $\sigma$-algebra containing all sets of the form $[a_1,b_1] \times \cdots \times [a_n,b_n]$. In formulas,
$$\mathcal{B}^n := \sigma\big([a_1,b_1] \times \cdots \times [a_n,b_n]\,;\ a_j \leq b_j \in \mathbb{R} \text{ for } 1 \leq j \leq n\big).$$

One can define the *Borel $\sigma$-algebra* on an interval or on a subset of $\mathbb{R}^n$ analogously to Definition 6.1.3. We use the notation $\mathcal{B}$ and $\mathcal{B}^n$ for those cases too as there is little danger of confusion. Using the fact that $\mathcal{B}$ is a $\sigma$-algebra, one can show that it also contains all open and half-open intervals as well as all open and closed sets. For example, we can write an open interval as

$$(a,b) = \bigcup_{j=1}^{\infty} [a + 1/n, b - 1/n].$$

In (6.1.1), we were able to give an explicit expression for $\mathbb{P}[A]$ for all events $A \in \mathcal{F}$, using the fact we could explicitly enumerate all the elements in the set $A$. Unfortunately, for most $\sigma$-algebras $\mathcal{F}$, including the Borel $\sigma$-algebra, it is not possible to give an explicit expression, or even a simple description, for all $A \in \mathcal{F}$. As a consequence, it is not possible in general to define $\mathbb{P}[A]$ as directly as in (6.1.1). Therefore, we have to take a few intermediate steps to construct a measure or a probability measure. We explain this here briefly and illustrate it with the Lebesgue measure. For this we need algebras, contents and Carathéodory's extension theorem. We begin with

**Definition 6.1.4** (Algebra). Let $\Omega$ be a non-empty set.

- A family $\mathcal{A}$ of subsets of $\Omega$ is called an *algebra* over $\Omega$ if
  - $\Omega \in \mathcal{A}$,
  - $A^c \in \mathcal{A}$ for all $A \in \mathcal{A}$ and
  - $\bigcup_{j=1}^{n} A_j \in \mathcal{A}$ for all $n \in \mathbb{N}$ and $A_1,\dots,A_n \in \mathcal{A}$.
- Let $\Omega$ be a set and $\mathcal{H}$ be a non-empty family of subsets of $\Omega$. We then define $\mathcal{A}(\mathcal{H})$ to be the smallest algebra containing $\mathcal{H}$. In formulas: If $\mathcal{A}$ is an algebra with $\mathcal{H} \subset \mathcal{A}$ then $\mathcal{A}(\mathcal{H}) \subset \mathcal{A}$.

The main difference between an algebra and a $\sigma$-algebra is that countable unions need not be contained in an algebra. The advantage of an algebra $\mathcal{A}$ is that it is in general much easier to give an explicit expression for the sets in $\mathcal{A}$.

**Definition 6.1.5.** Let $\mathcal{A}$ be an algebra over a set $\Omega$. A content is a function $\mu : \mathcal{A} \to [0, \infty]$ such that for all $n \in \mathbb{N}$ and all disjoint sets $A_1, \ldots, A_n \in \mathcal{A}$,

$$\mu\left(\bigcup_{i=1}^{n} A_i\right) = \sum_{i=1}^{n} \mu(A_i) \quad \text{and} \quad \mu(\emptyset) = 0. \tag{6.1.4}$$

A content is similar to a measure, but a content only needs to be finitely additive, while a measure must be $\sigma$-additive. A content looks almost like a measure, but not all contents can be extended to measures. Fortunately, if a content is $\sigma$-additive on the underlying algebra then it can be extended to a measure. This is the topic of the following theorem.

**Theorem 6.1.6** (Carathéodory's extension theorem). *Let $\mathcal{A}$ be an algebra and $\mu$ be a content on $\mathcal{A}$. Suppose that*

$$\mu\left(\bigcup_{j=1}^{\infty} A_j\right) = \sum_{j=1}^{\infty} \mu(A_j) \tag{6.1.5}$$

*for any sequence of disjoint sets $(A_j)_{j=1}^{\infty}$ in $\mathcal{A}$ with $\bigcup_{j=1}^{\infty} A_j \in \mathcal{A}$. Then $\mu$ can be extended to a measure on the $\sigma$-algebra $\sigma(\mathcal{A})$.*

We omit the proof of this theorem, but it can be found for instance in [1, Chapter 1.3]. As mentioned above, we can now use Theorem 6.1.6 to show the existence of the Lebesgue measure.

**Example 6.1.7.** *We show in this example the existence of the* Lebesgue measure *on the interval $[0, T]$ with $T > 0$. More precisely, we show the existence of a measure $\mathcal{L}$ such that*

$$\mathcal{L}\big[[a, b]\big] = b - a \ \text{for all}\ 0 \leq a \leq b \leq T, \tag{6.1.6}$$

*and which assigns to each $A \in \mathcal{B}$ a reasonable value. We do this in several steps. We give only the most important details.*

*Step 1: Define $\mathcal{H}$ to be a family of all closed intervals in $[0, T]$, that is*

$$\mathcal{H} = \{[a, b]\,;\ 0 \leq a \leq b \leq T\}.$$

*Then we can define $\mathcal{L}$ on $\mathcal{H}$ by (6.1.6). In particular, $\mathcal{L}(\{x\}) = 0$ for all $x \in [0, T]$.*

*Step 2: Next we determine $\mathcal{A} := \mathcal{A}(\mathcal{H})$. It is straightforward to see that each $A \in \mathcal{A}$ can be written as*

$$A = \bigcup_{j=1}^{n} I_j \tag{6.1.7}$$

*for some $n \in \mathbb{N}$, where each $I_j$ is an open, closed or half-open interval. The representation of $A$ in* (6.1.7) *is unique if we require dists$(I_i, I_j) > 0$ for all $i \neq j$, where*

$$\text{dists}(I,J) := \inf\{|x-y|\,;\, x \in I, y \in J\} \text{ for } I,J \subset \mathbb{R}.$$

*Step 3: We now extend $\mathcal{L}$ to $\mathcal{A}$. We begin with intervals. Since $\mathcal{L}(\{x\}) = 0$, we set*

$$\mathcal{L}\big[(a,b)\big] = \mathcal{L}\big[[a,b)\big] = \mathcal{L}\big[(a,b]\big] = b-a \text{ for } a \le b.$$

*Now let $A \in \mathcal{A}$ be arbitrary. Write $A$ as in* (6.1.7)*, assuming that dists$(I_i, I_j) > 0$ for all $i \neq j$. Then define*

$$\mathcal{L}[A] = \sum_{j=1}^{n} \mathcal{L}[I_j]. \tag{6.1.8}$$

*As the representation of $A$ is unique, the function $\mathcal{L}(A)$ is well-defined.*

*Step 4: We now have to show that $\mathcal{L}$ is a content. For this we have to show that $\mathcal{L}$ is additive. Note, for $s_0 < s_1 < s_2$,*

$$\mathcal{L}\big[(s_0,s_2)\big] = \mathcal{L}\big[(s_0,s_1)\big] + \mathcal{L}\big[[s_1,s_2)\big] = \mathcal{L}\big[(s_0,s_1]\big] + \mathcal{L}\big[(s_1,s_2)\big].$$

*Using this it is a straightforward inductive argument to see that $\mathcal{L}$ is additive.*

*Step 5: Finally, to show that $\mathcal{L}$ fulfils* (6.1.5)*, it is sufficient to verify that*

$$\mathcal{L}\big[(s_0,s_\infty)\big] = \sum_{j=1}^{\infty} \mathcal{L}\big[(s_{j-1},s_j]\big]$$

*for all $(s_j)_{j=0}^{\infty}$ with $s_j < s_{j+1}$ and $s_\infty = \sup_j s_j$.*

Example 6.1.7 can be extended a little further.

**Theorem 6.1.8** (Lebesgue–Stieltjes measure)**.** *Let $\Omega = \mathbb{R}$ and $F : \Omega \to [0,1]$ be a non-decreasing, right-continuous function such that*

$$\lim_{x\to-\infty} F(x) = 0 \text{ and } \lim_{x\to\infty} F(x) = 1.$$

*Then there exists a probability measure on $\mathbb{P}_F$ on $(\Omega, \mathcal{B})$ such that*

$$\mathbb{P}_F\big[(a,b]\big] = F(b) - F(a) \text{ for all } a \le b. \tag{6.1.9}$$

The proof of this theorem is similar to the computation in Example 6.1.7. The main difference is that one has to be more careful with the boundary points of intervals as a single point can get a non-zero probability.

We also have to work with measures and probability measures on product spaces $\Omega_1 \times \Omega_2$. The two main cases of interest are $\mathbb{R}^n$ and $\Omega \times [0,T]$. The case $\Omega \times [0,T]$ is in particular important for stochastic integration. The following proposition shows that we can combine two probability spaces.

**Proposition 6.1.9.** Let $(\Omega_1, \mathcal{F}_1, \mathbb{P}_1)$ and $(\Omega_2, \mathcal{F}_2, \mathbb{P}_2)$ be two probability spaces. Let $\mathcal{F}_1 \times \mathcal{F}_2$ be the $\sigma$-algebra over $\Omega_1 \times \Omega_2$ generated by the sets $A_1 \times A_2$ with $A_1 \in \mathcal{F}_1$ and $A_2 \in \mathcal{F}_2$.

Then there exists a unique probability measure $\mathbb{P}$ on $(\Omega_1 \times \Omega_2, \mathcal{F}_1 \times \mathcal{F}_2)$ with

$$\mathbb{P}[A_1 \times A_2] = \mathbb{P}_1[A_1] \cdot \mathbb{P}_2[A_2] \quad \text{for all } A_1 \in \mathcal{F}_1 \text{ and } A_2 \in \mathcal{F}_2.$$

This proposition follows again from Carathéodory's extension theorem. Proposition 6.1.9 is also valid for measures. Therefore, we can combine a probability measure $\mathbb{P}$ with the Lebesgue measure $\mathcal{L}$ and construct a measure on $\Omega \times [0, T]$ with

$$\mathbb{P} \times \mathcal{L}\big(A \times [a, b]\big) = \mathbb{P}[A] \cdot (b - a) \quad \text{for all } A \in \mathcal{F} \text{ and } a \leq b \in \mathbb{R}. \tag{6.1.10}$$

In a similar way, we can combine Example 6.1.7 and Proposition 6.1.9 to extend the Lebesgue measure $\mathcal{L}$ to $\mathbb{R}^n$. More precisely, there is a measure on $(\mathbb{R}^n, \mathcal{B}^n)$ with

$$\mathcal{L}\big([a_1, b_1] \times \cdots \times [a_n, b_n]\big) = \prod_{j=1}^{n} (b_j - a_j).$$

## 6.1.3 Random Variables

We are interested in random variables $X$ on a probability space $(\Omega, \mathcal{F}, \mathbb{P})$. Of course, we would like to study expressions like

$$\mathbb{P}[X = 0] \quad \text{or} \quad \mathbb{P}[1 \leq X \leq 2].$$

However, our probability measure can only assign a value to the events $A \in \mathcal{F}$. It is thus natural to require that $\{X \in [a, b]\} \in \mathcal{F}$. This leads to the following definition.

**Definition 6.1.10.** Let $(\Omega, \mathcal{F}, \mathbb{P})$ be a probability space. A *random variable* on $(\Omega, \mathcal{F}, \mathbb{P})$ is a function $X : \Omega \to \mathbb{R}$, such that

$$\{X \leq x\} \in \mathcal{F} \text{ for all } x \in \mathbb{R}. \tag{6.1.11}$$

In other words, $X$ is a $\mathcal{F}$-measureable function (see Definition 2.3.20).

Note that the condition (6.1.11) is equivalent to

$$\{X \in A\} \in \mathcal{F} \quad \text{for all } A \in \mathcal{B}. \tag{6.1.12}$$

This follows from the definition of the Borel $\sigma$-algebra $\mathcal{B}$ and the identities

$$\{X \in A\}^c = \{X \in A^c\} \quad \text{and} \quad \bigcup_{j=1}^{\infty} \{X \in A_j\} = \left\{ X \in \bigcup_{j=1}^{\infty} A_j \right\} \tag{6.1.13}$$

with $A, A_j \subset \mathbb{R}$ arbitrary. In a similar way, one can show that

$$\{(X_1, \ldots, X_n) \in A\} \in \mathcal{F} \quad \text{for all } A \in \mathcal{B}^n, \tag{6.1.14}$$

where $X_1, \ldots, X_n$ are random variables on a probability space. As already mentioned, in most cases we can choose $\mathcal{F}$ sufficiently rich that it contains all events we might be interested in. For this reason, in almost all practical applications, all functions considered on a probability space are random variables. In particular, all functions we are considering in this course are random variables. We also need the following notion.

**Definition 6.1.11.** A function $f : \mathbb{R}^n \to \mathbb{R}$ is called *Borel-measurable* if $f^{-1}(A) \in \mathcal{B}^n$ for all $A \in \mathcal{B}$.

If $X_1, \ldots, X_n$ are random variables on a probability space $(\Omega, \mathcal{F}, \mathbb{P})$ and $f$ is Borel-measurable then $f(X_1, \ldots, X_n)$ is a random variable too. This follows immediately from (6.1.14) and Definition 6.1.11. It is useful to keep in mind that all continuous functions and all *cadlag* functions are Borel-measurable. Recall, a function $f : \mathbb{R} \to \mathbb{R}$ is called *cadlag* if $f$ is right continuous with left limits. In formulas, we have

$$f(t) = \lim_{s \downarrow t} f(s) \quad \text{and the limit } \lim_{s \uparrow t} f(s) \text{ exists} \quad \text{for all } t \in \mathbb{R}.$$

The name *cadlag* originates from the French: 'continue á droite, limite á gauche'. An important quantity of a random variable is the distribution.

**Definition 6.1.12.** Let $X$ be a random variable. The *cumulative distribution function* $F_X : \mathbb{R} \to [0, 1]$ of $X$ is defined exactly as in the discrete case as

$$F_X(x) := \mathbb{P}[X \leq x].$$

We usually write $F(x) = F_X(x)$ if there is no likelihood of confusion.

The joint distribution of two or more random variables is defined similarly.

**Definition 6.1.13.** A random variable $X$ is called (absolutely) continuous if there exists an integrable function $f_X : \mathbb{R} \to [0, \infty]$ with

$$F_X(x) = \int_{-\infty}^{x} f_X(t)\, dt \quad \text{for all } x \in \mathbb{R}.$$

The function $f_X$ is called the density of $X$. We usually write $f(x) = f_X(x)$ if there is no likelihood of confusion.

The joint density of two or more continuous random variables is defined similarly. One of the most important continuous distributions is described as follows.

**Definition 6.1.14.** A random variable $X$ on probability space $(\Omega, \mathcal{F}, \mathbb{P})$ is said to have a *Gaussian* or *normal* distribution with parameters $\mu \in \mathbb{R}$ and $\sigma > 0$ if it is continuous and has the density $f(t) = \frac{1}{\sqrt{2\pi}\sigma} \exp\left(-\frac{(t-\mu)^2}{2\sigma}\right)$. We write in this case $X \sim \mathcal{N}(\mu, \sigma^2)$.

For convenience, we have summarised key properties of the normal distribution in Section A.1.2.

The following result tells us that, given a probability measure $\mathbb{P}$, we can define expectation with respect to $\mathbb{P}$.

**Theorem 6.1.15.** *Let $X$ be a non-negative random variable on a probability space $(\Omega, \mathcal{F}, \mathbb{P})$. Then $X$ has a well-defined integral with respect to $\mathbb{P}$, called the* expectation *of $X$ and denoted*

$$\mathbb{E}[X] = \int_{\Omega} X(\omega)\, d\mathbb{P}. \tag{6.1.15}$$

*Furthermore, for any random variable $X$, if $\mathbb{E}[|X|] < \infty$, we can define $\mathbb{E}[X] = \mathbb{E}[X^+] - \mathbb{E}[X^-]$ where $X^+ = \max\{X, 0\} = X \vee 0$ and $X^- = (-X) \vee 0$. In this case, we say that the expectation of $X$ exists or $X$ is integrable.*

*Expectation satisfies the following properties:*

*(i)* $\mathbb{E}[\mathbb{1}_A] = \mathbb{P}[A]$ *for all* $A \in \mathcal{F}$.
*(ii)* $\mathbb{E}[\lambda X + \mu Y] = \lambda \mathbb{E}[X] + \mu \mathbb{E}[Y]$, *for all integrable random variables $X, Y$ and scalars $\lambda, \mu \in \mathbb{R}$.*
*(iii) If $X$ is a positive random variable, then $\mathbb{E}[X] \geq 0$. Hence, if $X$ is an integrable random variable then $\mathbb{E}[|X|] \geq |\mathbb{E}[X]|$.*

*Proof.* The proof involves constructing the Lebesgue integral with respect to the probability measure $\mathbb{P}$. We just give an overview of the main ideas; the details of how to construct the Lebesgue integral can be found in any standard textbook on measure theory.

Firstly, one defines expectation for random variables of the form $X = \mathbb{1}_A$ for some $A \in \mathcal{F}$ by setting

$$\mathbb{E}[\mathbb{1}_A] = \mathbb{P}[A].$$

One then extends the definition to *simple random variables*, that is random variables which can be written in the form $X = \sum_{i=1}^{n} c_i \mathbb{1}_{B_i}$, for some $n \in \mathbb{N}$, $c_i \in \mathbb{R}$ and disjoint $B_i \in \mathcal{F}$, by setting

$$\mathbb{E}\left[\sum_{i=1}^{n} c_i \mathbb{1}_{B_i}\right] = \sum_{i=1}^{n} c_i \mathbb{P}[B_i].$$

Next, if $X$ is any non-negative random variable, define

$$\mathbb{E}[X] = \sup\{\mathbb{E}[Y] : Y \text{ is simple with } 0 \leq Y \leq X\}.$$

It can be shown that any non-negative random variable can be written as an increasing limit of simple random variables. One can then check that $\mathbb{E}[X]$ is well-defined (but may be infinite).

Lastly, suppose that $X$ is a general random variable $X$ for which $\mathbb{E}[|X|] < \infty$. Then $X^+$ and $X^-$ are non-negative, so $\mathbb{E}\left[X^+\right]$ and $\mathbb{E}\left[X^-\right]$ are well-defined and finite. One can therefore set $\mathbb{E}[X] = \mathbb{E}\left[X^+\right] - \mathbb{E}\left[X^-\right]$. (Note that the condition $\mathbb{E}[|X|] < \infty$ is needed, to avoid expressions of the form $\infty - \infty$.)

It is relatively straightforward to check that expectation defined in this way satisfies the stated properties. □

We give an alternative expression for $\mathbb{E}[X]$ in terms of the distribution

$$\mathbb{E}[X] = \int_{\mathbb{R}} x \, dF_X(x), \tag{6.1.16}$$

where $dF_X(x)$ denotes the Riemann–Stieltjes integral with respect to the cumulative distribution function of $X$. We will discuss the Riemann–Stieltjes integral in Section 8.2. If $X$ is a discrete random variable, then (6.1.16) reduces to the expression in Theorem 2.1.13. If $X$ is a continuous random variable with density $f_X$, then

$$\mathbb{E}[X] = \int_{\mathbb{R}} x \cdot f_X(x) \, dx. \tag{6.1.17}$$

### 6.1.4 Convergence of Random Variables

An important element of probability theory concerns the convergence of sequences of random variables. In contrast to the convergence of real numbers, there are many different types of convergence of random variables. We begin with the weakest one.

**Definition 6.1.16.** Let $X_n$ with $n \in \mathbb{N}$ and $X$ be random variables. We say that $X_n$ converges to $X$ in distribution (or in law, or converges weakly), abbreviated $X_n \xrightarrow{d} X$, if

$$\mathbb{P}[X_n \leq x] \to \mathbb{P}[X \leq x] \quad \text{as } n \to \infty \tag{6.1.18}$$

for all points $x$ of continuity of $\mathbb{P}[X \leq x]$ considered as a function in $x$.

We have to exclude all $x$ such that $\mathbb{P}[X \leq x]$ is not a continuous function in $x$ to avoid pathological cases. Consider, for instance, $X_n = (-1/2)^n$ and $X = 0$. In this case, $X_n$ and $X$ are just constants and therefore quite trivial random variables. Also, $X_n \to X$ as real numbers. We therefore would definitely like to have $X_n$ converge to $X$, regardless of which kind of convergence we use. On the other hand, $\mathbb{P}[X_n \leq 0] \nrightarrow \mathbb{P}[X \leq 0]$. The reason for this is that $\mathbb{P}[X = 0] > 0$ so $F_X(x)$ has a jump point at $x = 0$. In view of this, it is reasonable to exclude discontinuity points of $F_X(x)$ in the definition.

Convergence in distribution of a vector of random variables is defined similarly to Definition 6.1.16. We simply replace (6.1.18) by

$$\mathbb{P}\left[X_n^{(1)} \le x_1, \dots, X_n^{(k)} \le x_k\right] \to \mathbb{P}\left[X^{(1)} \le x_1, \dots, X^{(k)} \le x_k\right] \tag{6.1.19}$$

and exclude all $(x_1, \dots, x_k)$, where $\mathbb{P}\left[X^{(1)} \le x_1, \dots, X^{(k)} \le x_k\right]$ is not continuous.

Note that in Definition 6.1.16 only the distribution of $X$ matters. We therefore often omit the random variable $X$ and just mention the distribution. For instance, we say '*$X_n$ converges in distribution to a normal distribution*' instead of '*$X_n$ converges in distribution to a random variable with normal distribution*'. A classical result of convergence in distribution is the central limit theorem, see Theorem A.1.5.

In Definition 6.1.16, all $X_n$ and $X$ can be defined on different probability spaces. However, for most types of convergence, one needs that all $X_n$ and $X$ are defined on the same probability space. An example is convergence in probability. Recall, a sequence $(X_n)_{n\in\mathbb{N}}$ converges to $X$ in probability, abbreviated $X_n \stackrel{p}{\to} X$, if for all $\epsilon > 0$,

$$\mathbb{P}\left[|X_n - X| > \epsilon\right] \to 0 \ \text{ as } n \to \infty. \tag{6.1.20}$$

We will not need convergence in probability in this course so will not discuss it further. The main two types of convergence we use in this course are almost sure convergence and convergence in mean square.

**Definition 6.1.17.** Let $X_n$ with $n \in \mathbb{N}$ and $X$ be random variables defined on the same probability space $(\Omega, \mathcal{F}, \mathbb{P})$. We say that $(X_n)_{n\in\mathbb{N}}$ converges to $X$ *almost surely*, abbreviated $X_n \stackrel{as}{\to} X$, if

$$\mathbb{P}\left[\{\omega \in \Omega : X_n(\omega) \to X(\omega) \text{ as } n \to \infty\}\right] = 1.$$

Two random variables $X$ and $Y$ are called almost surely equal, abbreviated $X \stackrel{as}{=} Y$, if

$$\mathbb{P}\left[\{\omega \in \Omega : X(\omega) = Y(\omega)\}\right] = 1. \tag{6.1.21}$$

In probability theory one often wants to show that $\mathbb{E}[X_n] \to \mathbb{E}[X]$ but only knows that $X_n \stackrel{as}{\to} X$. In such situations, one can use the following theorem.

**Theorem 6.1.18** (Lebesgue's dominated convergence). *Let $(X_n)_{n\in\mathbb{N}}$ and $X$ be random variables such that $X_n \stackrel{as}{\to} X$. If there is a random variable $Z$ with $|X_n| < Z$ for all $n$ and $\mathbb{E}[Z] < \infty$, then $\mathbb{E}[X_n] \to \mathbb{E}[X]$.*

This is one of the most important theorems from measure theory and is widely used in probability theory. A proof of this theorem can be found, for instance, in [1, Theorem 16.4]. For completeness, we would like to mention that $X_n \stackrel{as}{\to} X$ does not imply in general $\mathbb{E}[X_n] \to \mathbb{E}[X]$. As an example, choose the uniform measure on $[0,1]$ and define $X_n(\omega) = n$ for $0 \le \omega \le 1/n$ and $X_n(\omega) = 0$ otherwise. In this case, $X_n \stackrel{as}{\to} 0$ but $\mathbb{E}[X_n] = 1$ for all $n$.

Next we introduce convergence in mean square. For this, we first define the $L^p$-norm.

**Definition 6.1.19.** Let $(\Omega, \mathcal{F}, \mathbb{P})$ be a probability space and $1 \leq p < \infty$. The $L^p$-norm of a random variable $X$ is defined as

$$\|X\|_{L^p(\Omega,\mathbb{P},\mathcal{F})} := \left(\mathbb{E}\left[|X|^p\right]\right)^{1/p} \tag{6.1.22}$$

if the expression in (6.1.22) is finite. Furthermore, we denote by $\mathcal{L}^p(\Omega, \mathbb{P}, \mathcal{F})$ the space of all random variables $X$ with $\|X\|_{L^p(\Omega,\mathbb{P},\mathcal{F})} < \infty$.

Note that in most situations, the probability space is clear from the context. We write $\|X\|_{L^p(\Omega)}$ or $\|X\|_p$ instead of $\|X\|_{L^p(\Omega,\mathbb{P},\mathcal{F})}$ if there is no danger of confusion. The $L^p$-norm with respect to measures is defined in the same way.

For the construction of the stochastic integral, we need the $L^2$-norm on the space $\Omega \times [0,t]$ with respect to the measure $\mathbb{P} \times \mathcal{L}$ (see (6.1.10)). If $X : \Omega \times [0,t]$ is given, then

$$\|X\|_{L^2(\Omega\times[0,t])} = \left(\int_0^t \mathbb{E}\left[X_s^2\right] ds\right)^{1/2}, \tag{6.1.23}$$

where $\mathbb{E}\left[X_s^2\right]$ is the expectation of $X_s(\omega) := X(\omega, s)$ with respect to $\mathbb{P}$ and $s$ is fixed. We will mainly work with the $L^2$-norm, and in some cases also with the $L^1$-norm. Thus we restrict our statements mainly to $L^2$. We now introduce convergence in mean square.

**Definition 6.1.20.** Let $X_n$ with $n \in \mathbb{N}$ and $X$ be random variables defined on the same probability space $(\Omega, \mathcal{F}, \mathbb{P})$. We say that $(X_n)_{n\in\mathbb{N}}$ converges to $X$ *in mean square* (or in $L^2$), abbreviated $X \overset{L^2}{\to} Y$, if

$$\|X_n - X\|_{L^2(\Omega)} \to 0 \text{ as } n \to \infty.$$

Let us now briefly compare the different types of convergence we have seen. We will only give a brief overview here, without going into detail.

- If a sequence of random variables converges almost surely and is also bounded by a square-integrable random variable, then this sequence also converges in mean square. In formulas:

$$X_n \overset{as}{\to} X \text{ and } |X_n| \leq Z \text{ with } \mathbb{E}\left[Z^2\right] < \infty \implies X_n \overset{L^2}{\to} X.$$

This follows immediately from the dominated convergence theorem, see above.

- Mean square convergence implies convergence in probability, that is

$$X_n \overset{L^2}{\to} X \implies X_n \overset{p}{\to} X.$$

This follows immediately from Markov's inequality.

- Convergence in probability implies convergence in distribution, that is

$$X_n \overset{p}{\to} X \implies X_n \overset{d}{\to} X.$$

To justify this, one can use that we have for all $\epsilon > 0$ the inequality

$$\mathbb{P}[X_n \le x] \le \mathbb{P}[X \le x + \epsilon] + \mathbb{P}[|X_n - X| > \epsilon].$$

Our main application of the $L^2$-norm and mean square convergence is the construction of the stochastic integral in Chapter 8. We look at the properties of the $L^2$-norm.

**Proposition 6.1.21.** Let $X, Y \in \mathcal{L}^2(\Omega, \mathbb{P}, \mathcal{F})$. Then

(a) $\|X\|_1 \le \|X\|_2$,
(b) $\|XY\|_1 \le \|X\|_2 \|Y\|_2$,
(c) $\|X + Y\|_2 \le \|X\|_2 + \|Y\|_2$,
(d) $\|\lambda X\|_2 = |\lambda| \|X\|_2$ for all $\lambda \in \mathbb{R}$ and
(e) $\|X\|_2 = 0$ implies that $X \overset{as}{=} 0$.

*Proof.* We give here only a brief overview of the proof.

(a) This follows immediately from Jensen's inequality with the function $f(x) = x^2$.
(b) For each $\lambda > 0$, $(|X| - \lambda|Y|)^2 = X^2 - 2\lambda|XY| + \lambda^2 Y^2 \ge 0$. This implies that

$$2\mathbb{E}[|XY|] \le \frac{\mathbb{E}[X^2]}{\lambda} + \lambda \mathbb{E}\left[Y^2\right].$$

Choosing $\lambda = \sqrt{\frac{\mathbb{E}[X^2]}{\mathbb{E}[Y^2]}}$ completes the proof.
(c) Using the definitions of $\|\cdot\|_1$ and $\|\cdot\|_2$ and point (b) gives

$$\begin{aligned}(\|X + Y\|_2)^2 &= \mathbb{E}\left[(X + Y)^2\right] = \mathbb{E}\left[X^2\right] + 2\|XY\|_1 + \mathbb{E}\left[X^2\right] \\ &\le (\|X\|_2)^2 + 2\|X\|_2\|Y\|_2 + (\|Y\|_2)^2 = (\|X\|_2 + \|Y\|_2)^2.\end{aligned}$$

(d) This follows immediately from the definition of $\|X\|_2$.
(e) We argue by contraposition. Let $X$ be a random variable, which is not almost surely equal to 0. Then there exists a $\lambda > 0$ such that $\mathbb{P}[|X| \ge \lambda] > 0$. This implies that $|X| \ge \lambda \mathbb{1}_A(\omega)$, where $A := \{X \ge \lambda\}$. We get

$$\mathbb{E}\left[X^2\right] \ge \mathbb{E}\left[(\lambda \mathbb{1}_A(\omega))^2\right] = \lambda^2 \mathbb{P}[|X| \ge \lambda] > 0.$$

This completes the proof. □

Proposition 6.1.21 shows that the $L^2$-norm is *a priori* just a seminorm as it is not positive definite. However, we can turn $\|\cdot\|_2$ into a norm. For this, let us make some observations.

- $\|X\|_2 = \|Y\|_2$ if $X \overset{as}{=} Y$,
- $X \overset{as}{=} X$ and
- if $X \overset{as}{=} Y$ and $Y \overset{as}{=} Z$ then $X \overset{as}{=} Z$.

Therefore $\overset{as}{=}$ is an equivalence relation. Also, if $X \overset{as}{=} Y$ then we cannot distinguish between $X$ and $Y$ in practice. Even if we repeat an experiment a billion times, all realisations of $X$ and $Y$ would be equal with probability one. For this reason, we no longer distinguish between two random variables if they are almost surely equal. Formally, define

$$L^2(\Omega, \mathbb{P}, \mathcal{F}) := \mathcal{L}^2(\Omega, \mathbb{P}, \mathcal{F})/\sim, \tag{6.1.24}$$

where $\sim$ denotes the equivalence relation given by $\overset{as}{=}$. In view of the above observations, $\|\cdot\|_2$ is well-defined on $L^2(\Omega, \mathbb{P}, \mathcal{F})$ and is also a norm on $L^2(\Omega, \mathbb{P}, \mathcal{F})$.

**Definition 6.1.22.** A sequence $(X_n)_{n\in\mathbb{N}}$ in $L^2(\Omega, \mathbb{P}, \mathcal{F})$ is called a Cauchy sequence if, for every $\epsilon$, there is an $n_0 \in \mathbb{N}$ such that $\|X_n - X_m\| \leq \epsilon$ for all $m, n \geq n_0$.

We now have the following result.

**Theorem 6.1.23.** *Let $(X_n)_{n\in\mathbb{N}}$ be a Cauchy sequence in $L^2(\Omega, \mathbb{P}, \mathcal{F})$.*

- *Then there exists an $X \in L^2(\Omega, \mathbb{P}, \mathcal{F})$ such that $X_n \overset{L^2}{\to} X$ as $n \to \infty$.*
- *The limit is almost surely unique, that is if $X_n \overset{L^2}{\to} X$ and $X_n \overset{L^2}{\to} Y$, then $X \overset{as}{=} Y$.*
- *There exists a subsequence $(X_{n_j})_{j\in\mathbb{N}}$ such that $X_{n_j} \overset{as}{\to} X$ as $j \to \infty$.*

For completeness, we give the proof of this theorem in Section A.4.2. It is important to keep in mind that mean square convergence does not imply almost sure convergence.

## 6.2 Review of Stochastic Processes

We will study some stochastic models for stock prices over a given time period. Mathematically, we model these prices by a collection of random variables indexed by time. Such collections of random variables are called *stochastic processes*.

**Definition 6.2.1** (Stochastic process). Let $(\Omega, \mathcal{F}, \mathbb{P})$ be a probability space and $\mathcal{T}$ be a non-empty set. Suppose that for each $t \in \mathcal{T}$, there is a real-valued random variable $X_t$ on $(\Omega, \mathcal{F}, \mathbb{P})$. Then the function

$$X : \mathcal{T} \times \Omega \longrightarrow \mathbb{R}$$

with $X(t, \omega) := X_t(\omega)$ is called a *stochastic process* on $(\Omega, \mathcal{F}, \mathbb{P})$ with index set $\mathcal{T}$. We use the notation $X$ as well as $\{X_t, t \in \mathcal{T}\}$ for a stochastic process.

The index set $\mathcal{T}$ typically has a heuristic interpretation such as time or spatial location. Here we only use the interpretation of $\mathcal{T}$ as time and therefore we always take $\mathcal{T} \subset \mathbb{R}$. The random variable

$X_t$ represents the state of the process at time $t$. For instance, if $X$ models the price of a stock then $X_t$ is the price of the stock at time $t$. We will mainly work with $\mathcal{T} = \{0, 1, \dots, T\}$ or $\mathcal{T} = [0, T]$ for some $0 < T \leq \infty$.

**Definition 6.2.2.** Let $X$ be a *stochastic process* with index set $\mathcal{T}$.

- If $\mathcal{T} = \mathbb{N}_0$ (or a subset of $\mathbb{N}_0$), we say that $X$ is a *discrete-time* stochastic process.
- If $\mathcal{T} = [0, \infty)$ or any interval in $\mathbb{R}$, we say that $X$ is a *continuous-time* stochastic process.

In the definition of a stochastic process $X$, we allow $X$ to take all values in $\mathbb{R}$. However, not all stochastic processes $X$ take every value in $\mathbb{R}$. It is useful to introduce the following definition.

**Definition 6.2.3** (State space). The *state space* of a stochastic process $X$ is the set of possible values of $X$, taken over all $t \in \mathcal{T}$ and $\omega \in \Omega$.

The state space can be a discrete set, such as the integers $\mathbb{Z}$. In this case, $X_t$ is a discrete random variable for all $t \in \mathcal{T}$. The state space can also be a continuous set such as an interval, or the set of real numbers $\mathbb{R}$. Let us consider some examples.

**Example 6.2.4** (Simple random walk). *The stochastic process* $\{X_n, n \in \mathbb{N}\}$ *is a simple random walk if* $X_0 = 0$ *and*

$$\mathbb{P}[X_{n+1} = X_n + 1] = p, \qquad \mathbb{P}[X_{n+1} = X_n - 1] = 1 - p$$

*for some* $p \in [0, 1]$. *Each step in the walk corresponds to the outcome of a Bernoulli trial and is independent of all other steps. This is a discrete-time stochastic process with a discrete state space. A realisation of this process is illustrated in Figure 6.1. Note that a realisation of a stochastic process is typically called a sample path, see also the comments just after Example 6.2.5.*

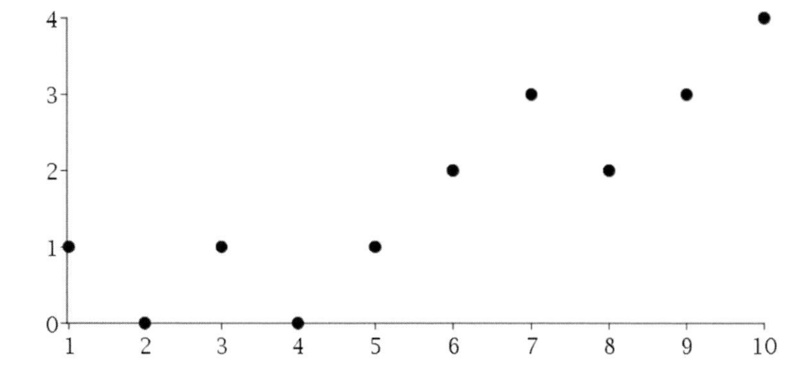


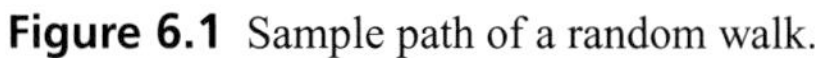
**Figure 6.1** Sample path of a random walk.

**Example 6.2.5** (Counting and Poisson processes). *A process $\{X_t, t \geq 0\}$ is called a counting process if $X_t \in \mathbb{N}_0$ for all t and $X_s \leq X_t$ for $s \leq t$. Heuristically, a counting process is used to count the number of incidents (e.g. volcanic eruptions) which occur up to time t. An important example of a counting process is a* Poisson process. *This is a stochastic process dependent on a parameter $\lambda > 0$ (the rate). It is a counting process satisfying the following properties:*

- $X_0 = 0$.
- *For $s \leq t$, $X_t - X_s$ is independent of all $X_u$ with $u \leq s$.*
- *For $s \leq t$, $X_t - X_s \sim Pois(\lambda(t-s))$.*

*For a Poisson process,*

$$\mathbb{P}[X_t - X_s = m] = \mathbb{P}[X_{t-s} = m] = e^{-\lambda(t-s)}\frac{(\lambda(t-s))^m}{m!} \qquad \text{with } m \in \mathbb{N}_0.$$

*Counting and Poisson processes are both continuous-time stochastic processes with a discrete state space. A sample path of a Poisson process is illustrated in Figure 6.2.*

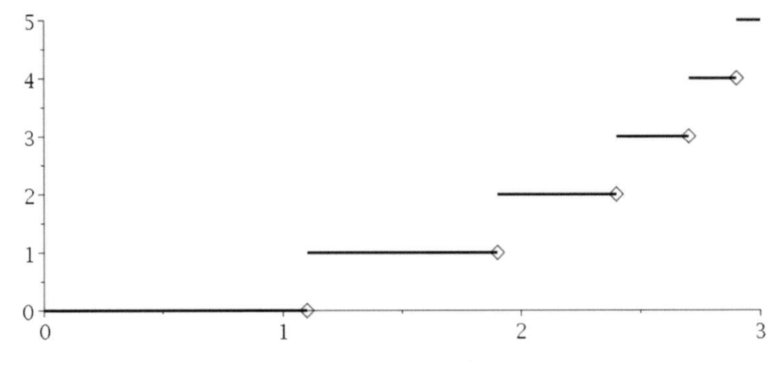


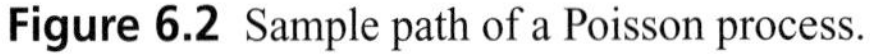

**Figure 6.2** Sample path of a Poisson process.

A stochastic process is by definition a collection of random variables, which we can view as a function $X : \mathcal{T} \times \Omega \to \mathbb{R}$. If we now choose a fixed $\omega \in \Omega$ and insert this $\omega$ into $X$, then we obtain a function $X_\omega : \mathcal{T} \to \mathbb{R}$ with $X_\omega(t) := X(t, \omega)$. This is a deterministic function and is called the *sample path* (or *trajectory*) of $X$ at $\omega$. We can therefore interpret a stochastic process $X$ as a function $X : \Omega \to E$, where $E$ is a set of functions from $\mathcal{T} \to \mathbb{R}$. In other words, we assign to each $\omega \in \Omega$ a function from $\mathcal{T} \to \mathbb{R}$ and can view a stochastic process as randomly choosing a function. In certain situations this view of a stochastic process can be more convenient than viewing it as a collection of random variables. Although the set $E$ can be any set of functions $\mathcal{T} \to \mathbb{R}$, that is there is no need to assume that the functions in $E$ are continuous or have any other nice properties, in most applications it is usual for the functions in $E$ to have some regularities. For instance, one might wish to require that all sample paths are continuous. Alternatively, it often occurs (or is convenient to assume) that the sample paths are right-continuous functions with left limits (i.e. cadlag functions). For example, the sample paths of a Poisson process are *cadlag* functions.

**Definition 6.2.6.** A stochastic process $X = \{X_t, t \in \mathcal{T}\}$ is called *continuous* if all sample paths are continuous functions in $t$. Similarly, we call a stochastic process *cadlag* if all sample paths are *cadlag* functions in $t$.

Note that (almost) all continuous-time stochastic processes $X$ that we will consider are continuous or cadlag.

Studying continuous-time stochastic processes requires one to work with random functions. This is not particularly easy, so one way to extract some information is to study the joint behaviour at some given times $t_1, \ldots, t_n \in \mathcal{T}$.

**Definition 6.2.7** (Finite-dimensional distributions)**.** The *finite-dimensional distributions* of the stochastic process $X$ are the distributions of the vectors

$$(X_{t_1}, \ldots, X_{t_n}), \quad t_1 < \cdots < t_n \in \mathcal{T},$$

for all possible choices of times $t_1 < t_2 < \cdots < t_n \in \mathcal{T}$ and for every $n \geq 1$.

The finite-dimensional distributions clearly determine a discrete-time stochastic process. It can be shown that the finite-dimensional distributions determine a continuous-time stochastic process $X$ if the process is continuous or cadlag, see for instance [2]. However, in general, the finite-dimensional distributions do not determine a stochastic process $X$.

**Example 6.2.8.** *Let $\mathcal{T} = [0,1]$ and $(\Omega, \mathcal{F}, \mathbb{P})$ be a probability space and let $U$ be a random variable on this space with $U \sim \text{Uniform}(0,1)$. Define two stochastic processes $X$ and $Y$ on $(\Omega, \mathcal{F}, \mathbb{P})$ by*

$$X(t,\omega) := 0 \text{ and } Y(t,\omega) := \begin{cases} 1, & \text{if } t = U(\omega), \\ 0, & \text{otherwise.} \end{cases}$$

*We now take a look at finite-dimensional distributions of $X$ and $Y$. For simplicity, assume $n = 1$. Let $t \in [0,1]$ be fixed. Then*

$$\mathbb{P}[Y_t = 1] = \mathbb{P}[\{\omega \in \Omega;\ Y(t,\omega) = 1\}] = \mathbb{P}[\{\omega \in \Omega;\ U(\omega) = t\}] = \mathbb{P}[U = t] = 0$$

*because $U$ is uniformly distributed. Since $Y_t$ can only be $0$ or $1$, we get*

$$\mathbb{P}[X_t = 0] = 1 \text{ and } \mathbb{P}[Y_t = 0] = 1.$$

*The computations for $n > 1$ are similar. Therefore, $X$ and $Y$ have the same finite-dimensional distributions, but are different processes. Indeed, the sample paths of $X$ and $Y$ are different for all $\omega \in \Omega$.*

A natural question at this point is: Suppose that some finite-dimensional distributions are given, does there exist a process with these finite-dimensional distributions? Kolmogorov's extension theorem states that the answer is yes if these finite-dimensional distributions fulfil some natural consistency conditions. We will not study this question here and instead refer the interested reader to [2].

Let us next look at another important example of a stochastic process. Recall the definition of the multivariate Gaussian distribution from Definition 5.2.1.

**Definition 6.2.9** (Gaussian process). A *Gaussian process* is any stochastic process whose finite-dimensional distributions are multivariate Gaussians.

Since an $n$-dimensional multivariate Gaussian random vector is determined by its expectation vector $\mu$ and covariance matrix $\Sigma$, a Gaussian process is determined by the collection of expectations, $\mathbb{E}[X_t]$, and covariances, $\mathrm{Cov}(X_s, X_t)$, for all $s, t \in \mathcal{T}$.

**Example 6.2.10** (Discrete Gaussian white noise). *The* discrete Gaussian white noise *is a stochastic process* $\{X_t,\ t \in \mathbb{N}_0\}$, *where* $X_0, X_1, \ldots$ *are i.i.d.* $N(0, \sigma^2)$ *random variables. A sample path of a* discrete Gaussian white noise *is illustrated in Figure 6.3.*

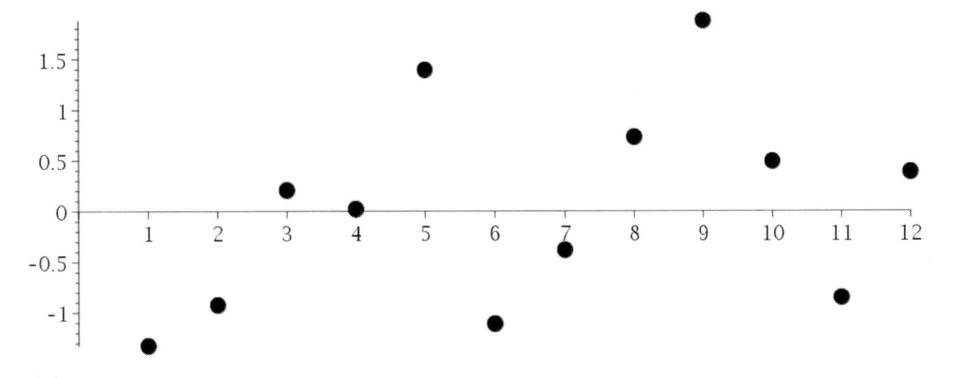


**Figure 6.3** Sample path of discrete white noise.

**Example 6.2.11** (Gaussian random walk). *In the* Gaussian random walk, $\mathcal{T} = \mathbb{N}_0$ *and each* step $X_n - X_{n-1}$ *is an independent* $N(\mu, \sigma^2)$ *random variable. The process* $\{W_t^{(n)}, t \in [0, T; n]\}$ *from Chapter 5 is an example of a Gaussian random walk.*

The stochastic processes in Examples 6.2.10 and 6.2.11 are both discrete-time stochastic processes on continuous state spaces. The notions in the following definition are useful.

**Definition 6.2.12.** Let $X$ be a stochastic process.

- $X$ is called *stationary* if the finite-dimensional distributions are invariant under shifts in time. More precisely, $(X_{t_1}, \ldots, X_{t_n})$ has the same distribution as $(X_{t_1+h}, \ldots, X_{t_n+h})$ for all possible choices of $n \geq 1$, $t_1 < \cdots < t_n \in \mathcal{T}$ and $h$ such that $t_1 + h, \ldots, t_n + h \in \mathcal{T}$.

- $X$ has *stationary increments* if $(X_{t_2} - X_{t_1}, \ldots, X_{t_n} - X_{t_{n-1}})$ has the same distribution as $(X_{t_2+h} - X_{t_1+h}, \ldots, X_{t_n+h} - X_{t_{n-1}+h})$ for all possible choices of $n \geq 1$, $t_1 < \cdots < t_n \in \mathcal{T}$ and $h$ such that $t_1 + h, \ldots, t_n + h \in \mathcal{T}$.
- $X$ has *independent increments* if $(X_{t_2} - X_{t_1}, \ldots, X_{t_n} - X_{t_{n-1}})$ are independent random variables for every choice of $n \geq 1$ and $t_1 < \cdots < t_n \in \mathcal{T}$.

Discrete Gaussian white noise is an example of a stationary process. The Poisson process is an example of a stochastic process with stationary and independent increments but it is not a stationary process.

## 6.3 Filtrations and Conditional Expectations

In this section we look at filtrations and conditional expectations and introduce stopping times. Since we work with an underlying probability space $(\Omega, \mathcal{F}, \mathbb{P})$, we will assume that any $\sigma$-algebra $\mathcal{G}$ is contained in $\mathcal{F}$, that is $\mathcal{G} \subset \mathcal{F}$.

**Definition 6.3.1** (Measurable). Let $(\Omega, \mathcal{F}, \mathbb{P})$ be a probability space and $X$ a random variable. Given a $\sigma$-algebra $\mathcal{G} \subseteq \mathcal{F}$, we say that $X$ is *$\mathcal{G}$-measurable* or *measurable with respect to the $\sigma$-algebra $\mathcal{G}$* if

$$\{X \in [a, b]\} \in \mathcal{G} \text{ for all } a \leq b \text{ with } a, b \in \mathbb{R}.$$

One defines a $\sigma$-algebra generated by a random variable $X$ in the general case in exactly the same way as in the discrete case, see Definition 2.3.11. The main difference is that in general we cannot illustrate $\sigma(X)$ as in Section 2.3.3 nor explicitly determine $\sigma(X)$ as in Lemma 2.3.15. We automatically get that $\sigma(X) \subseteq \mathcal{F}$ since all random variables on $(\Omega, \mathcal{F}, \mathbb{P})$ are by definition $\mathcal{F}$-measurable. As in the discrete case, we have to model the available information at different times. We do this by extending the definition of a filtration in Definition 2.3.18.

**Definition 6.3.2** (Filtration). Given an indexing set $\mathcal{T} \subseteq [0, \infty)$, a *filtration* is a collection of $\sigma$-algebras $\mathbb{F} = \{\mathcal{F}_t : t \in \mathcal{T}\}$ satisfying $\mathcal{F}_s \subseteq \mathcal{F}_t$ for all $0 \leq s \leq t$.

We also extend the definition of an adapted process from Definition 3.2.2 and give the definition of a natural filtration.

**Definition 6.3.3** (Adapted process). A stochastic process $\{X_t, t \in \mathcal{T}\}$ is *adapted* to a filtration $\mathbb{F} = \{\mathcal{F}_t, t \in \mathcal{T}\}$ if $X_t$ is $\mathcal{F}_t$-measurable for all $t \in \mathcal{T}$.

**Definition 6.3.4** (Natural filtration). Given a stochastic process $X$, the *natural filtration* is the smallest filtration $\mathbb{F}$ for which $\{X_t, t \in \mathcal{T}\}$ is adapted to $\mathbb{F}$.

If $\mathbb{F}$ is the natural filtration of a stochastic process and an event $A \in \mathcal{F}_t$, then it is possible to tell whether or not the event $A$ has occurred just by observing the stochastic process up to time $t$. It can be shown that the natural filtration is given by $\mathcal{F}_t = \sigma(\{X_s, 0 \leq s \leq t\})$. Observe that $\sigma(\{X_s, 0 \leq s \leq t\})$ is generated by events of the form

$$\{X_{s_1} \in [a_1, b_1], \ldots, X_{s_k} \in [a_k, b_k]\}$$

with $0 \leq s_1 < \cdots < s_k \leq t$, $k \in \mathbb{N}$, $a_j \leq b_j$ and $a_j, b_j \in \mathbb{R}$ and all arbitrary. Therefore, only events involving finite or countably many $X_s$ are automatically contained in this $\sigma$-algebra. Events that involve uncountably many $X_s$ do not need to be contained. For instance, the event

$$\{X_s \leq x;\ 0 \leq s \leq t\} = \bigcap_{0 \leq s \leq t} \{X_s \leq x\} \tag{6.3.1}$$

can be contained in $\sigma(\{X_s, 0 \leq s \leq t\})$, but is not necessarily so. If we encounter an event as in (6.3.1) and would like to show that this event is contained, then we have to express this event using at most countably many $X_s$.

Let us now make two useful observations.

**Proposition 6.3.5.** Let $(X_n)_{n \in \mathbb{N}}$ and $X$ be given with $X_n \xrightarrow{as} X$. If $\mathcal{G}$ is a $\sigma$-algebra for which all $X_n$ are $\mathcal{G}$ measurable, then $X$ is also $\mathcal{G}$ measurable.

*Proof.* For simplicity, assume that $X(\omega) = \lim_{n \to \infty} X_n(\omega)$ for all $\omega$ in the sample space. Then for all $a \leq b$,

$$\{X \in [a, b]\} = \bigcap_{j \in \mathbb{N}} \bigcup_{n_0 \in \mathbb{N}} \bigcap_{n \geq n_0} \{X_n \in [a - 1/j, b + 1/j]\}. \tag{6.3.2}$$

By assumption, $\{X_n \in [a - 1/j, b + 1/j]\} \in \mathcal{G}$ for all $a$, $b$, $n$ and $j$. Furthermore, the RHS of (6.3.2) involves only countable unions and intersections. Thus $\{X \in [a, b]\} \in \mathcal{G}$. This completes the proof of this case. The proof in the general case is similar and we thus omit it. □

We now extend Dynkin's lemma (see Lemma 2.3.22) to the general case.

**Lemma 6.3.6** (Dynkin's lemma, general version). Let $(\Omega, \mathcal{F}, \mathbb{P})$ be a probability space and $X, Y_1, \ldots, Y_T$ be random variables on this space. Then the following are equivalent.

- $X$ is $\sigma(Y_1, \ldots, Y_T)$-measurable.
- There exists a Borel-measurable function $f$ such that $X = f(Y_1, \ldots, Y_T)$.

This lemma is also known as Doob–Dynkin's lemma, but for simplicity we just call it Dynkin's lemma. We give the proof of this lemma in the case $T = 1$ in Section A.4.3 and the proof in the general situation can be found in [6, Lemma 1.13].

We next look at conditional expectations and martingales in the general situation. First, recall the general definition of conditional expectation from Section 2.3.5. Note that we have to do some minor adjustments since we are working with general probability spaces.

**Definition 6.3.7** (Conditional expectation with respect to a $\sigma$-algebra). Let $(\Omega, \mathcal{F}, \mathbb{P})$ be a probability space, $X$ be a random variable with $\mathbb{E}[|X|] < \infty$ and $\mathcal{G} \subset \mathcal{F}$ be a $\sigma$-algebra over $\Omega$. Then the *conditional expectation* $\mathbb{E}[X|\mathcal{G}]$ *of* $X$ *given* $\mathcal{G}$ is a random variable which is $\mathcal{G}$-measurable and fulfils

$$\mathbb{E}[X \cdot \mathbb{1}_A(\omega)] = \mathbb{E}\Big[\mathbb{E}[X|\mathcal{G}] \cdot \mathbb{1}_A(\omega)\Big] \quad \text{for all } A \in \mathcal{G}. \tag{6.3.3}$$

If $Y$ is a random variable then the *conditional expectation of* $X$ *given* $Y$ is defined by

$$\mathbb{E}[X|Y] = \mathbb{E}[X|\sigma(Y)].$$

As already mentioned in Section 2.3.5, $\mathbb{E}[X|\mathcal{G}]$ always exists. In addition, $\mathbb{E}[X|\mathcal{G}]$ is almost surely unique. In other words, if we have two random variables $Z_1$ and $Z_2$ fulfilling the conditions of Definition 2.2.14, then $Z_1 \stackrel{as}{=} Z_2$. Therefore, we can just write $\mathbb{E}[X|\mathcal{G}]$ without any danger of confusion. Let us consider an example.

**Example 6.3.8.** *Suppose that* $\mathcal{G} = \{\emptyset, B, B^c, \Omega\}$ *for some* $B \in \mathcal{F}$ *with* $0 < \mathbb{P}[B] < 1$. *Given an integrable random variable* $X$, *define the random variable* $Z$ *by*

$$Z(\omega) = \begin{cases} \frac{\mathbb{E}[X\mathbb{1}_B]}{\mathbb{P}[B]}, & \text{if } \omega \in B, \\ \frac{\mathbb{E}[X\mathbb{1}_{B^c}]}{\mathbb{P}[B^c]}, & \text{if } \omega \in B^c, \end{cases}$$

*where* $\mathbb{1}_A(\omega)$ *denotes the indicator function, see* (2.2.6). *Then* $Z$ *is* $\mathcal{G}$*-measurable since it is constant on* $B$ *and* $B^c$. *Furthermore,*

$$\begin{aligned}
\mathbb{E}[Z\mathbb{1}_\emptyset] &= \mathbb{E}[X\mathbb{1}_\emptyset] = 0, \\
\mathbb{E}[Z\mathbb{1}_B] &= \mathbb{E}\left[\frac{\mathbb{E}[X\mathbb{1}_B]}{\mathbb{P}[B]} \cdot \mathbb{1}_B\right] = \frac{\mathbb{E}[X\mathbb{1}_B]}{\mathbb{P}[B]} \cdot \mathbb{E}[\mathbb{1}_B] = \frac{\mathbb{E}[X\mathbb{1}_B]}{\mathbb{P}[B]} \cdot \mathbb{P}[B] = \mathbb{E}[X\mathbb{1}_B], \\
\mathbb{E}[Z\mathbb{1}_{B^c}] &= \mathbb{E}\left[\frac{\mathbb{E}[X\mathbb{1}_{B^c}]}{\mathbb{P}[B^c]} \cdot \mathbb{1}_{B^c}\right] = \frac{\mathbb{E}[X\mathbb{1}_{B^c}]}{\mathbb{P}[B^c]} \cdot \mathbb{E}[\mathbb{1}_{B^c}] = \frac{\mathbb{E}[X\mathbb{1}_{B^c}]}{\mathbb{P}[B^c]} \cdot \mathbb{P}\left[B^c\right] = \mathbb{E}[X\mathbb{1}_{B^c}], \\
\mathbb{E}[Z\mathbb{1}_\Omega] &= \mathbb{E}[Z(\mathbb{1}_B + \mathbb{1}_{B^c})] = \mathbb{E}[Z\mathbb{1}_B] + \mathbb{E}[Z\mathbb{1}_{B^c}] \\
&= \mathbb{E}[X\mathbb{1}_B] + \mathbb{E}[X\mathbb{1}_{B^c}] = \mathbb{E}[X(\mathbb{1}_B + \mathbb{1}_{B^c})] = \mathbb{E}[X\mathbb{1}_\Omega].
\end{aligned}$$

*Therefore* $\mathbb{E}[X\mathbb{1}_A] = \mathbb{E}[Z\mathbb{1}_A]$ *for all* $A \in \mathcal{G}$, *which implies* $Z = \mathbb{E}[X|\mathcal{G}]$.

*Observe that if we take $X = \mathbb{1}_A$ for some $A \in \mathcal{F}$, then for $\omega \in B$,*

$$\mathbb{E}\left[\mathbb{1}_A|\mathcal{G}\right](\omega) = \frac{\mathbb{E}\left[\mathbb{1}_A \mathbb{1}_B\right]}{\mathbb{P}\left[B\right]} = \frac{\mathbb{P}\left[A \cap B\right]}{\mathbb{P}\left[B\right]} = \mathbb{P}\left[A|B\right].$$

In the above example, it was easy to check condition (6.3.3) since we have an explicit expression for all events $A \in \mathcal{G}$. In the general situation, however, this is no longer possible. In this case, one can often use the following proposition.

**Proposition 6.3.9.** Let $(\Omega, \mathcal{F}, \mathbb{P})$ be a probability space, $X$ be a random variable with $\mathbb{E}\left[|X|\right] < \infty$ and $\mathcal{G} \subset \mathcal{F}$ be a $\sigma$-algebra over $\Omega$. Assume that $\mathcal{G}$ is generated by a family $\mathcal{I}$, which is stable under intersection and contains $\Omega$. In formulas,

$$\mathcal{G} = \sigma(\mathcal{I}), \ \Omega \in \mathcal{I} \text{ and } A, B \in \mathcal{I} \Longrightarrow A \cap B \in \mathcal{I}.$$

If $Z$ is a $\mathcal{G}$-measurable random variable for which $\mathbb{E}\left[X \cdot \mathbb{1}_A(\omega)\right] = \mathbb{E}\left[Z \cdot \mathbb{1}_A(\omega)\right]$ for all $A \in \mathcal{I}$, then $Z = \mathbb{E}\left[X|\mathcal{G}\right]$.

The proof of this proposition is straightforward, but requires the $\pi$–$\lambda$ theorem. For completeness, we give the proof of Proposition 6.3.9 in Section A.4.4.

One of the challenges of working with Definition 6.3.7 is that it does not provide a method for computing $\mathbb{E}\left[X|\mathcal{G}\right]$; rather, it states the properties that need to be satisfied by some random variable $Z$ in order to have $Z = \mathbb{E}\left[X|\mathcal{G}\right]$. The procedure for computing $\mathbb{E}\left[X|\mathcal{G}\right]$ is therefore to 'guess' the solution $Z$, and then check that the required properties hold. Since most $\sigma$-algebras we consider here are generated by random variables, it is useful to keep in mind the heuristic interpretation

$\mathbb{E}\left[X|Y\right]$ *is the average of the random variable $X$ when we know the random variable $Y$.*

We have studied the conditional expectation of discrete random variables in Section 2.2 in detail and in Definition 2.2.11 gave an explicit expression in the discrete case. In the remaining part of this book, we work mainly with continuous random variables, so we need to have a closer look at this case.

**Example 6.3.10.** *Suppose that $X$ and $Y$ are continuous random variables with joint density function $f_{X,Y}(x,y)$ and suppose that $X$ is integrable. For simplicity we assume that $f_{X,Y}$ is a continuous function and $f_{X,Y}(x,y) > 0$ for all $x, y$. The marginal distributions of $X$ and $Y$ are given by*

$$f_X(x) = \int_{-\infty}^{\infty} f_{X,Y}(x,w)dw \quad \text{and} \quad f_Y(y) = \int_{-\infty}^{\infty} f_{X,Y}(z,y)dz.$$

*Furthermore, the conditional density of $X$ given $Y$ is*

$$f_{X|Y}(x|y) = \frac{f_{X,Y}(x,y)}{f_Y(y)}.$$

*We can interpret* $f_{X|Y}(\cdot|y)$ *as the density of* $X$ *if we know that* $Y = y$. *If we know that* $Y = y$, *then the average of* $X$ *is*

$$h(y) = \int_{-\infty}^{\infty} x f_{X|Y}(x|y) dx.$$

*Heuristically, we can interpret* $h(y)$ *as* $\mathbb{E}[X|Y = y]$, *but of course* $\mathbb{E}[X|Y = y]$ *is not defined in this case since* $\mathbb{P}[Y = y] = 0$. *In view of the heuristic interpretation of* $\mathbb{E}[X|Y]$ *as well as the expression for* $\mathbb{E}[X|Y]$ *in* (2.2.12) *in the discrete case, we can guess that* $\mathbb{E}[X|Y] = h(Y)$. *Let us now check that this is correct. First,* $h(y)$ *is a continuous function in* $y$ *since by assumption* $f_{X,Y}$ *is continuous. Lemma 6.3.6 implies that* $h(Y)$ *is* $\sigma(Y)$*-measurable. It remains to check* (6.3.3). *We begin with the event* $A = \{Y \in [a, b]\}$ *with* $a \leq b$. *We have*

$$\begin{aligned}\mathbb{E}[h(Y)\mathbb{1}_A(\omega)] &= \int_{-\infty}^{\infty} h(y)\mathbb{1}_{\{y\in[a,b]\}} f_Y(y)\, dy = \int_{-\infty}^{\infty}\int_{-\infty}^{\infty} x f_{X|Y}(x|y)\, \mathbb{1}_{\{y\in[a,b]\}} f_Y(y)\, dxdy \\ &= \int_{-\infty}^{\infty}\int_{-\infty}^{\infty} x\mathbb{1}_{\{y\in[a,b]\}} f_{X,Y}(x,y)\, dxdy = \mathbb{E}[X\mathbb{1}_A(\omega)].\end{aligned}$$

*Therefore* (6.3.3) *is fulfilled for* $A = \{Y \in [a, b]\}$. *Since these events generate* $\sigma(Y)$, *we can apply Proposition 6.3.9 and immediately get that* $\mathbb{E}[X|Y] = h(Y)$.

When $\mathcal{G}$ is neither a finite set nor equal to $\sigma(Y)$ for some random variable $Y$, it is difficult, if not impossible, to calculate $\mathbb{E}[X|\mathcal{G}]$ explicitly. For this reason, it is important to be able to manipulate expressions involving conditional expectation without knowing their particular form. Many of the properties we saw in Section 2.2.2 also hold in the general situation. Before we state them, we introduce independent $\sigma$-algebras.

**Definition 6.3.11.** Let $(\Omega, \mathcal{F}, \mathbb{P})$ be a probability space and $\mathcal{G}, \mathcal{G}' \subset \mathcal{F}$ be two $\sigma$-algebras over $\Omega$. The $\sigma$-algebras $\mathcal{G}$ and $\mathcal{G}'$ are called independent if $\mathbb{P}[A \cap B] = \mathbb{P}[A]\,\mathbb{P}[B]$ for all $A \in \mathcal{G}$ and $B \in \mathcal{G}'$.

Using the $\pi$–$\lambda$ theorem, one can show that $\mathcal{G}$ and $\mathcal{G}'$ are independent if they are generated by independent random variables. We omit the proof and instead refer the reader to [1, Chapter 3]. The following theorem gives the main properties of conditional expectations which we require.

**Theorem 6.3.12** (Properties of conditional expectation)**.** *Let* $X$ *and* $Y$ *be integrable random variables and let* $\mathcal{G} \subseteq \mathcal{F}$ *be a* $\sigma$*-algebra. Then the following properties hold.*

*(i)* $\mathbb{E}[\lambda X + \mu Y|\mathcal{G}] = \lambda\mathbb{E}[X|\mathcal{G}] + \mu\mathbb{E}[Y|\mathcal{G}]$, *for all scalars* $\lambda, \mu \in \mathbb{R}$.

*(ii)* $|\mathbb{E}[X|\mathcal{G}]| \leq \mathbb{E}[|X|\,|\mathcal{G}]$.

(iii) *Let $\varphi : \mathbb{R} \to \mathbb{R}$ be a convex function with $\mathbb{E}[|\varphi(X)|] < \infty$. Then*

$$\varphi(\mathbb{E}[X|\mathcal{G}]) \leq \mathbb{E}[|\varphi(X)|\mathcal{G}].$$

(iv) *If $\mathcal{G} = \{\emptyset, \Omega\}$, then $\mathbb{E}[X|\mathcal{G}] = \mathbb{E}[X]$.*

(v) *If $Y$ is $\mathcal{G}$-measurable and $XY$ is integrable, then $\mathbb{E}[XY|\mathcal{G}] = Y\mathbb{E}[X|\mathcal{G}]$.*

(vi) *Tower property: If $\mathcal{G}'$ is another $\sigma$-algebra with $\mathcal{G}' \subseteq \mathcal{G}$, then $\mathbb{E}\left[\mathbb{E}[X|\mathcal{G}]|\mathcal{G}'\right] = \mathbb{E}\left[X|\mathcal{G}'\right]$. In particular, taking $\mathcal{G}' = \{\emptyset, \Omega\}$ gives $\mathbb{E}[\mathbb{E}[X|\mathcal{G}]] = \mathbb{E}[X]$.*

(vii) *If $\sigma(X)$ and $\mathcal{G}$ are independent, then $\mathbb{E}[X|\mathcal{G}] = \mathbb{E}[X]$.*

*Proof.* We give only the proofs of (ii) and (vii) here. The proofs of the other properties are left as Exercise 6.4. We begin with (ii). Consider the event $A = \{\mathbb{E}[X|\mathcal{G}] > \mathbb{E}[|X|\,|\mathcal{G}]\}$. Then $A \in \mathcal{G}$, since $\mathbb{E}[X|\mathcal{G}]$ and $\mathbb{E}[|X|\,|\mathcal{G}]$ are both $\mathcal{G}$-measurable. Suppose that $\mathbb{P}[A] > 0$. This, together with the definition of conditional expectations, implies

$$\begin{aligned} 0 &< \mathbb{E}\left[\left(\mathbb{E}[X|\mathcal{G}] - \mathbb{E}[|X|\,|\mathcal{G}]\right) \cdot \mathbb{1}_A\right] = \mathbb{E}\left[\mathbb{E}[X|\mathcal{G}] \cdot \mathbb{1}_A\right] - \mathbb{E}\left[\mathbb{E}[|X|\,|\mathcal{G}] \cdot \mathbb{1}_A\right] \\ &= \mathbb{E}\left[X \cdot \mathbb{1}_A\right] - \mathbb{E}\left[|X| \cdot \mathbb{1}_A\right] = \mathbb{E}\left[(X - |X|) \cdot \mathbb{1}_A\right] \leq 0, \end{aligned}$$

which is a contradiction. Therefore $\mathbb{P}[A] = 0$. Similarly, $\mathbb{P}[-\mathbb{E}[X|\mathcal{G}] > \mathbb{E}[|X|\,|\mathcal{G}]] = 0$. This completes the proof of (ii). For (iv), first observe that since $\mathbb{E}[X]$ is a constant, it is $\mathcal{G}$-measurable. Now let $A \in \mathcal{G}$. Then $\mathbb{1}_A$ and $X$ are independent. Hence

$$\mathbb{E}[X\mathbb{1}_A] = \mathbb{E}[X]\,\mathbb{E}[\mathbb{1}_A] = \mathbb{E}[\mathbb{E}[X]\,\mathbb{1}_A],$$

where the last equality used that $\mathbb{E}[X]$ is a constant, and so $\mathbb{E}[X] = \mathbb{E}[X|\mathcal{G}]$. □

We saw in Chapters 3 and 4 that martingales play an important role in the pricing of a contingent claim. We therefore have to generalise the definition of a martingale in Definition 3.3.15 to continuous time.

**Definition 6.3.13** (Martingale). A stochastic process $\{X_t, t \geq 0\}$ is a *martingale* with respect to a filtration $\mathbb{F} = \{\mathcal{F}_t, t \geq 0\}$ if

(a) $\{X_t, t \geq 0\}$ is adapted to $\mathbb{F}$, that is $X_t$ is $\mathcal{F}_t$-measurable for all $t \geq 0$,

(b) $X_t$ is integrable for all $t \geq 0$, that is $\mathbb{E}[|X_t|] < \infty$ and

(c) $\mathbb{E}[X_t|\mathcal{F}_s] = X_s$ for each $0 \leq s \leq t$.

Property (c) embodies the idea of a fair game. If $X_t$ denotes the wealth at time $t$ of a gambler playing a fair game, then his expected fortune at time $t > s$, given the outcomes up to time $s$, is just his wealth at time $s$. In particular,

$$\mathbb{E}[X_t] = \mathbb{E}[X_0] \quad \text{for all} \quad t \geq 0. \tag{6.3.4}$$

The proof of this equation is almost the same as in discrete time, see Lemma 3.3.18.

Many examples of martingales in continuous time, especially those we need in the Black–Scholes model, involve Brownian motion. However, as we have not yet discussed Brownian motion, we postpone those examples to Chapter 7 (see Example 7.3.6). Instead, we will now give the compensated Poisson process as an example of a continuous-time martingale.

**Example 6.3.14** (Compensated Poisson process). *Let $N = \{N_t, t \geq 0\}$ be a Poisson process with rate $\lambda > 0$, see Example 6.2.5. Let $\mathbb{F}$ be the natural filtration of the Poisson process $N$. Define $X_t := N_t - \lambda t$. Then the process $\{X_t, t \geq 0\}$ is a martingale. We have to check the properties in Definition 6.3.13.*

*(a) $X_t = N_t - \lambda t$ is $\mathcal{F}_t$-measurable since $N_t$ is $\mathcal{F}_t$-measurable and $\lambda t$ is a constant.*
*(b) $N_t$ is Poisson distributed so $\mathbb{E}[|N_t|] = \mathbb{E}[N_t] = \lambda t$. Therefore, $\mathbb{E}[|X_t|] \leq 2\lambda t < \infty$.*
*(c) Since $N_t - N_s$ is independent of $N_s$ for all $0 \leq s \leq t$,*

$$\begin{aligned}\mathbb{E}[X_t \mid N_s] &= \mathbb{E}[N_t - N_s + N_s \mid N_s] - \lambda t \\ &= \mathbb{E}[N_t - N_s \mid N_s] + \mathbb{E}[N_s \mid N_s] - \lambda t \\ &= \mathbb{E}[N_t - N_s] + N_s - \lambda t = N_s - \lambda s = X_s.\end{aligned}$$

*The tower property then implies*

$$\mathbb{E}[X_t \mid \mathcal{F}_s] = \mathbb{E}\big[\mathbb{E}[X_t \mid N_s] \bigm| \mathcal{F}_s\big] = \mathbb{E}[X_s \mid \mathcal{F}_s] = X_s.$$

*Therefore, the process $\{X_t, t \geq 0\}$ is indeed a martingale.*

## 6.4 Exercises

**Exercise 6.1.** *Let $U$ be uniformly distributed on $[0,1]$ (i.e. with probability density function $f_U(x) = \mathbb{1}_{[0,1]}(x)$), and let $U_j$ be mutually independent copies of $U$ on $(\Omega, \mathcal{F}, \mathbb{P})$.*

*(a) Find the cumulative distribution function $F_U$ of $U$.*
*(b) Show that the cumulative distribution function $F_{Z_n}$ of the random variable $Z_n = \max\{U_1, U_2, \ldots, U_n\}$ satisfies $F_{Z_n} = F_U^n$, and hence find $F_{Z_n}$.*
*(c) Show that $Z_n \to 1$ in probability as $n \to \infty$.*
*(d) Show that*

$$\mathbb{P}[n(1 - Z_n) \leq t] \to \begin{cases} 1 - e^{-t}, & t > 0, \\ 0, & t \leq 0, \end{cases}$$

*as $n \to \infty$. (Therefore one can obtain an exponential distribution from a sequence of uniform random variables.)*

**Exercise 6.2.** *For each of the following Gaussian processes $X$, obtain the two-dimensional distributions $(X_{t_1}, X_{t_2})$ where $t_1, t_2 \in \mathcal{T}$.*

*(a) $X$ is a discrete Gaussian white noise.*
*(b) $X$ is a Gaussian random walk.*
*(c) $X$ is defined by $X(t) = tZ + b$ for $t \in [0, \infty)$, where $Z \sim N(\mu, \sigma^2)$.*

**Exercise 6.3.** *Let $\Omega$ be the set of infinite sequences of heads and tails. Define $X$ to be the simple random walk starting from 0 that goes up if a head occurs and down if a tail occurs, so, for example, if $\omega = HHTHTTTHT\ldots$, then*

$$(X_0(\omega), X_1(\omega), \ldots) = (0, 1, 2, 1, 2, 1, 0, -1, 0, -1, \ldots).$$

*In the natural filtration $\mathbb{F} = \{\mathcal{F}_n, n \in \mathbb{N}\}$, show that $\mathcal{F}_n$ contains $2^{2^n}$ elements.*

**Exercise 6.4.** *Give the proofs of properties (i), (iii), (iv), (v) and (vi) from Theorem 6.3.12.*

**Exercise 6.5.** *Let $N = \{N_t, t \geq 0\}$ be a Poisson process with rate $\lambda > 0$ and let $\mathbb{F}$ be the natural filtration of the Poisson process $N$. Define $X_t := (N_t - \lambda t)^2 - \lambda t$. Show that the process $\{X_t, t \geq 0\}$ is a martingale.*

# 7 Brownian Motion

The computations in Section 5.6 suggest that it is reasonable to (try to) extend the discrete Black–Scholes model from Section 5.2 to a continuous-time model on $[0, T]$. The main difference is that we have to assign a stock price to each $t \in [0, T]$ instead of only to each $t \in [0, T; n]$. It is natural to request that the stock price is a continuous function in $t$. The first step in doing this is to extend the process $\{W_t^{(n)}, t \in [0, T; n]\}$ to a continuous-time process on $[0, T]$ and study its properties. This is the main topic of this chapter.

## 7.1 Definition of Brownian Motion

The generalisation of the Gaussian process $\{W_t^{(n)}, t \in [0, T; n]\}$ to a continuous stochastic process is called Brownian motion.

**Definition 7.1.1** (Brownian motion, Version 1). Let $x \in \mathbb{R}$ and $\sigma > 0$ be given. A continuous-time stochastic process $\{W_t, t \geq 0\}$ is called a *Brownian motion* (or *Wiener process*) if

(a) $W_0 = x$.
(b) For every $0 \leq s \leq t$, $W_t - W_s \sim \mathcal{N}(0, \sigma^2(t - s))$.
(c) $\{W_t, t \geq 0\}$ has (stationary and) independent increments, see Definition 6.2.12.
(d) $W_t$ is continuous in $t \geq 0$, see Definition 6.2.6.

The parameter $\sigma^2$ is called the *diffusivity* and $x$ the starting point. The case $x = 0$ and $\sigma = 1$ is called *standard Brownian motion*. Unless we specify otherwise, we will assume that $W = \{W_t, t \geq 0\}$ is a standard Brownian motion starting from 0.

Brownian motion $W$ is a continuous-time stochastic process. It has continuous state space, continuous sample paths, but it is not a stationary process.

Conditions (b) and (c) in Definition 7.1.1 together determine the finite-dimensional distributions of $\{W_t, t \geq 0\}$ and in particular tell us that $W$ is a Gaussian process. Since Gaussian processes are determined by the collection of expectations and covariance matrices of the finite-dimensional distributions, we have the following equivalent definition of a standard Brownian motion, which is often easier to work with.

**Definition 7.1.2** (Brownian motion, Version 2). A continuous-time Gaussian process $\{W_t, t \geq 0\}$ is a *standard Brownian motion* if

(a) $W_0 = 0$.
(b) $\mathbb{E}[W_t] = 0$ for all $t$.
(c) For every $s, t > 0$, $\mathrm{Cov}(W_s, W_t) = s \wedge t$, where $s \wedge t = \min(s, t)$.
(d) $W_t$ is continuous in $t \geq 0$.

The following questions naturally arise on encountering Brownian motion for the first time:

- How should one visualise Brownian motion physically? Is there a natural driving force or mechanism behind Brownian motion?
- On which probability space can $W$ be realised?
- Does a process $W$ exist?

For comparison, consider a Poisson process $\{P_t, t \geq 0\}$. The mechanism behind a Poisson process (see Example 6.2.5) can be described as follows. At time $t = 0$ start to count some incidents, for instance arriving customers. The waiting time to the first incident is exponentially distributed. The waiting time from the first to the second incident is also exponentially distributed and independent of the first waiting time. Similarly for the other waiting times. Then $P_t$ is the number of incidents that have happened by time $t$. Mathematically, we can construct a Poisson process $\{P_t, t \geq 0\}$ as follows. Let $(T_j)_{j=1}^{\infty}$ be an i.i.d. sequence of random variables with $T_j \sim \mathrm{Exp}(\lambda)$ for some $\lambda > 0$. Define $I_0 := 0$ and for $n \in \mathbb{N}$ and $t \geq 0$ set

$$I_n := \sum_{j=1}^{n} T_j \quad \text{and} \quad P_t := \max\{n \in \mathbb{N}_0,\, I_n \leq t\}. \tag{7.1.1}$$

The random variable $I_n$ can be interpreted as the time at which the $n$th incident occurs, for instance the $n$th customer arrives. One can now show that $\{P_t, t \geq 0\}$ is indeed a Poisson process. However, we omit this computation here as it can be found in many textbooks on stochastic processes.

Let us now return to Brownian motion. The mechanism behind a Brownian motion is unfortunately not as simple as for a Poisson process. Heuristically, one can try to adjust the argument used in Chapter 5 for $\{W_t^{(n)}, t \in [0, T; n]\}$ and replace the time step $h = T/n$ by an infinitesimally small time step $dt$. This results in an infinitesimally small Gaussian jump at each infinitesimally small time step $dt$. This heuristic interpretation can be a little bit confusing. Therefore, it is often better to view a Brownian motion as a random curve, that is randomly selecting a curve in $C(0, \infty)$. We will not prove the existence of Brownian motion here, but we would like to mention at this point that one can use the sample space $\Omega = C(0, \infty)$ to construct Brownian motion. Further details can be found, for instance, in [2].

Brownian motion arises naturally as the limit of discrete-time processes. Recall that the central limit theorem (CLT) states that the Gaussian distribution occurs as the limit of rescaled sums of i.i.d. random variables, see Theorem A.1.5. Brownian motion arises from a generalisation of this result as the limit of random walks scaled in both time and space.

**Theorem 7.1.3.** *Let $(Y_i)_{i=1}^{\infty}$ be a sequence of i.i.d. random variables with $\mathbb{E}[Y_i] = 0$ and $\mathrm{Var}(Y_i) = 1$. For each $n \geq 1$, define the stochastic process $\{Z_t^{(n)}, t \geq 0\}$ by*

$$Z_t^{(n)} = \frac{1}{\sqrt{n}} \sum_{1 \leq i \leq nt} Y_i, \quad t \geq 0. \tag{7.1.2}$$

*Then for any $k \geq 1$, $0 \leq t_1 \leq \cdots \leq t_k$,*

$$\left(Z_{t_1}^{(n)}, \ldots, Z_{t_k}^{(n)}\right) \xrightarrow{d} (W_{t_1}, \ldots, W_{t_k}), \tag{7.1.3}$$

*where $\{W_t, t \geq 0\}$ is a Brownian motion.*

Note that $Z_t^{(n)} = 0$ for $0 \leq t < 1/n$ since the empty sum is interpreted as 0. Theorem 7.1.3 shows that the finite-dimensional distributions of $\{Z_t^{(n)}, t \geq 0\}$ converge in distribution to the finite-dimensional distributions of the Brownian motion. Unfortunately, the finite-dimensional distributions do not determine a process uniquely, see Example 6.2.8. Therefore one has to request a little bit more for the convergence of continuous-time stochastic processes. However, we will not discuss this here as it goes beyond the scope of this book and we do not need it for our purposes. Instead, we refer the interested reader to [2]. For completeness, the process $\{Z_t^{(n)}, t \geq 0\}$ does indeed converge to Brownian motion.

*Proof of Theorem 7.1.3.* We give the proof only in the case when $k = 2$; the general case is similar. Suppose $0 \leq s \leq t$. By linearity properties of the multivariate Gaussian distribution, in order to show that $(Z_s^{(n)}, Z_t^{(n)}) \xrightarrow{d} (W_s, W_t)$, it is sufficient to show that

$$(Z_s^{(n)}, Z_t^{(n)} - Z_s^{(n)}) \xrightarrow{d} (W_s, W_t - W_s). \tag{7.1.4}$$

We now have

$$Z_t^{(n)} - Z_s^{(n)} = \frac{1}{\sqrt{n}} \sum_{i=\lfloor ns \rfloor+1}^{\lfloor nt \rfloor} Y_i$$

with $\lfloor x \rfloor = \max\{n \in \mathbb{Z}; n \leq x\}$. Since the $Y_i$ are independent, $Z_t^{(n)} - Z_s^{(n)}$ and $Z_s^{(n)}$ are independent. Therefore

$$\mathbb{P}\left[Z_s^{(n)} \leq z, Z_t^{(n)} - Z_s^{(n)} \leq w\right] = \mathbb{P}\left[Z_s^{(n)} \leq z\right] \mathbb{P}\left[Z_t^{(n)} - Z_s^{(n)} \leq w\right].$$

We can write

$$Z_t^{(n)} - Z_s^{(n)} = \frac{\sum_{i=\lfloor ns \rfloor+1}^{\lfloor nt \rfloor} Y_i}{\sqrt{\lfloor nt \rfloor - \lfloor ns \rfloor}} \frac{\sqrt{\lfloor nt \rfloor - \lfloor ns \rfloor}}{\sqrt{n}}.$$

By the central limit theorem, see Theorem A.1.5, the first ratio converges in distribution to an $\mathcal{N}(0, 1)$ random variable. By basic analysis the second ratio converges to $\sqrt{t-s}$. Hence

$$\lim_{n \to \infty} \mathbb{P}\left[Z_t^{(n)} - Z_s^{(n)} \leq w\right] = \Phi(w/\sqrt{t-s}) = \mathbb{P}\left[W_t - W_s \leq w\right].$$

Using the fact that $Z_0^{(n)} = W_0 = 0$, we get a similar result for $Z_s^{(n)}$. Therefore

$$\lim_{n\to\infty} \mathbb{P}\left[Z_s^{(n)} \le z, Z_t^{(n)} - Z_s^{(n)} \le w\right] = \mathbb{P}\left[W_s \le z\right]\mathbb{P}\left[W_t - W_s \le w\right]$$
$$= \mathbb{P}\left[W_s \le z, W_t - W_s \le w\right],$$

where the final step used the fact that Brownian motion has independent increments. □

Theorem 7.1.3 shows that the process $\{W_t^{(n)}, t \ge 0\}$ from Chapter 5 is a good approximation of Brownian motion for $n$ large. This theorem also suggests a method for simulating Brownian motion with a computer: generate a realisation of a large number of i.i.d. random variables $Y_i$ and then use (7.1.2). This leads to a graph as in Figure 7.1, which is an approximation to a sample path of Brownian motion. Note that many mathematical software packages have a built-in tool to generate Brownian motions.

## 7.2 Properties of Sample Paths of Brownian Motion

We know that Brownian motion is continuous, that is each sample path is continuous. A natural question at this point is to ask which other properties a typical sample path has. These properties have been studied quite intensively and a lot is known about them.

Let us start with a simple observation of Figure 7.1. This figure suggests that a sample path of a Brownian motion is very rough. In fact, a typical sample path is not differentiable at any point.

**Theorem 7.2.1.** *Let $t_0 \ge 0$ be arbitrary. Then, with probability 1, Brownian motion is not differentiable at $t_0$. Furthermore,*

$$\mathbb{P}\left[\limsup_{t\to t_0} \left|\frac{W_t - W_{t_0}}{t - t_0}\right| = \infty\right] = 1. \tag{7.2.1}$$

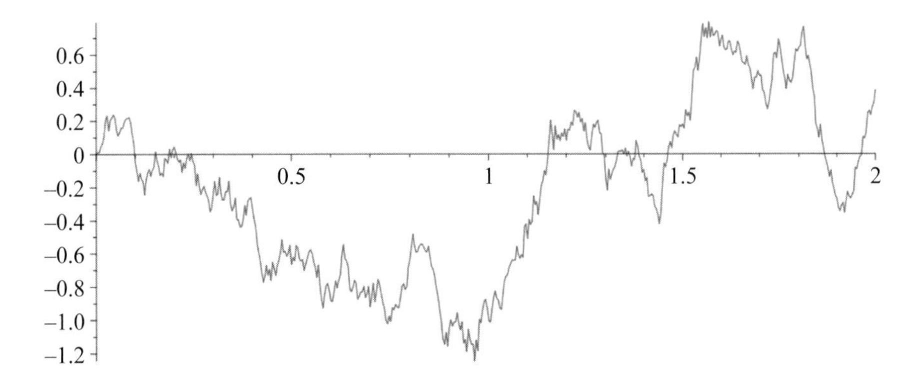


**Figure 7.1** Sample path of Brownian motion.

*Proof.* We have to show that

$$\mathbb{P}\left[\{\omega \in \Omega; W(t,\omega) \text{ is differentiable at } t_0\}\right] = 0,$$

where $\Omega$ is the sample space on which the Brownian motion is realised. Suppose now that we select randomly an $\omega \in \Omega$ and look at the sample path of the Brownian motion corresponding to this $\omega$. If this sample path is differentiable at $t_0$, then we must have

$$\limsup_{t\to t_0} \left| \frac{W_t(\omega) - W_{t_0}(\omega)}{t - t_0} \right| < \infty.$$

Therefore it is sufficient to establish (7.2.1) to prove the theorem. Let $(t_j)_{j=1}^{\infty}$ be a sequence with $t_j \to t_0$. For each $M > 0$,

$$\mathbb{P}\left[\limsup_{t\to t_0} \left| \frac{W_t - W_{t_0}}{t - t_0} \right| = \infty\right] \geq \lim_{M\to\infty} \mathbb{P}\left[\limsup_{j\to\infty} \left| \frac{W_{t_j} - W_{t_0}}{t_j - t_0} \right| > M\right].$$

Furthermore,

$$\mathbb{P}\left[\limsup_{j\to\infty} \left| \frac{W_{t_j} - W_{t_0}}{t_j - t_0} \right| > M\right] = \mathbb{P}\left[\bigcap_{j=1}^{\infty}\bigcup_{n=j}^{\infty}\left\{\left| \frac{W_{t_n} - W_{t_0}}{t_n - t_0} \right| > M\right\}\right]$$

$$= \lim_{j\to\infty} \mathbb{P}\left[\bigcup_{n=j}^{\infty}\left\{\left| \frac{W_{t_n} - W_{t_0}}{t_n - t_0} \right| > M\right\}\right] \geq \limsup_{j\to\infty} \mathbb{P}\left[\left| \frac{W_{t_j} - W_{t_0}}{t_j - t_0} \right| > M\right].$$

Now $W_{t_j} - W_{t_0} \sim \mathcal{N}(0, |t_j - t_0|)$. So, as $j \to \infty$,

$$\mathbb{P}\left[\left| \frac{W_{t_j} - W_{t_0}}{t_j - t_0} \right| > M\right] = \mathbb{P}\left[|Z| > M\sqrt{t_j - t_0}\right] \longrightarrow \mathbb{P}\left[|Z| > 0\right] = 1, \tag{7.2.2}$$

where $Z \sim \mathcal{N}(0,1)$. This completes the proof. □

However, the sample paths of a Brownian motion are not arbitrary rough. Let us recall the following definition.

**Definition 7.2.2.** Let $\alpha > 0$. A function $f : [0,\infty) \to \mathbb{R}$ is said to be locally $\alpha$-Hölder continuous if for every $0 \leq a < b < \infty$ there exists a $c = c(f,\alpha,a,b) > 0$ such that

$$|f(t) - f(s)| \leq c|t - s|^{\alpha} \qquad \text{for all } t, s \in [a,b]. \tag{7.2.3}$$

One can now show that

**Proposition 7.2.3.** For $\alpha > \frac{1}{2}$, almost all sample paths of Brownian motion are nowhere locally $\alpha$-Hölder continuous. For $\alpha < \frac{1}{2}$, almost all sample paths of Brownian motion are locally $\alpha$-Hölder continuous.

We do not give the proof of this proposition here and refer the interested reader to [14, Corollary 9.3 and 10.2]. The behaviour of the Brownian motion as $t \to \infty$ has also been studied carefully.

**Proposition 7.2.4.** With probability 1, Brownian motion $\{W_t, t \geq 0\}$ is unbounded.

*Proof.* Let $M \in \mathbb{N}$ be arbitrary. Then

$$\begin{aligned}\mathbb{P}[|W_t| \leq M \text{ for all } t \geq 0] &\leq \mathbb{P}[|W_n| \leq M \text{ for all } n \in \mathbb{N}] \\ &\leq \mathbb{P}[|W_1| \leq M \text{ and } |W_{n+1} - W_n| \leq 2M \text{ for all } n \in \mathbb{N}].\end{aligned}$$

Since the increments of Brownian motion are independent and stationary,

$$\mathbb{P}[|W_t| \leq M \text{ for all } t \geq 0] \leq \mathbb{P}[|W_1| \leq M] \cdot \lim_{N\to\infty} \left(\mathbb{P}[|W_1| \leq 2M]\right)^N.$$

Using that $W_1 \sim \mathcal{N}(0,1)$, $\mathbb{P}[|W_1| \leq 2M] < 1$ and so $\mathbb{P}[|W_t| \leq M \text{ for all } t \geq 0] = 0$. Then

$$\begin{aligned}\mathbb{P}[W_t \text{ is bounded for all } t \geq 0] &= \mathbb{P}\left[\bigcup_{M\in\mathbb{N}} \{|W_t| \leq M \text{ for all } t \geq 0\}\right] \\ &= \lim_{M\to\infty} \mathbb{P}[|W_t| \leq M \text{ for all } t \geq 0] = 0.\end{aligned}$$

This completes the proof. □

## 7.3 Transformations of Brownian Motion

A useful tool when working with Brownian motion is that it is invariant under certain transformations.

**Theorem 7.3.1.** *Let $\{W_t, t \geq 0\}$ be a Brownian motion. Then each of the following processes is also a Brownian motion.*

(i) *Reflection: $X_t = -W_t$, $t \geq 0$.*
(ii) *Markov property: $X_t = W_{\tau+t} - W_\tau$, $t \geq 0$, for any fixed time $\tau \in \mathbb{R}$.*
(iii) *Time-reversal: $X_t = W_\tau - W_{\tau-t}$, $0 \leq t \leq \tau$, for any fixed time $\tau \in \mathbb{R}$.*
(iv) *Rescaling: $X_t = c^{-1/2} W_{ct}$, $t \geq 0$ where $c > 0$.*
(v) *Inversion: $X_t = tW_{1/t}$, $t > 0$ and $W_0$ is taken to be 0.*

*Proof.* We start with the proof of (ii). We check the conditions of Definition 7.1.2. Since $\{W_t, t \geq 0\}$ is a (continuous-time) Gaussian process, so is $\{W_{\tau+t}, t \geq 0\}$. Since linear combinations of multivariate Gaussian vectors are still multivariate Gaussians, we see that $\{W_{\tau+t} - W_\tau, t \geq 0\}$ is a Gaussian process. Furthermore

(a) $X_0 = W_\tau - W_\tau = 0$.

(b) $\mathbb{E}[X_t] = \mathbb{E}[W_{\tau+t}] - \mathbb{E}[W_\tau] = 0$.

(c) We have

$$\begin{aligned}\operatorname{Cov}(X_s, X_t) &= \operatorname{Cov}(W_{\tau+s} - W_\tau, W_{\tau+t} - W_\tau)\\ &= \operatorname{Cov}(W_{\tau+s}, W_{\tau+t}) - \operatorname{Cov}(W_\tau, W_{\tau+t}) - \operatorname{Cov}(W_{\tau+s}, W_\tau) + \operatorname{Cov}(W_\tau, W_\tau)\\ &= (\tau+s)\wedge(\tau+t) - \tau - \tau + \tau\\ &= s \wedge t.\end{aligned}$$

(d) The sample path $W_t(\omega)$ is continuous in $t \geq 0$ for all $\omega \in \Omega$. Thus the sample path $W_{\tau+t}(\omega)$ is also continuous for all $\omega \in \Omega$ and hence so is $X_t$ a continuous process.

This completes the proof of (ii). The proofs of the other cases are similar so (with the exception of (v)) they are left as Exercise 7.2. The proof of the continuity at $t = 0$ in case (v) is a little bit more tricky. The details for this computation can be found for instance in [14, Section 2.13]. □

Transformation (iv) tells us that Brownian motion looks the same on different scales. Processes with this property are called *fractal processes*. Transformation (v) enables us to relate properties of $W_t$ as $t \to \infty$ to those of $W_t$ as $t \to 0$ and vice versa.

There are many processes related to Brownian motion. We give some examples.

**Definition 7.3.2** (Brownian motion with drift). Let $\{W_t, t \geq 0\}$ be a Brownian motion. A *Brownian motion with drift* is a process of the form

$$X_t = \mu t + \sigma W_t, \quad t \geq 0,$$

where $\mu \in \mathbb{R}$ and $\sigma > 0$ are constants. We call $\mu$ the *drift* of the process and $\sigma$ the *diffusivity*.

A *Brownian motion with drift* is a continuous Gaussian process with

$$\mathbb{E}[X_t] = \mu t \quad \text{and} \quad \operatorname{Cov}(X_s, X_t) = \sigma^2 (s \wedge t).$$

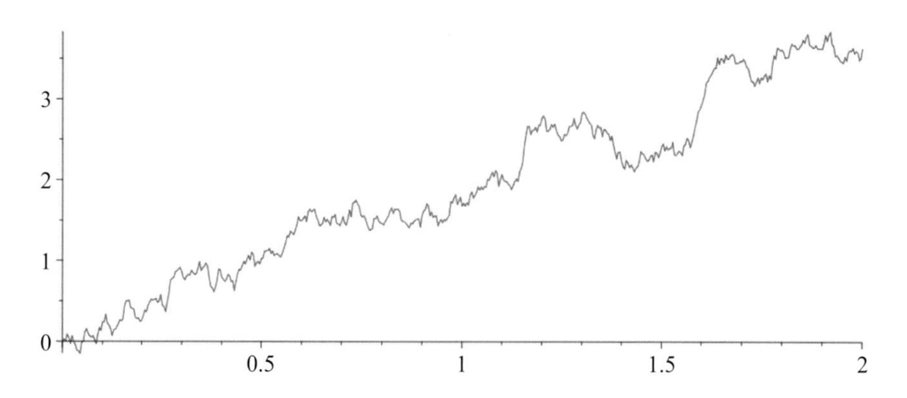


**Figure 7.2** Sample path of a Brownian motion with drift $\mu = 2$ and $\sigma = 1$.

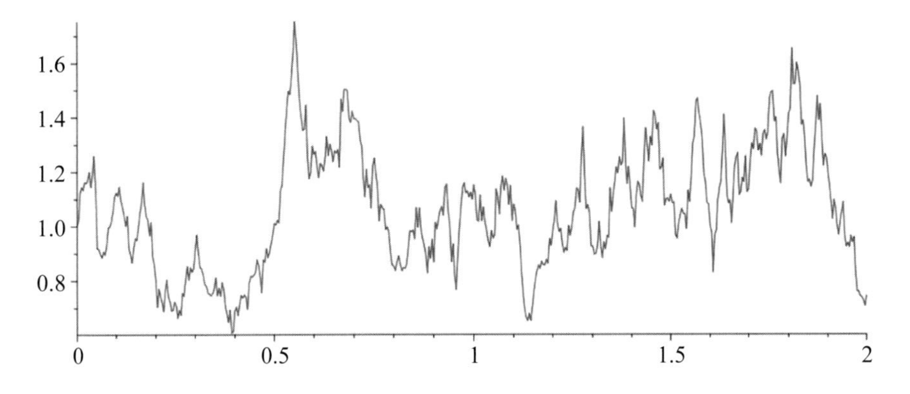


**Figure 7.3** Sample path of a geometric Brownian motion.

A sample path is given in Figure 7.2.

**Definition 7.3.3** (Geometric Brownian motion)**.** Let $\{W_t, t \geq 0\}$ be a Brownian motion. A *geometric Brownian motion* is a process of the form

$$X_t = e^{\mu t + \sigma W_t}, \quad t \geq 0,$$

where $\mu \in \mathbb{R}$ and $\sigma > 0$ are constants again called the *drift* and *diffusivity*, respectively.

Observe that $X_t$ is the exponential of a Brownian motion with drift. This is *not* a Gaussian process. A sample path is illustrated in Figure 7.3. To find the expectation and variance of $X_t$, we need to compute the *moment generating function* of an $\mathcal{N}(0, 1)$ random variable $Z$:

$$\mathbb{E}\left[e^{\lambda Z}\right] = \frac{1}{\sqrt{2\pi}} \int_{-\infty}^{\infty} e^{\lambda z} e^{-z^2/2} dz = e^{\lambda^2/2} \frac{1}{\sqrt{2\pi}} \int_{-\infty}^{\infty} e^{-(z-\lambda)^2/2} dz = e^{\lambda^2/2}.$$

Then

$$\mathbb{E}\left[X_t\right] = e^{\mu t} \mathbb{E}\left[e^{\sigma W_t}\right] = e^{\mu t} \mathbb{E}\left[e^{\sigma \sqrt{t} Z}\right] = e^{\mu t + \sigma^2 t/2}$$

and

$$\mathrm{Var}(X_t) = \mathbb{E}\left[X_t^2\right] - \mathbb{E}\left[X_t\right]^2 = e^{2\mu t + 2\sigma^2 t} - e^{2\mu t + \sigma^2 t} = e^{2\mu t + \sigma^2 t}(e^{\sigma^2 t} - 1).$$

The geometric Brownian motion occurs as a scaling limit of the binomial model defined in Chapter 3. For this, however, the parameters must be chosen correctly and this choice may seem a little strange at first glance. The details can be found in the following example.

**Example 7.3.4.** *For each $n \geq 1$, let $\xi_i^n$, $i = 1, 2, \ldots$ be a sequence of i.i.d. random variables where*

$$\mathbb{P}\left[\xi_i^n = u_n\right] = p_n \quad \text{and} \quad \mathbb{P}\left[\xi_i^n = d_n\right] = 1 - p_n.$$

*In the binomial model, the stock price after $j$ steps is given by*

$$S_j^n = S_0 \prod_{i=1}^{j} \xi_i^n.$$

*Suppose that*

$$u_n = e^{\sigma/\sqrt{n}}, \quad d_n = e^{-\sigma/\sqrt{n}} \quad \text{and} \quad p_n = \frac{1}{2}\left(1 + \frac{\mu}{\sigma\sqrt{n}}\right).$$

*We shall show that, as $n \to \infty$, $S_{\lfloor nt \rfloor}^n$ approximates a geometric Brownian motion with drift $\mu$ and diffusivity $\sigma$ starting from $S_0$, so that*

$$\{\log(S_{\lfloor nt \rfloor}^n / S_0), t \geq 0\} \xrightarrow{d} \{\mu t + \sigma W_t, t \geq 0\}.$$

*Let $Z_t^n = \log(S_{\lfloor nt \rfloor}^n / S_0)$. We need to show that for any $k \geq 1$, $0 \leq t_1 \leq \cdots \leq t_k$ and $z_1, \ldots, z_k$,*

$$\lim_{n \to \infty} \mathbb{P}\left[Z_{t_i}^n \leq z_i, 1 \leq i \leq k\right] = \mathbb{P}\left[\mu t_i + \sigma W_{t_i} \leq z_i, 1 \leq i \leq k\right],$$

*where $\{W_t, t \geq 0\}$ is a Brownian motion. The proof is similar to the proof of Theorem 7.1.3, therefore we do the computation only in the case $k = 1$. For $t \geq 0$*

$$Z_t^n = \sum_{i=1}^{\lfloor nt \rfloor} \log(\xi_i^n).$$

*Now*

$$\begin{aligned}
\mathbb{E}\left[\log(\xi_i^n)\right] &= p_n \log u_n + (1 - p_n) \log d_n \\
&= \frac{1}{2}\left(1 + \frac{\mu}{\sigma\sqrt{n}}\right)\frac{\sigma}{\sqrt{n}} - \frac{1}{2}\left(1 - \frac{\mu}{\sigma\sqrt{n}}\right)\frac{\sigma}{\sqrt{n}} \\
&= \frac{\mu}{n}
\end{aligned}$$

*and*

$$\begin{aligned}
\mathbb{E}\left[(\log(\xi_i^n))^2\right] &= p_n(\log u_n)^2 + (1 - p_n)(\log d_n)^2 \\
&= \frac{1}{2}\left(1 + \frac{\mu}{\sigma\sqrt{n}}\right)\frac{\sigma^2}{n} + \frac{1}{2}\left(1 - \frac{\mu}{\sigma\sqrt{n}}\right)\frac{\sigma^2}{n} \\
&= \frac{\sigma^2}{n}.
\end{aligned}$$

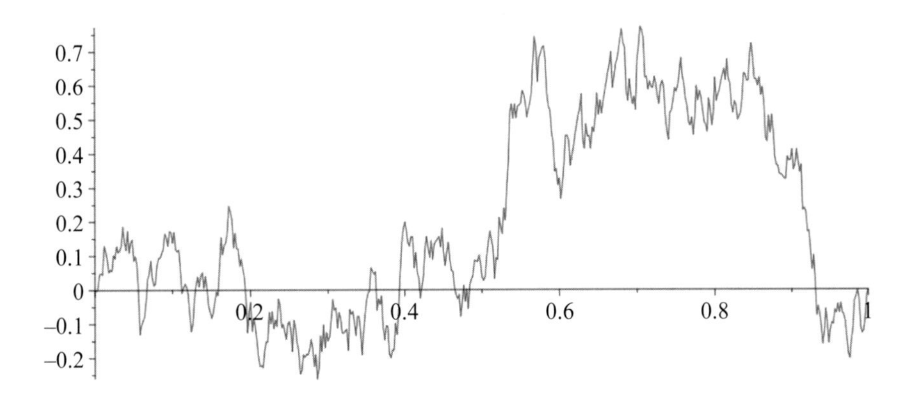


**Figure 7.4** Sample path of a Brownian bridge.

*Therefore* $\mathrm{Var}(\log(\xi_i^n)) = \frac{\sigma^2}{n} - \frac{\mu^2}{n^2}$. *Define for* $i \in \mathbb{N}$

$$Y_i^n := \frac{\log(\xi_i^n) - \mu/n}{\sqrt{\sigma^2/n - \mu^2/n^2}}.$$

*Then* $\mathbb{E}\left[Y_i^n\right] = 0$ *and* $\mathrm{Var}(Y_i^n) = 1$. *Applying the central limit theorem to* $(Y_i^n)_{i=1}^n$ *gives*

$$\begin{aligned} Z_t^n &= \sum_{i=1}^{\lfloor nt \rfloor} \log(\xi_i^n) = \sqrt{\sigma^2/n - \mu^2/n^2} \sum_{i=1}^{\lfloor nt \rfloor} Y_i^n + \frac{\mu \lfloor nt \rfloor}{n} \\ &= \sqrt{\sigma^2 - \mu^2/n} \frac{\sqrt{\lfloor nt \rfloor}}{\sqrt{n}} \left( \frac{1}{\sqrt{\lfloor nt \rfloor}} \sum_{i=1}^{\lfloor nt \rfloor} Y_i^n \right) + \frac{\mu \lfloor nt \rfloor}{n} \\ &\xrightarrow{d} \sigma \sqrt{t} Z + \mu t \end{aligned}$$

*as* $n \to \infty$, *where* $Z \sim \mathcal{N}(0, 1)$ *as required.*

**Definition 7.3.5** (Brownian bridge). Let $\{W_t, 0 \le t \le 1\}$ be a Brownian motion on the interval $[0, 1]$. A *Brownian bridge* is a process of the form

$$X_t = W_t - tW_1, \quad 0 \le t \le 1.$$

Observe that $X_0 = W_0 - 0W_1 = 0$ and $X_1 = W_1 - 1W_1 = 0$, so $X_t$ is ‘pinned down’ at $t = 0$ and $t = 1$. An example of a sample path of $X_t$ is given in Figure 7.4. This is a Gaussian process with $\mathbb{E}[X_t] = 0$ and

$$\begin{aligned} \mathrm{Cov}(X_t, X_s) &= \mathrm{Cov}\left((W_t - tW_1)(W_s - sW_1)\right) \\ &= \mathrm{Cov}(W_t, W_s) - t\,\mathrm{Cov}(W_1, W_s) - s\,\mathrm{Cov}(W_t, W_1) + ts\,\mathrm{Cov}(W_1, W_1) \end{aligned}$$

$$= t \wedge s - ts - ts + ts$$
$$= t \wedge s - ts.$$

Recall the definition of a martingale from Definition 6.3.13. Several martingales arise naturally as processes constructed from Brownian motion. Let us consider some examples.

**Example 7.3.6.** *Let $\{W_t, t \geq 0\}$ be a Brownian motion and let $\mathbb{F}$ be the natural filtration for $\{W_t, t \geq 0\}$, see Definition 6.3.4. The following processes are martingales with respect to $\mathbb{F}$.*

*(i)* $X_t = W_t$.
*(ii)* $X_t = W_t^2 - t$.
*(iii)* $X_t = e^{\sigma W_t - \sigma^2 t/2}$, *for any* $\sigma \in \mathbb{R}$.

*To verify that each of these is a martingale it is necessary to check conditions (a), (b) and (c) of Definition 6.3.13. We illustrate this only for (ii); the computations for (i) are straightforward and the computations for (iii) are similar to the computations in Section 5.4, so these are both left to Exercise 7.5.*

*(a)* *Let* $g(x) = x^2 - t$. *This is a continuous function of* $x$ *and hence* $X_t = g(W_t)$ *is* $\sigma(W_t)$*-measurable by Lemma 6.3.6.*
*(b)* *Since* $W_t \sim \mathcal{N}(0, t)$,
$$\mathbb{E}\left[|X_t|\right] \leq \mathbb{E}\left[(W_t)^2\right] + t = 2t < \infty.$$
*(c)* *For* $0 \leq s \leq t$, *we use that* $W_s$ *is* $\mathcal{F}_s$*-measurable and* $W_t - W_s$ *is independent of* $\mathcal{F}_s$. *By Theorem 6.3.12,*
$$\begin{aligned}
\mathbb{E}\left[X_t|\mathcal{F}_s\right] &= \mathbb{E}\left[(W_t)^2 - t\,|\mathcal{F}_s\right] = \mathbb{E}\left[\left((W_t - W_s) + W_s\right)^2 - t\,|\mathcal{F}_s\right] \\
&= \mathbb{E}\left[(W_t - W_s)^2\,|\mathcal{F}_s\right] + 2\mathbb{E}\left[W_s(W_t - W_s)\,|\mathcal{F}_s\right] + \mathbb{E}\left[(W_s)^2\,|\mathcal{F}_s\right] - t \\
&= \mathbb{E}\left[(W_t - W_s)^2\right] + 2W_s\mathbb{E}\left[W_t - W_s\right] + (W_s)^2 - t \\
&= (t - s) + 0 + (W_s)^2 - t \quad (\textit{since } W_t - W_s \sim \mathcal{N}(0, t - s)) \\
&= X_s.
\end{aligned}$$

In the remainder of this book, we will only use the natural filtration $\mathbb{F}$ of a Brownian motion $\{W_t, t \geq 0\}$, unless explicitly stated otherwise.

## 7.4 Exercises

**Exercise 7.1.** *Check that Definitions 7.1.1 and 7.1.2 are indeed equivalent.*

**Exercise 7.2.** *Prove properties (i), (iii) and (iv) from Theorem 7.3.1.*

**Exercise 7.3.** *Let $\{W_t, t \geq 0\}$ be a Brownian motion. Fix $\tau > 0$ and set*

$$X_t = \begin{cases} W_t, & \text{for } t \leq \tau \\ 2W_\tau - W_t, & \text{for } t > \tau. \end{cases}$$

*Show that the process $\{X_t, t \geq 0\}$ is a Brownian motion.*

**Exercise 7.4.** *Suppose that $\{X_t, 0 \leq t \leq 1\}$ is a Brownian bridge and let $Z \sim N(0,1)$ be independent of $\{X_t, 0 \leq t \leq 1\}$. Set $W_t = X_t + tZ$. Show that $\{W_t, 0 \leq t \leq 1\}$ is a Brownian motion (restricted to $[0,1]$).*

**Exercise 7.5.** *Let $\{W_t, t \geq 0\}$ be a Brownian motion and let $\mathbb{F}$ be the natural filtration.*

(a) *Show that $X_t = W_t$ is a martingale with respect to $\mathbb{F}$.*
(b) *Let $X_t = e^{\mu t + \sigma W_t}$ be a geometric Brownian motion. Show that $X_t$ is a martingale if and only if $\mu = -\sigma^2/2$.*
(c) *Let $M_t = e^{\sigma W_t - \sigma^2 t/2}$. By expanding $M_t$ as a power series in $\sigma$, give an alternative proof that $W_t$ and $W_t^2 - t$ are martingales.*
(d) *Write down a martingale which involves the term $W_t^3$.*

# 8 Stochastic Integration

The goal of this chapter is to define stochastic integrals of the form

$$\int_0^T X_s\, dW_s$$

where $\{W_s, s \geq 0\}$ is a Brownian motion. In general, one can define a stochastic integral with respect to any martingale or semimartingale. However, for financial applications we only require integrals with respect to Brownian motion, so our focus will be on this case. An introduction to more general stochastic integrals can be found for instance in [14].

In Chapter 7 we saw that the sample paths of Brownian motion are very rough and in fact nowhere differentiable. This has major consequences for the definition of the integral. Before we attempt to define integration with respect to Brownian motion, we shall discuss how to define integration with respect to processes which have much smoother sample paths.

## 8.1 The Riemann Integral

We recall the Riemann integral which is used to define the ordinary integral in any course on calculus. We begin with some notation.

- A partition of the interval $[0, T]$ is a finite sequence $\mathbf{t} = (t_i)_{i=0}^n$ with $n \in \mathbb{N}$ and
$$0 = t_0 < t_1 < \cdots < t_n = T.$$
- An *intermediate partition* of a partition $\mathbf{t} = (t_i)_{i=0}^n$ is a sequence $\mathbf{y} = (y_i)_{i=1}^n$ satisfying
$$t_{i-1} \leq y_i \leq t_i \text{ for } i = 1, \ldots, n.$$
- The norm of a partition $\mathbf{t}$ is defined as $\|\mathbf{t}\| := \max\{|t_i - t_{i-1}|\,;\, 1 \leq i \leq n\}$.
- A partition $\mathbf{t}'$ is called a *refinement* of $\mathbf{t}$ if $\mathbf{t}'$ contains all the points from $\mathbf{t}$ and some additional points, again sorted by order of magnitude.

Note that a partition of an arbitrary set is defined differently, see Definition 2.1.3. However, we can easily identify the above definition with Definition 2.1.3 by splitting the interval $[0, T]$ into the subsets $[t_0, t_1)$, ..., $[t_{n-2}, t_{n-1})$ and $[t_{n-1}, T]$. It does not matter for our purposes whether the intervals are open, half-open or closed. Therefore, it is sufficient to consider the sequence $\mathbf{t} = (t_i)_{i=0}^n$. We can now recall the definition of the Riemann integral.

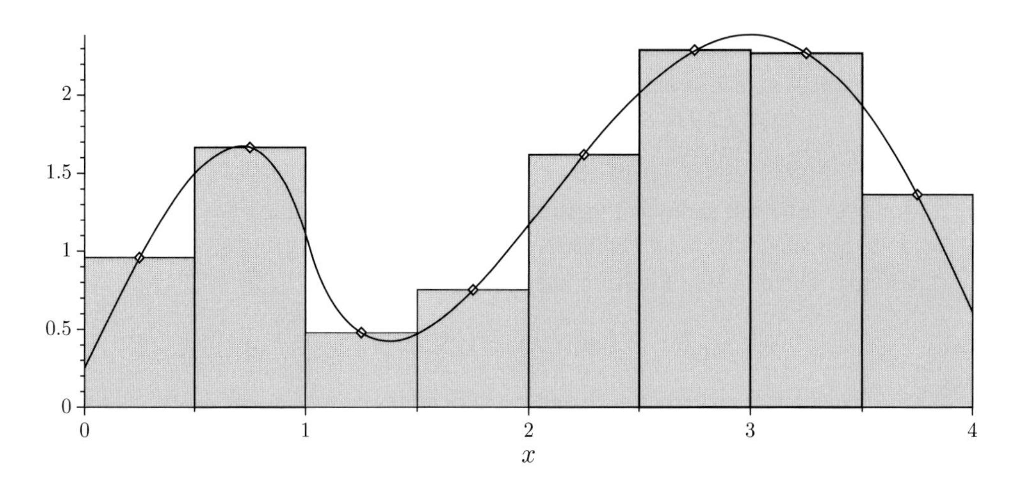


**Figure 8.1** Illustration of a Riemann sum.

**Definition 8.1.1** (Riemann integral). Let $f : [0, \infty) \to \mathbb{R}$ and $T > 0$ be given. Let $\mathbf{t} = (t_i)_{i=0}^n$ be a partition of $[0, T]$ and $\mathbf{y} = (y_i)_{i=1}^n$ an intermediate partition. The *Riemann sum* is defined by

$$S(f, \mathbf{t}, \mathbf{y}) = \sum_{i=1}^{n} f(y_i)(t_i - t_{i-1}). \tag{8.1.1}$$

The *Riemann integral of* $f$ is defined as the limit

$$\int_0^T f(s)ds = \lim_{\|\mathbf{t}\| \to 0} S(f, \mathbf{t}, \mathbf{y}) = \lim_{\|\mathbf{t}\| \to 0} \sum_{i=1}^{n} f(y_i)(t_i - t_{i-1}) \tag{8.1.2}$$

if this limit exists and does not depend on the choice of partition and intermediate partition.

An illustration of (8.1.1) can be found in Figure 8.1. The Riemann integral exists if $f$ is sufficiently smooth. For example, suppose $f$ is continuous. For a given partition, $\mathbf{t} = (t_i)_{i=0}^n$, pick $y_i^-, y_i^+ \in [t_{i-1}, t_i]$ so that

$$f(y_i^-) = \inf_{t_{i-1} \le y \le t_i} f(y) \quad \text{and} \quad f(y_i^+) = \sup_{t_{i-1} \le y \le t_i} f(y).$$

Then set

$$S^-(f, \mathbf{t}) = \sum_{i=1}^{n} f(y_i^-)(t_i - t_{i-1}) \quad \text{and} \quad S^+(f, \mathbf{t}) = \sum_{i=1}^{n} f(y_i^+)(t_i - t_{i-1}). \tag{8.1.3}$$

If $\mathbf{t}'$ is a *refinement* of $\mathbf{t}$ then it is straightforward to see that

$$S^-(f, \mathbf{t}) \le S^-(f, \mathbf{t}') \le S^+(f, \mathbf{t}') \le S^+(f, \mathbf{t}). \tag{8.1.4}$$

Suppose now we have a sequence $(\mathbf{t}_n)_{n\in\mathbb{N}}$ of partitions such that $\|\mathbf{t}_n\| \to 0$ as $n \to \infty$ and $\mathbf{t}_n$ is a refinement of $\mathbf{t}_m$ if $n > m$. Then (8.1.4) implies that the following limits exist:

$$S^-(f,T) = \lim_{n\to\infty} S^-(f,\mathbf{t}_n) \quad \text{and} \quad S^+(f,T) = \lim_{n\to\infty} S^+(f,\mathbf{t}_n).$$

Moreover, one can show that these limits do not depend on the sequence of $(\mathbf{t}_n)_{n\in\mathbb{N}}$. However, we do not show this here. Furthermore, $S^-(f,\mathbf{t}) \le S(f,\mathbf{t},\mathbf{y}) \le S^+(f,\mathbf{t})$ for each intermediate partition. Therefore the Riemann integral exists provided $S^-(f,t) = S^+(f,t)$. For $\delta > 0$ set

$$w_f(\delta) := \sup\{|f(y) - f(y')| : 0 \le y, y' \le T, |y - y'| < \delta\}. \tag{8.1.5}$$

Since $f$ is continuous, $w_f(\delta) < \infty$ and $w_f(\delta) \to 0$ as $\delta \to 0$. Suppose now that $\|\mathbf{t}\| \le \delta$ for some $\delta > 0$, then

$$S^+(f,\mathbf{t}) - S^-(f,\mathbf{t}) \le \sum_{i=1}^n |f(y_i^+) - f(y_i^-)|(t_i - t_{i-1}) \le \sum_{i=1}^n w_f(\delta)(t_i - t_{i-1}) = w_f(\delta)\cdot T.$$

Since $f$ is continuous, $w_f(\delta)\cdot T$ converges to 0 as $\delta \to 0$ and so $S^-(f,T) = S^+(f,T)$ as required.

**Example 8.1.2.** *Let $\{W_s, s \ge 0\}$ be a Brownian motion realised on the probability space $(\Omega, \mathcal{F}, \mathbb{P})$ and $\mathbb{F}$ be a filtration such that $\{W_s, s \ge 0\}$ is $\mathbb{F}$-adapted. Since $\{W_s, s \ge 0\}$ has continuous sample paths, we can use Riemann integration to define*

$$I_t(\omega) = \int_0^t W_s(\omega)ds$$

*for all $t \ge 0$ and $\omega \in \Omega$. In other words, we consider each sample path separately and assign the Riemann integral to each as we would for an ordinary function. Indeed, if a partition $\mathbf{t}$ of $[0,t]$ and an intermediate partition $\mathbf{y}$ is given then, for each $\omega \in \Omega$,*

$$S(W(\omega),\mathbf{t},\mathbf{y}) = \sum_{i=1}^n W_{y_i}(\omega)(t_i - t_{i-1}).$$

*Since $t \mapsto W_t(\omega)$ is continuous, $S(W(\omega),\mathbf{t},\mathbf{y}) \overset{as}{\to} I_t(\omega)$, and the limit does not depend on the choice of partition and intermediate partition.*

*We shall show that $\{I_t, t \ge 0\}$ defines an $\mathbb{F}$-adapted Gaussian process with*

$$\mathbb{E}[I_t] = 0 \quad \textit{and} \quad \mathrm{Cov}(I_s, I_t) = \int_0^s\int_0^t (u \wedge v)dudv,$$

*where we denote $u \wedge v = \min(u,v)$.*

*We are free to choose the most convenient partition and intermediate partition for doing the analysis. Therefore, fix $n \in \mathbb{N}$ and take the partition $\mathbf{t} = (t_i)_{i=0}^{\lfloor nt\rfloor+1}$, where $t_i = i/n$ for $i = 0, \ldots, \lfloor nt\rfloor$ and $t_{\lfloor nt\rfloor+1} = t$. Take the intermediate partition defined by $y_i = (i-1)/n$. Then*

$$I_t = \lim_{n\to\infty} S_n(t),$$

*where*

$$S_n(t) = \frac{1}{n}\sum_{i=1}^{\lfloor nt \rfloor} W_{(i-1)/n} + W_{\lfloor nt \rfloor/n}(t - \lfloor nt \rfloor/n). \tag{8.1.6}$$

*By construction, $S_n(t)$ is $\mathcal{F}_t$-measurable. Therefore, by Proposition 6.3.5, $I_t$ is $\mathcal{F}_t$-measurable for all $t$. To show that $\{I_t, t \geq 0\}$ is a Gaussian process, we need to show that the finite-dimensional distributions are multivariate Gaussians. Suppose that $0 \leq s_1 < \cdots < s_k \leq t$. Then $(S_n(s_1), \ldots, S_n(s_k))$ is a linear transformation of the multivariate Gaussian random variable $(W_{1/n}, \ldots, W_{\lfloor nt \rfloor/n})$ and hence $(S_n(s_1), \ldots, S_n(s_k))$ is multivariate Gaussian. But*

$$(I_{s_1}, \ldots, I_{s_k}) = \lim_{n\to\infty}(S_n(s_1), \ldots, S_n(s_k)),$$

*and the limit of a sequence of multivariate Gaussian random variables is a multivariate Gaussian random variable. Therefore, $\{I_t, t \geq 0\}$ is a Gaussian process. Since $\mathbb{E}[W_t] = 0$ for all $t$, it follows from (8.1.6) that*

$$\mathbb{E}[I_t] = \lim_{n\to\infty}\mathbb{E}[S_n(t)] = 0.$$

*The final term in (8.1.6) converges to zero almost surely. Therefore*

$$\begin{aligned}
\mathrm{Cov}(I_s, I_t) &= \mathrm{Cov}\left(\lim_{m\to\infty} S_m(s), \lim_{n\to\infty} S_n(t)\right) \\
&= \lim_{m\to\infty}\lim_{n\to\infty}\frac{1}{mn}\sum_{i=1}^{\lfloor ms \rfloor}\sum_{j=1}^{\lfloor nt \rfloor}\mathrm{Cov}(W_{t_{i-1}}, W_{t_{j-1}}) \\
&= \lim_{m\to\infty}\lim_{n\to\infty}\frac{1}{mn}\sum_{i=1}^{\lfloor ms \rfloor}\sum_{j=1}^{\lfloor nt \rfloor}\frac{(i-1)s}{m}\wedge\frac{(j-1)t}{n} \\
&= \lim_{m\to\infty}\frac{1}{m}\sum_{i=1}^{\lfloor ms \rfloor}\lim_{n\to\infty}\frac{1}{n}\sum_{j=1}^{\lfloor nt \rfloor}\frac{(i-1)s}{m}\wedge\frac{(j-1)t}{n} \\
&= \lim_{m\to\infty}\frac{1}{m}\sum_{i=1}^{\lfloor ms \rfloor}\int_0^t\left(\frac{(i-1)s}{m}\wedge u\right)du \\
&= \int_0^s\int_0^t (u\wedge v)dudv.
\end{aligned}$$

*The last two lines used the Riemann sum definitions of the respective integrals.*

## 8.2 The Riemann–Stieltjes Integral

We now generalise this approach. One of the motivations for this is that we want to give meaning to expressions like

$$\int_{\mathbb{R}} x\,dF_X(x) \text{ or } \int_{\mathbb{R}} x^2\,dF_X(x),$$

where $X$ is an arbitrary random variable. This is important since there are random variables that are neither discrete nor continuous nor a mix of both. Considerations similar to the construction of the Riemann integral lead to the following definition.

**Definition 8.2.1** (Riemann–Stieltjes integral). Suppose that we have two functions $f : [0, \infty) \to \mathbb{R}$ and $g : [0, \infty) \to \mathbb{R}$ and $T > 0$. Let $\mathbf{t}$ and $\mathbf{y}$ be as in Definition 8.1.1. The *Riemann–Stieltjes sum* is defined by

$$S(f, g, \mathbf{t}, \mathbf{y}) = \sum_{i=1}^{n} f(y_i)(g(t_i) - g(t_{i-1})). \tag{8.2.1}$$

The *Riemann–Stieltjes integral of f with respect to g* is defined as the limit

$$\int_0^T f(s)dg(s) = \lim_{\|\mathbf{t}\| \to 0} S(f, g, \mathbf{t}, \mathbf{y}) = \lim_{\|\mathbf{t}\| \to 0} \sum_{i=1}^{n} f(y_i)(g(t_i) - g(t_{i-1})) \tag{8.2.2}$$

if this limit exists and does not depend on the choice of partition and intermediate partition.

The Riemann–Stieltjes integral plays an important role for the expectation of general random variables, see (6.1.16). Also, the Riemann–Stieltjes sum has a natural interpretation in finance. Suppose that a stock price is given by $\{S_t, t \geq 0\}$. Consider a trading strategy and denote by $\alpha_t$ the number of stocks we hold at time $t$. Suppose we only trade the stocks at the times $0 = t_0 < t_1 < \cdots < t_n = T$. Therefore, we hold $\alpha_{t_{i-1}}$ stocks in the time interval $[t_{i-1}, t_i]$. The term $\alpha_{t_{i-1}}(S_{t_i} - S_{t_{i-1}})$ represents the amount of profit made during the time interval $[t_{i-1}, t_i]$. The total profit made up to time $T$ is therefore given by

$$\sum_{i=1}^{n} \alpha_{t_{i-1}}(S_{t_i} - S_{t_{i-1}}). \tag{8.2.3}$$

We have deduced a similar expression in Lemma 4.1.4 for the value of the portfolio of a self-financing trading strategy in the finite market model. Note that the trading strategy in (8.2.3) is not self-financing in general and mainly serves here as a motivation.

We are interested in the conditions needed on $f$ and $g$ in order for the Riemann–Stieltjes integral to exist.

**Example 8.2.2.** *Suppose that $f : [0, \infty) \to \mathbb{R}$ is continuous and $g : [0, \infty) \to \mathbb{R}$ is differentiable with continuous derivative $g'$. Given a partition $\mathbf{t} = (t_i)_{i=0}^n$, by the mean value theorem, there exists an intermediate partition $z_1, \ldots, z_n$ such that*

$$g(t_i) - g(t_{i-1}) = g'(z_i)(t_i - t_{i-1}) \quad \text{for} \quad i = 1, \ldots, n.$$

*But then, for any intermediate partition $\mathbf{y} = (y_i)_{i=1}^n$, the Riemann–Stieltjes sum is given by*

$$S(f, g, \mathbf{t}, \mathbf{y}) = \sum_{i=1}^{n} f(y_i)g'(z_i)(t_i - t_{i-1}).$$

*Now define*

$$\widetilde{S}(f,g,\mathbf{t},\mathbf{y}) := \sum_{i=1}^{n} f(y_i)g'(y_i)(t_i - t_{i-1}).$$

*Observe that since $f$ is bounded and $g'$ is continuous on $[0,T]$,*

$$S(f,g,\mathbf{t},\mathbf{y}) - \widetilde{S}(f,g,\mathbf{t},\mathbf{y}) \to 0 \quad as \quad \|\mathbf{t}\| \to 0.$$

*But $\widetilde{S}(f,g,\mathbf{t},\mathbf{y})$ is just the Riemann sum for the function $fg'$. Since $f$ and $g'$ are both continuous, the Riemann integral of $fg'$ exists. Therefore, the Riemann–Stieltjes integral of $f$ with respect to $g$ exists and is given by*

$$\int_0^T f(s)dg(s) = \int_0^T f(s)g'(s)ds. \tag{8.2.4}$$

Suppose now that $X$ is a continuous variable with density $f_X$. If we choose $f(s) = s$ and $g(s) = F_X(s)$ in (8.2.4), then we recover the expression for $\mathbb{E}[X]$ in (6.1.17) and get

$$\mathbb{E}[X] = \int_{\mathbb{R}} s\, dF_X(s) = \int_{\mathbb{R}} s \cdot f_X(x)\, ds. \tag{8.2.5}$$

It is natural to ask what happens if we replace $X$ by a discrete random variable. In this case, $F_X$ is piecewise constant, non-decreasing and right continuous. Let us consider a simple example.

**Example 8.2.3.** *Suppose that $T > u > 0$, $f : [0,\infty) \to \mathbb{R}$ is continuous. Let $a, b \in \mathbb{R}$ be given and define $g : [0,\infty) \to \mathbb{R}$ as $g(s) = b$ if $s \geq u$ and $g(s) = a$ if $0 \leq s < u$. Consider a partition of $\mathbf{t} = (t_i)_{i=1}^n$ and let $\mathbf{y} = (y_i)_{i=1}^n$ be an intermediate partition. Then*

$$g(t_i) - g(t_{i-1}) = \begin{cases} b - a, & \text{if } t_{i-1} < u \leq t_i, \\ 0, & \text{otherwise.} \end{cases}$$

*If we substitute this into (8.2.1), then we get*

$$S(f,g,\mathbf{t},\mathbf{y}) = (b-a)f(y_j), \tag{8.2.6}$$

*where $[t_{j-1}, t_j]$ is the interval that contains $u$ (i.e. $y_j \in [t_{j-1}, t_j]$ and $u \in [t_{j-1}, t_j]$). If $\|t\| \to 0$ then the $y_i$ in (8.2.6) converge to $u$. Since $f$ is continuous, we get*

$$\int_{\mathbb{R}} f(s)\, dg(s) = (b-a)f(u).$$

*Therefore, the Riemann–Stieltjes integral evaluates the function $f$ at the point $u$ and multiplies it by the size of the jump of $g$ at the point $u$. This computation can easily be generalised to all functions $g$ which are piecewise constant. This includes the cumulative distribution functions of discrete random variables. Suppose that $X$ is a discrete random variable taking only the values $x_1, x_2, \ldots$. Then the size of the jump of $F_X$ at the point $x_j$ is equal to the probability $\mathbb{P}[X = x_j]$. It immediately follows that*

$$\int_{\mathbb{R}} f(s)\, dF_X(s) = \sum_{j=1}^{\infty} f(x_j)\mathbb{P}\left[X = x_j\right] = \mathbb{E}\left[f(X)\right],$$

*provided that* $\mathbb{E}\left[|f(X)|\right] < \infty$.

The requirement that $g$ is differentiable or piecewise constant is not necessary for the existence of the Riemann–Stieltjes integral. The integral also exists if $g$ has finite variation.

**Definition 8.2.4** (Finite variation). The *variation* of a function $g : [0, \infty) \to \mathbb{R}$ over the interval $[0, T]$ is defined as

$$V_g(T) = \sup_{\mathbf{t}} \sum_{i=1}^{n} |g(t_i) - g(t_{i-1})|,$$

where the supremum is taken over all partitions $\mathbf{t} = (t_i)_{i=0}^n$ of $[0, T]$. A function $g$ has *finite variation* if $V_g(T) < \infty$.

We now show that the Riemann–Stieltjes integral exists if $f$ is continuous and $g$ has finite variation. We use a similar argument as for the Riemann integral in Section 8.1. Our first step is to construct an upper and a lower bound for the Riemann–Stieltjes sum, similar to the ones in (8.1.3). However, as $g(t_i)-g(t_{i-1})$ can be positive or negative, we have to choose the intermediate partition more carefully. Explicitly, for a given partition, $\mathbf{t} = (t_i)_{i=0}^n$, pick $y_i^- \in [t_{i-1}, t_i]$ so that

$$f(y_i^-) = \begin{cases} \inf_{t_{i-1} \le y \le t_i} f(y), & \text{if } g(t_i) - g(t_{i-1}) \ge 0, \\ \sup_{t_{i-1} \le y \le t_i} f(y), & \text{if } g(t_i) - g(t_{i-1}) < 0. \end{cases}$$

Similarly, pick $y_i^+ \in [t_{i-1}, t_i]$ so that

$$f(y_i^+) = \begin{cases} \sup_{t_{i-1} \le y \le t_i} f(y), & \text{if } g(t_i) - g(t_{i-1}) \ge 0, \\ \inf_{t_{i-1} \le y \le t_i} f(y), & \text{if } g(t_i) - g(t_{i-1}) < 0. \end{cases}$$

Then set

$$S^-(f, g, \mathbf{t}) = \sum_{i=1}^{n} f(y_i^-)(g(t_i) - g(t_{i-1})) \quad \text{and} \quad S^+(f, g, \mathbf{t}) = \sum_{i=1}^{n} f(y_i^+)(g(t_i) - g(t_{i-1})).$$

As for the Riemann integral, let $(\mathbf{t}_n)_{n\in\mathbb{N}}$ be a sequence of partitions such that $\mathbf{t}_n$ is a refinement of $\mathbf{t}_m$ if $n > m$ and $\|\mathbf{t}_n\| \to 0$ as $n \to \infty$. Then the following limits exist:

$$S^-(f, g, T) = \lim_{n\to\infty} S^-(f, g, \mathbf{t}_n) \quad \text{and} \quad S^+(f, g, T) = \lim_{n\to\infty} S^+(f, g, \mathbf{t}_n).$$

Moreover, one can show that these limits do not depend on the sequence of $(\mathbf{t}_n)_{n\in\mathbb{N}}$. Again, we do not show this here. Also, $S^-(f, g, \mathbf{t}) \le S(f, g, \mathbf{t}, \mathbf{y}) \le S^-(f, g, \mathbf{t})$ for any intermediate partition $\mathbf{y}$. Therefore, the Riemann–Stieltjes integral will exist provided $S^+(f, g, T) = S^-(f, g, T)$. However, if $\|\mathbf{t}\| \le \delta$ for some $\delta > 0$ then, with $w_f(\delta)$ as in (8.1.5),

$$S^+(f,g,\mathbf{t}) - S^-(f,g,\mathbf{t}) \le \sum_{i=1}^{n} |f(y_i^+) - f(y_i^-)||g(t_i) - g(t_{i-1})|$$
$$\le w_f(\delta) \sum_{i=1}^{n} |g(t_i) - g(t_{i-1})| \le w_f(\delta) V_g(T).$$

Now $w_f(\delta)V_g(T) < \infty$ as by assumption $g$ has finite variation. Since $f$ is continuous, we immediately get $w_f(\delta)V_g(T) \to 0$ as $\delta \to 0$ and so $S^-(f,g,T) = S^+(f,g,T)$ as required.

Many functions have bounded variation. For instance, if a function $g$ is non-decreasing and bounded then it has finite variation $V_g(T)$. Indeed,

$$V_g(T) = \sup_{\mathbf{t}} \sum_{i=1}^{n} |g(t_i) - g(t_{i-1})| = \sup_{\mathbf{t}} \sum_{i=1}^{n} g(t_i) - g(t_{i-1}) = g(T) - g(0).$$

It follows that the cumulative distribution function $F_X$ of a random variable $X$ always has bounded variation. Therefore, expressions of the form $\int_{\mathbb{R}} f(x)\,dF_X(x)$ are meaningfully defined, provided $f$ is continuous.

A useful property of the Riemann–Stieltjes integral is that it admits integration by parts.

**Proposition 8.2.5.** Suppose that $f : [0,\infty) \to \mathbb{R}$ and $g : [0,\infty) \to \mathbb{R}$ and $T > 0$ are given. If the Riemann–Stieltjes integral $\int_0^T g(s)\,df(s)$ exists, then so does the Riemann–Stieltjes integral $\int_0^T f(s)\,dg(s)$ and it satisfies

$$\int_0^T f(s)\,dg(s) = f(T)g(T) - f(0)g(0) - \int_0^T g(s)\,df(s). \tag{8.2.7}$$

The main step in the proof of this proposition is to apply partial summation to the Riemann–Stieltjes sum in (8.2.1). We leave the details of this proof as Exercise 8.1.

## 8.3 Quadratic Variation

We would like to be able to make sense of integrals such as $\int_0^T W_s dW_s$, where $\{W_s, s \ge 0\}$ is a Brownian motion. The natural approach is to try to argue as in Example 8.1.2. More precisely, one is tempted to try to consider $\int_0^T W_s dW_s$ as a Riemann–Stieltjes integral separately for each sample path. However, the following example shows that we quickly run into difficulties with this approach.

**Example 8.3.1.** *Let $\{W_s, s \ge 0\}$ be a Brownian motion and $\mathbf{t} = (t_i)_{i=0}^n$ be a partition of $[0,T]$. Consider the two intermediate partitions*

$$\mathbf{y}^d = (y_i^d)_{i=1}^n \text{ with } y_i^d = t_{i-1}, \text{ and } \mathbf{y}^u = (y_i^u)_{i=1}^n \text{ with } y_i^u = t_i \text{ for } i = 1,\dots,n.$$

*The corresponding Riemann–Stieltjes sums are*

$$S(W,\mathbf{t},\mathbf{y}^d) = \sum_{i=1}^{n} W_{t_{i-1}}(W_{t_i} - W_{t_{i-1}}) \quad \textit{and} \quad S(W,\mathbf{t},\mathbf{y}^u) = \sum_{i=1}^{n} W_{t_i}(W_{t_i} - W_{t_{i-1}}). \tag{8.3.1}$$

*If it is possible to define $\int_0^T W_s dW_s$ as a Riemann–Stieltjes integral for a sample path, then both sums in* (8.3.1) *converge to this integral as* $\|t\| \to 0$. *However,*

$$S(W,\mathbf{t},\mathbf{y}^u) - S(W,\mathbf{t},\mathbf{y}^d) = \sum_{i=1}^{n}(W_{t_i} - W_{t_{i-1}})^2 > 0 \tag{8.3.2}$$

*and furthermore,*

$$\mathbb{E}\left[S(W,\mathbf{t},\mathbf{y}^u) - S(W,\mathbf{t},\mathbf{y}^d)\right] = \sum_{i=1}^{n} \mathbb{E}\left[(W_{t_i} - W_{t_{i-1}})^2\right] = \sum_{i=1}^{n}(t_i - t_{i-1}) = T. \tag{8.3.3}$$

*This computation suggests that $S(W,\mathbf{t},\mathbf{y}^u) - S(W,\mathbf{t},\mathbf{y}^d)$ is not converging to* 0. *Unfortunately, the above calculation is not sufficient to show this; see comment after Theorem 6.1.18. The remaining argument is postponed until after the proof of Theorem 8.3.3. At this point, we just state that the Riemann–Stieltjes integral does not exist for Brownian motion integrated with respect to Brownian motion with probability* 1.

We use Example 8.3.1 in the argument below to show that almost all sample paths of Brownian motion do not have finite variation. However, Brownian motion does have finite quadratic variation.

**Definition 8.3.2** (Quadratic variation). Let $g : [0,\infty) \to \mathbb{R}$ be a function. The *quadratic variation* of $g$ over the interval $[0,T]$ is defined as the limit

$$Q_g(T) = \lim_{\|\mathbf{t}\|\to 0} \sum_{i=1}^{n} |g(t_i) - g(t_{i-1})|^2, \tag{8.3.4}$$

if this limit exists and does not depend on the choice of partitions $\mathbf{t} = (t_i)_{i=0}^n$ of $[0,T]$. We often write $[g]_T$ for $Q_g(T)$.

Note that $V_g(T)$ in Definition 8.3.2 is always defined, even though it might be infinite. The quantity $Q_g(T)$, on the other hand, does not need to exist. For instance, consider the function $g$ with $g(s) = 0$ if $s$ is rational and $g(s) = 1$ if $s$ is irrational. Then, for each $m \in \mathbb{N}$, it is possible to construct a sequence of partitions such that the expression in (8.3.4) converges to $m$. At this point, we would like to compute the variation and the quadratic variation of stochastic processes, especially for Brownian motion. We therefore have to substitute a stochastic process into Definition 8.2.4 and Definition 8.3.2. For the variation, this is unambiguous and we can assign the variation directly to each sample path. On the other hand, we have to be more careful with the quadratic variation. More precisely, if we substitute a stochastic process into the sum on the RHS of (8.3.4) then this sum becomes a random variable. As the quadratic variation is then defined as a limit, we have to state in which sense this random variable converges. Mostly, one would prefer almost sure convergence; this is not always possible and it is often easier to prove $L^2$-convergence.

**Theorem 8.3.3.** *The quadratic variation of Brownian motion exists as a mean square limit for any sequence $(\mathbf{t}_j)_{j=1}^{\infty}$ of partitions of $[0,T]$ with $\|\mathbf{t}_j\| \to 0$ and satisfies $[W]_T = T$.*

*Proof.* Let $\mathbf{t} = (t_i)_{i=0}^n$ be a partition of $[0,T]$ and let

$$Q(\mathbf{t}) = \sum_{i=1}^n (W_{t_i} - W_{t_{i-1}})^2.$$

By definition of mean square convergence in Definition 6.1.20, we have to show that

$$\|Q(\mathbf{t}) - T\|_{L^2(\Omega)} \to 0 \quad \text{as} \quad \|\mathbf{t}\| \to 0.$$

We know from Example 8.3.1 that $\mathbb{E}\left[Q(\mathbf{t})\right] = T$. Furthermore, Brownian motion has independent increments. Hence

$$\|Q(\mathbf{t}) - T\|_{L^2(\Omega)}^2 = \mathbb{E}\left[(Q(\mathbf{t}) - T)^2\right] = \mathrm{Var}(Q(\mathbf{t})) = \mathbb{E}\left[\left(\sum_{i=1}^n (W_{t_i} - W_{t_{i-1}})^2\right)^2\right] - T^2$$

$$= \sum_{i=1}^n \sum_{j=1}^n \mathbb{E}\left[(W_{t_i} - W_{t_{i-1}})^2 (W_{t_j} - W_{t_{j-1}})^2\right] - T^2$$

$$= \sum_{i=1}^n \mathbb{E}\left[(W_{t_i} - W_{t_{i-1}})^4\right] + \sum_{i \neq j} \mathbb{E}\left[(W_{t_i} - W_{t_{i-1}})^2\right] \mathbb{E}\left[(W_{t_j} - W_{t_{j-1}})^2\right] - T^2.$$

Now, using the fact that $\mathbb{E}\left[Z^4\right] = 3$ if $Z \sim \mathcal{N}(0,1)$,

$$\sum_{i=1}^n \mathbb{E}\left[(W_{t_i} - W_{t_{i-1}})^4\right] = \sum_{i=1}^n (t_i - t_{i-1})^2 \mathbb{E}\left[Z^4\right] = 3 \sum_{i=1}^n (t_i - t_{i-1})^2.$$

Also,

$$\sum_{i \neq j} \mathbb{E}\left[(W_{t_i} - W_{t_{i-1}})^2\right] \mathbb{E}\left[(W_{t_j} - W_{t_{j-1}})^2\right] = \sum_{i \neq j} (t_i - t_{i-1})(t_j - t_{j-1})$$

$$= \sum_{i=1}^n \sum_{j=1}^n (t_i - t_{i-1})(t_j - t_{j-1}) - \sum_{i=1}^n (t_i - t_{i-1})^2$$

$$= \left(\sum_{i=1}^n (t_i - t_{i-1})\right)^2 - \sum_{i=1}^n (t_i - t_{i-1})^2 = T^2 - \sum_{i=1}^n (t_i - t_{i-1})^2.$$

Hence

$$\|Q(\mathbf{t}) - T\|_{L^2(\Omega)}^2 = 2 \sum_{i=1}^n (t_i - t_{i-1})^2 \leq 2 \max_{1 \leq i \leq n} (t_i - t_{i-1}) \sum_{i=1}^n (t_i - t_{i-1}) \leq 2T \|\mathbf{t}\|.$$

Therefore $\|Q(\mathbf{t}) - T\|_{L^2(\Omega)} \to 0$ as $\|\mathbf{t}\| \to 0$ and hence $[W]_T = \lim Q(\mathbf{t}) = T$. □

We can now give the remaining argument needed in Example 8.3.1. We clearly have

$$Q(\mathbf{t}) = S^+(f, g, \mathbf{t}) - S^-(f, g, \mathbf{t}).$$

The proof of Theorem 8.3.3 shows that $Q(\mathbf{t}_j) \overset{L^2}{\to} T$ for any sequence $(\mathbf{t}_j)_{j=1}^{\infty}$ with $\|\mathbf{t}_j\| \to 0$. Theorem 6.1.23 then implies that there exists a subsequence $(\mathbf{t}_{j_m})_{m=1}^{\infty}$ such that $Q(\mathbf{t}_{j_m}) \overset{as}{\to} T$. It follows that $Q(\mathbf{t}_{j_m})$ cannot converge to 0, which completes the proof. An immediate consequence is the following corollary.

**Corollary 8.3.4.** Almost all sample paths of Brownian motion have infinite variation.

*Proof.* If a sample path has finite variation then the Riemann–Stieltjes integral $\int_0^T W_s dW_s$ exists for this sample path. This implies that $Q(\mathbf{t}) \to 0$ for this sample path. On the other hand, $Q(\mathbf{t}_j) \overset{L^2}{\to} T$ and so $Q(\mathbf{t}) \to 0$ with probability 0. □

## 8.4 Construction of the Stochastic Integral

In this section, we introduce the stochastic integral with respect to Brownian motion $W$. We assume for the rest of this chapter that each process, including the Brownian motion $W$, is realised on the probability space $(\Omega, \mathcal{F}, \mathbb{P})$ and that we are working with the natural filtration $\mathbb{F}$ of Brownian motion. We will focus on integration over the time interval $[0, T]$ with $T > 0$ fixed and therefore assume that each process is defined over the time interval $[0, T]$.

Suppose we have a process $X$ and we would like to define its integral with respect to Brownian motion. In other words, we want to define the expression $\int_0^T X_s \, dW_s$. A naive approach would be to try and argue as for the Riemann–Stieltjes integral. More precisely, one could consider the Riemann–Stieltjes sum for each sample path separately and hope that this sum converges for almost all sample paths. However, Example 8.3.1 shows that two problems occur.

- The first is that almost surely the sample paths of Brownian motion do not have finite variation. Therefore, we cannot expect that the Riemann–Stieltjes integral with respect to Brownian motion will exist for almost all sample paths. The solution to overcome this problem goes back to Itô and is to replace almost sure convergence by convergence in mean square with respect to the underlying probability measure $\mathbb{P}$.
- The second problem relates to the intermediate partition. More precisely, if we look at (8.3.1) then we see that the limit of a Riemann–Stieltjes sum depends on the choice of the intermediate partition. To overcome this problem, we need to make a fixed choice. In view of expressions occurring in finance, see for instance (8.2.3), it is natural to choose the point on the left. This

was the choice made by Itô and we will see below that it results in an integral which has a number of useful properties. Other choices are possible and result in integrals with different properties. The most common alternative to the Itô integral is the Stratonovich integral (see e.g. [13]).

Taking into account these two points, we consider the sums

$$S(X, W, \mathbf{t}) := \sum_{i=1}^{n} X_{t_{i-1}}(W_{t_i} - W_{t_{i-1}}) \tag{8.4.1}$$

and study their behaviour as $\|\mathbf{t}\| \to 0$ with respect to $\|\cdot\|_{L^2(\Omega)}$. Specifically, we propose the following definition and then investigate whether this definition gives rise to a process with reasonable properties.

**Definition 8.4.1** (Itô integral). Let $X = \{X_s, 0 \leq s \leq T\}$ be an adapted, cadlag process. Further, let $\mathbf{t} = (t_i)_{i=0}^n$ be a partition of $[0, T]$. The *stochastic integral of $X$ with respect to $W$* is defined as the mean square limit

$$\int_0^T X_s \, dW_s := \lim_{\|\mathbf{t}\| \to 0} S(X, W, \mathbf{t}) = \lim_{\|\mathbf{t}\| \to 0} \sum_{i=0}^{n} X_{t_i}(W_{t_i} - W_{t_{i-1}}) \tag{8.4.2}$$

if this limit exists and does not depend on the choice of partition.

It is important to keep in mind that (8.4.2) has to be understood as a mean square limit and not as an almost sure limit. In other words, (8.4.2) has to be interpreted as

$$\left\| \int_0^T X_s \, dW_s - S(X, W, \mathbf{t}) \right\|_{L^2(\Omega)} \to 0 \text{ as } \|\mathbf{t}\| \to 0.$$

The question we would like to resolve is for which processes $X$ the above expression converges and whether such a limit behaves as desired. Certainly, we would want any reasonable definition of an integral to satisfy

$$\int_0^T \mathbb{1}_{[a,b)}(s) \, dW_s = W_b - W_a \text{ for } 0 \leq a \leq b \leq T. \tag{8.4.3}$$

Let us check that (8.4.1) converges for $X = \mathbb{1}_{[a,b)}$ and that the limit satisfies (8.4.3). We substitute the process $X = \mathbb{1}_{[a,b)}$ into (8.4.1) and show that $S(\mathbb{1}_{[a,b)}, W, \mathbf{t})$ converges in $L^2$ to $W_b - W_a$ as $\|\mathbf{t}\| \to 0$. Assume that $a < b$ since this clearly holds if $a = b$. Suppose that $\mathbf{t} = (t_i)_{i=0}^n$ is a partition of $[0, T]$ and $\|\mathbf{t}\| \leq 4(b - a)$. Then

$$S(\mathbb{1}_{[a,b)}, W, \mathbf{t}) = \sum_{i=1}^{n} \mathbb{1}_{[a,b)}(t_{i-1})(W_{t_i} - W_{t_{i-1}}) = W_{t_u} - W_{t_d},$$

with $0 \leq d \leq u \leq n$ such that $t_{d-1} < a \leq t_d$ and $t_{u-1} < b \leq t_u$. Using that $\mathbb{E}[W_s] = 0$ for all $s$ and that Brownian motion has independent increments,

$$\left\|S(\mathbb{1}_{[a,b)}, W, \mathbf{t}) - (W_b - W_a)\right\|_{L^2(\Omega)}^2 = \mathbb{E}\left[\left((W_{t_u} - W_{t_d}) - (W_b - W_a)\right)^2\right]$$
$$= \mathbb{E}\left[\left((W_{t_u} - W_b) - (W_{t_d} - W_a)\right)^2\right] = \mathrm{Var}\left((W_{t_u} - W_b) - (W_{t_d} - W_a)\right)$$
$$= \mathrm{Var}\left(W_{t_u} - W_b\right) + \mathrm{Var}\left(W_{t_d} - W_a\right) = (t_u - b) + (t_u - a) \leq 2\|\mathbf{t}\|. \tag{8.4.4}$$

We therefore get, as required, that $S(X, W, \mathbf{t}) \xrightarrow{L^2} W_b - W_a$ as $\|\mathbf{t}\| \to 0$.

The next desirable property is that the stochastic integral should be linear. Consider processes $X$ which can be written as

$$X_s = \sum_{i=1}^{n} c_{i-1} \mathbb{1}_{[t_{i-1}, t_i)}(s) \text{ for all } s \geq 0, \tag{8.4.5}$$

where $c_0, c_1, \ldots, c_n$ are constants and $0 = t_0 < t_1 < \cdots < t_n = T$ are some times. We again need to check that (8.4.1) converges in this case and that the limit satisfies

$$\int_0^T X_s \, dW_s = \sum_{i=1}^{n} c_{i-1}(W_{t_i} - W_{t_{i-1}}).$$

Since $S(X, W, \mathbf{t})$ is by definition linear in $X$, this follows immediately from (8.4.3).

Processes of the form (8.4.5) occur naturally in finance. For instance, consider a trading strategy in which we trade only at the times $0 = t_0 < t_1 < \cdots < t_n = T$. Then the process giving either the number of bonds or the number of stocks held over the period $[0, T]$ has the same form as (8.4.5), provided the trading strategy is deterministic. In other words, we decide at time $t = 0$ how many bonds or stocks we will buy/sell at each of the times $t_1, \ldots, t_n$, regardless of how the market evolves. This is, of course, not likely to be a good strategy. We would therefore like to replace the constants $c_i$ in (8.4.5) by random variables $C_i$. Since a trading strategy can only depend on information that we have seen in the past, not on future outcomes, we should require that each $C_i$ is $\mathcal{F}_{t_i}$-measurable. This is equivalent to saying that $X$ is adapted. This leads to the following definition.

**Definition 8.4.2** (Simple process). We call an adapted process $X = \{X_s, 0 \leq s \leq T\}$ a *simple process* if it can be written as

$$X_s = \sum_{i=1}^{n} C_{i-1}(\omega) \mathbb{1}_{[t_{i-1}, t_i)}(s), \text{ for all } s \geq 0,\ \omega \in \Omega, \tag{8.4.6}$$

where $0 = t_0 < t_1 < \cdots \leq t_n = T$ are some times and $C_i$ are $\mathcal{F}_{t_i}$-measurable random variables with $\mathbb{E}\left[C_i^2\right] < \infty$ for $i = 0, 1, \ldots$.

The times $t_i$ in Definition 8.4.2 are not random and do not depend on the sample path of the Brownian motion $W$. Substituting (8.4.6) into (8.4.1) and doing a similar computation to that in (8.4.4) gives the following result.

**Proposition 8.4.3.** Let $X$ be a simple process satisfying (8.4.6). Then the stochastic integral $\int_0^T X_s\, dW_s$ exists and satisfies

$$\int_0^T X_s\, dW_s = \sum_{i=1}^{n} C_{i-1}(W_{t_i} - W_{t_{i-1}}). \tag{8.4.7}$$

Comparing the RHS of (8.4.7) to (8.2.3), we see that both have exactly the same form. In other words, if we think of a simple process as a trading strategy, then we can interpret the RHS of (8.4.7) as giving our total profit up to time $T$. In other words, we get again a desirable result. The stochastic integral is in general a random variable and not a constant. Furthermore, the stochastic integral is automatically $\mathcal{F}_T$-measurable since each $S(X, W, \mathbf{t})$ in (8.4.2) is $\mathcal{F}_T$-measurable and limits of measurable functions are measurable by Proposition 6.3.5.

We are interested in the conditions needed on $X$ so that the stochastic integral exists. We have shown above that it exists for all simple processes, but need to widen this class.

**Example 8.4.4.** *Suppose that $X$ is a process satisfying $\mathbb{E}\left[X_s^2\right] < \infty$ for all $s \geq 0$ and*

$$\lim_{t \to s} \mathbb{E}\left[(X_t - X_s)^2\right] = 0 \text{ for all } s \geq 0. \tag{8.4.8}$$

*A process with this property is called mean square continuous. We now show that the stochastic integral of a mean square continuous process $X$ exists with respect to Brownian motion. For this, we will use Theorem 6.1.23, but we first need some preliminaries. Define*

$$f(s,t) := \mathbb{E}\left[(X_t - X_s)^2\right].$$

*Condition (8.4.8) implies that $f$ is a continuous function. Also, for $\delta > 0$ set*

$$w_f(\delta) := \sup\{|f(x,y) - f(x',y')| : 0 \leq x, x', y, y' \leq T, |x - x'| < \delta, |y - y'| < \delta\}. \tag{8.4.9}$$

*Since $f$ is continuous, we have that $w_f(\delta) < \infty$ and $w_f(\delta) \to 0$ as $\delta \to 0$.*

*Now let $\mathbf{t} = (t_j)_{j=0}^n$ be a partition of $[0, T]$ and $\mathbf{t}'$ be the refined partition obtained by adding the point $u$ with $u \in (t_i, t_{i+1})$ to $\mathbf{t}$. Then*

$$\begin{aligned} S(X, W, \mathbf{t}) - S(X, W, \mathbf{t}') &= X_{t_i}\left(W_{t_{i+1}} - W_{t_i}\right) - \left(X_{t_i}\left(W_u - W_{t_i}\right) + X_u\left(W_{t_{i+1}} - W_u\right)\right) \\ &= (X_{t_i} - X_u)(W_{t_{i+1}} - W_u). \end{aligned}$$

*The random variables $X_{t_i} - X_u$ and $W_{t_{i+1}} - W_u$ are independent since we are working with the natural filtration of the Brownian motion. Therefore*

$$\begin{aligned} \|S(X, W, \mathbf{t}) - S(X, W, \mathbf{t}')\|_{L^2(\Omega)}^2 &= \mathbb{E}\left[(X_{t_i} - X_u)^2\right] \mathbb{E}\left[(W_{t_{i+1}} - W_u)^2\right] \\ &= \mathbb{E}\left[(X_{t_i} - X_u)^2\right](t_{i+1} - u) \\ &\leq w_f(\|\mathbf{t}\|)(t_{i+1} - u). \end{aligned} \tag{8.4.10}$$

*Suppose now that* $\mathbf{t}$ *and* $\mathbf{t}'$ *are partitions, where* $\mathbf{t}'$ *is a refined partition of* $\mathbf{t}$. *Using a similar computation to that in* (8.4.10)*, it is straightforward to see that*

$$\|S(X,W,\mathbf{t}) - S(X,W,\mathbf{t}')\|^2_{L^2(\Omega)} \le w_f(\|\mathbf{t}\|)T. \tag{8.4.11}$$

*This calculation is somewhat tedious and similar to the calculations in the proof of Theorem 8.4.6 below. For this reason, we omit this calculation. If* $\mathbf{t}$ *and* $\mathbf{t}'$ *are two arbitrary partitions of* $[0,T]$*, then denote by* $\mathbf{t}\cup\mathbf{t}'$ *the partition that contains all points of* $\mathbf{t}$ *and* $\mathbf{t}'$. *Clearly,* $\mathbf{t}\cup\mathbf{t}'$ *is a refinement of* $\mathbf{t}$ *and* $\mathbf{t}'$. *Using* (8.4.11) *with* $\mathbf{t}\cup\mathbf{t}'$ *and the triangle inequality, we obtain*

$$\|S(X,W,\mathbf{t}) - S(X,W,\mathbf{t}')\|_{L^2(\Omega)} \le \sqrt{w_f(\|\mathbf{t}\|)T} + \sqrt{w_f(\|\mathbf{t}'\|)T}. \tag{8.4.12}$$

*We can now show that the stochastic integral of* $X$ *exists. Let* $(\mathbf{t}_n)_{n=1}^\infty$ *be a sequence of partitions with* $\|\mathbf{t}_n\| \to 0$ *as* $n\to\infty$. *Equation* (8.4.12) *shows that* $S(X,W,\mathbf{t}_n)$ *is a Cauchy sequence in* $L^2(\Omega,\mathbb{P},\mathcal{F})$. *By Theorem 6.1.23 there exists a random variable* $Z$ *such that* $S(X,W,\mathbf{t}_n)$ *converges in mean square to this random variable. This random variable* $Z$ *is of course a candidate for the stochastic integral* $\int_0^T X_s\,dW_s$. *It remains to show that this limit is the same for each sequence of partitions. Suppose we have two sequences* $(\mathbf{t}_n)_{n=1}^\infty$ *and* $(\mathbf{t}'_n)_{n=1}^\infty$ *with* $\|\mathbf{t}_n\|\to 0$ *and* $\|\mathbf{t}'_n\|\to 0$ *as* $n\to\infty$. *Define*

$$\mathbf{t}''_n = \begin{cases} \mathbf{t}_m, & \text{if } n = 2m, \\ \mathbf{t}'_m, & \text{if } n = 2m-1. \end{cases}$$

*Clearly,* $\|\mathbf{t}''_n\|\to 0$ *as* $n\to\infty$*, so there exists a random variable* $Z''$ *such that* $S(X,W,\mathbf{t}''_n)$ *converges in mean square to* $Z''$. *Since* $(\mathbf{t}_n)_{n=1}^\infty$ *and* $(\mathbf{t}'_n)_{n=1}^\infty$ *are subsequences of* $(\mathbf{t}''_n)_{n=1}^\infty$*, it follows that*

$$S(X,W,\mathbf{t}_n) \xrightarrow{L^2} Z'' \text{ and } S(X,W,\mathbf{t}'_n) \xrightarrow{L^2} Z''.$$

*Since mean square limits are unique, we get that* $Z = Z''$*, so* $S(X,W,\mathbf{t}'_n) \xrightarrow{L^2} Z$. *Therefore the limit is independent of the sequence and the stochastic integral exists.*

Let us now use Example 8.4.4 in an explicit case.

**Example 8.4.5.** *Compute* $\int_0^T W_s dW_s$. *First observe that for all* $s,t\ge 0$,

$$\mathbb{E}\left[W_s^2\right] = s \text{ and } \mathbb{E}\left[(W_t - W_s)^2\right] = |t-s|.$$

*Therefore the condition* (8.4.8) *is fulfilled and* $\int_0^T W_s dW_s$ *exists. Choose the partition* $\mathbf{t} = (t_i)_{i=0}^n$ *with* $t_i = T/n$. *Now*

$$S(X,W,\mathbf{t}) = \sum_{i=1}^n W_{t_{i-1}}(W_{t_i} - W_{t_{i-1}}).$$

*Note that*

$$W_{t_{i-1}}(W_{t_i} - W_{t_{i-1}}) = \frac{1}{2}\left(W_{t_i}^2 - W_{t_{i-1}}^2 - (W_{t_i} - W_{t_{i-1}})^2\right).$$

*Using this, we obtain*

$$S(X,W,\mathbf{t}) = \frac{1}{2}\left(\sum_{i=1}^{n}(W_{t_i}^2 - W_{t_{i-1}}^2) - \sum_{i=1}^{n}(W_{t_i} - W_{t_{i-1}})^2\right)$$
$$= \frac{1}{2}(W_T^2 - W_0^2) - \frac{1}{2}\left(\sum_{i=1}^{n}(W_{t_i} - W_{t_{i-1}})^2\right).$$

*By the definition of the quadratic variation and Theorem 8.3.3,*

$$S(X,W,\mathbf{t}) \xrightarrow{L^2} \frac{1}{2}W_T^2 - \frac{1}{2}[W]_t = \frac{1}{2}W_T^2 - \frac{T}{2} \quad \textit{as } \|\mathbf{t}\| \to 0.$$

*Therefore*

$$\int_0^T W_s\, dW_s = \frac{1}{2}W_T^2 - \frac{T}{2}.$$

We cannot expect the stochastic integral to exist for every process $X$, so we return to the question of what conditions on the process will ensure existence of the limit in (8.4.2). If we consider for comparison Figure 8.1, then we see that a function can only be Riemann integrable if the approximating simple function converges (in $L^1([0,T])$) to this function as $\|\mathbf{t}\| \to 0$. It is therefore natural to expect that a similar condition needs to be fulfilled in the case of the stochastic integral. Define

$$R(X,\mathbf{t}) := \sum_{j=1}^{n} X_{t_{j-1}} \mathbb{1}_{[t_{j-1},t_j)}$$

and observe that, by Proposition 8.4.3,

$$S(X,W,\mathbf{t}) = \int_0^T R(X,\mathbf{t})_s dW_s.$$

We therefore anticipate that a condition such as

$$R(X,\mathbf{t}) \longrightarrow X \text{ as } \|\mathbf{t}\| \to 0 \tag{8.4.13}$$

will ensure that

$$\left\| \int_0^T R(X,\mathbf{t})_s dW_s - \int_0^T X_s dW_s \right\|_{L^2(\Omega)} \to 0 \text{ as } \|\mathbf{t}\| \to 0. \tag{8.4.14}$$

We need to specify which kind of convergence we mean in (8.4.13). Since the convergence in (8.4.14) is mean square convergence, mean square convergence of the approximating processes turns out to be the most suitable choice. As $X$ is a stochastic process, we have to use mean square convergence on the space $\Omega \times [0,T]$ with respect to the measure $\mathbb{P} \times \mathcal{L}$, see (6.1.10) and (6.1.23). Explicitly, we require

$$\|X - R(X,\mathbf{t})\|_{L^2(\Omega\times[0,T])} \to 0 \text{ as } \|\mathbf{t}\| \to 0, \tag{8.4.15}$$

where $\mathbf{t} = (t_i)_{i=0}^n$ is a partition of $[0, T]$. At this point, we have to assume that the process $X$ is $\mathcal{F}_T \times \mathcal{B}$-measurable, otherwise the expression in (8.4.15) is not well-defined. However, as we are working with adapted, cadlag processes (see Definition 8.4.1), this holds automatically, although we will not prove it here.

One can easily check that a process $X$ which fulfils the condition (8.4.8) in Example 8.4.4 is a special case of a process satisfying (8.4.15). Let $\mathbf{t}$ be a partition with $\|\mathbf{t}\| \leq \delta$ and let $w_f(\delta)$ be defined as in (8.4.9). Then

$$\begin{aligned}
\|X - R(X,\mathbf{t})\|^2_{L^2(\Omega\times[0,T])} &= \int_0^T \mathbb{E}\left[\left(X_s - R(X,\mathbf{t})_s\right)^2\right] ds \\
&= \sum_{j=1}^n \int_{t_{j-1}}^{t_j} \mathbb{E}\left[\left(X_s - X_{t_{j-1}}\right)^2\right] ds \\
&\leq \sum_{j=1}^n \int_{t_{j-1}}^{t_j} w_f(\delta)\, ds \\
&= T w_f(\delta).
\end{aligned} \tag{8.4.16}$$

Since $w_f(\delta) \to 0$ as $\delta \to 0$, (8.4.15) is fulfilled.

The argument in Example 8.4.4 can be generalised to show that the stochastic integral does indeed exist if (8.4.15) holds for all partitions on $[0, T]$, see Theorem 8.4.6 below. Unfortunately, it is often hard to check directly that (8.4.15) is fulfilled for all partitions for a general process $X$. However, in this book, we work only with cadlag processes and this allows us to establish (8.4.15) under a mild assumption. For any process $X$ defined on $\Omega \times [0, T]$, set

$$\|X\|_{L^{2,\infty}(\Omega\times[0,T])} := \left(\mathbb{E}\left[\sup_{s\in[0,T]} |X_s|^2\right]\right)^{1/2}. \tag{8.4.17}$$

Then $\|\cdot\|_{L^{2,\infty}(\Omega\times[0,T])}$ is a norm. In other words, for all processes $X$ and $Y$ defined on $\Omega \times [0, T]$ and $\lambda \in \mathbb{R}$:

- $\|\lambda X\|_{L^{2,\infty}(\Omega\times[0,T])} = |\lambda|\|X\|_{L^{2,\infty}(\Omega\times[0,T])}$,
- $\|X + Y\|_{L^{2,\infty}(\Omega\times[0,T])} \leq \|X\|_{L^{2,\infty}(\Omega\times[0,T])} + \|Y\|_{L^{2,\infty}(\Omega\times[0,T])}$,
- $\|\lambda X\|_{L^{2,\infty}(\Omega\times[0,T])} = 0$ implies that $X = 0$ a.s.

Furthermore, $\|X\|_{L^2(\Omega\times[0,T])} \leq \|X\|_{L^{2,\infty}(\Omega\times[0,T])}$. We claim that a sufficient condition for (8.4.15) to hold when $X$ is cadlag is that $\|X\|_{L^{2,\infty}(\Omega\times[0,T])} < \infty$. To see this, observe that if a function $f$ is cadlag then it has at most countably many discontinuity points. A proof of this fact can be found for the convenience of the reader in Section A.3. Therefore, the set of discontinuity points of $f$ has Lebesgue measure 0. If a process $X$ is cadlag then this implies that $R(X,\mathbf{t})(\omega, s) \to X(\omega, s)$ as $\|\mathbf{t}\| \to 0$ for all pairs $(\omega, s) \in \Omega \times [0, T]$, excluding a set of measure zero with respect to $\mathbb{P} \times \mathcal{L}$ which contains those points $(\omega, s)$ for which the sample path $t \mapsto X_t(\omega)$ has a discontinuity at $s$. By the dominated convergence theorem (Theorem 6.1.18), (8.4.15) will hold provided there exists some process $Z$ with $\|Z\|_{L^2(\Omega\times[0,T])} < \infty$ for which $|R(X,\mathbf{t})(\omega, s) - X(\omega, s)| \leq |Z(\omega, s)|$

for all $\mathbf{t}$. But this is immediate under the assumption that $\|X\|_{L^{2,\infty}(\Omega\times[0,T])} < \infty$. This leads to the following result.

**Theorem 8.4.6.** *Let $X$ be an adapted cadlag process satisfying $\|X\|_{L^{2,\infty}(\Omega\times[0,T])} < \infty$. The stochastic integral $\int_0^T X_s\,dW_s$ exists and satisfies the isometry*

$$\left\| \int_0^T X_s\,dW_s \right\|_{L^2(\Omega)} = \|X\|_{L^2(\Omega\times[0,T])}\,. \tag{8.4.18}$$

*Proof.* We first prove that (8.4.18) holds for simple processes. We already know by Proposition 8.4.3 that the stochastic integral exists in this case. Let $X$ be as in (8.4.6). Then

$$\left\| \int_0^T X_s\,dW_s \right\|^2_{L^2(\Omega)} = \mathbb{E}\left[\left(\int_0^T X_s\,dW_s\right)^2\right] = \mathbb{E}\left[\left(\sum_{i=1}^n C_{i-1}(W_{t_i} - W_{t_{i-1}})\right)^2\right]$$

$$= \sum_{i=1}^n \mathbb{E}\left[C_{i-1}^2(W_{t_i} - W_{t_{i-1}})^2\right] + \sum_{i\neq j} \mathbb{E}\left[\left(C_{i-1}(W_{t_i} - W_{t_{i-1}})\right)\left(C_{j-1}(W_{t_j} - W_{t_{j-1}})\right)\right].$$

By assumption, $C_i$ is $\mathcal{F}_{t_i}$-measurable. Therefore, $C_{i-1}$ and $W_{t_i} - W_{t_{i-1}}$ are independent since we are working with the natural filtration of Brownian motion. It follows that

$$\mathbb{E}\left[C_{i-1}^2(W_{t_i} - W_{t_{i-1}})^2\right] = \mathbb{E}\left[C_{i-1}^2\right]\mathbb{E}\left[(W_{t_i} - W_{t_{i-1}})^2\right] = \mathbb{E}\left[C_{i-1}^2\right](t_i - t_{i-1}).$$

Now suppose that $i > j$. Then $C_{i-1}C_{j-1}(W_{t_j} - W_{t_{j-1}})$ is $\mathcal{F}_{t_{i-1}}$-measurable and therefore independent of $W_{t_i} - W_{t_{i-1}}$. Since $\mathbb{E}\left[W_{t_i} - W_{t_{i-1}}\right] = 0$,

$$\mathbb{E}\left[\left(C_{i-1}(W_{t_i} - W_{t_{i-1}})\right)\left(C_{j-1}(W_{t_j} - W_{t_{j-1}})\right)\right] = 0.$$

Combining these two computations and using that the intervals $[t_{i-1}, t_i)$ are disjoint, we get

$$\left\| \int_0^T X_s\,dW_s \right\|^2_{L^2(\Omega)} = \sum_{i=1}^n \mathbb{E}\left[C_{i-1}^2\right](t_i - t_{i-1}) = \sum_{i=1}^n \mathbb{E}\left[C_{i-1}^2\right]\int_0^T \mathbb{1}_{[t_{i-1},t_i)}(s)\,ds$$

$$= \int_0^T \mathbb{E}\left[\sum_{i=1}^n C_{i-1}^2 \mathbb{1}_{[t_{i-1},t_i)}(s)\right] ds = \int_0^T \mathbb{E}\left[\left(\sum_{i=1}^n C_{i-1}\mathbb{1}_{[t_{i-1},t_i)}(s)\right)^2\right] ds$$

$$= \|X\|^2_{L^2(\Omega\times[0,T])}\,.$$

This completes the proof of (8.4.18) for simple functions.

We now show that the stochastic integral exists for general processes satisfying the assumptions. As in Example 8.4.4, we use Theorem 6.1.23. Let $X$ be given and $(\mathbf{t}^n)_{n=1}^\infty$ be a sequence of partitions with $\|\mathbf{t}^n\| \to 0$. Then the assumptions imply that

$$\left\|X - R(X, \mathbf{t}^n)\right\|_{L^2(\Omega\times[0,T])} \to 0 \text{ as } n \to \infty.$$

Note that, for each $n, m \in \mathbb{N}$, $R(X, \mathbf{t}^n) - R(X, \mathbf{t}^m)$ is a simple process and

$$\int_0^T R(X, \mathbf{t}^n)_s \, dW_s - \int_0^T R(X, \mathbf{t}^m)_s \, dW_s = \int_0^T \left(R(X, \mathbf{t}^n) - R(X, \mathbf{t}^m)\right)_s dW_s.$$

Using this and (8.4.18), we get for all $n, m \in \mathbb{N}$ that

$$\left\| \int_0^T \left(R(X, \mathbf{t}^n) - R(X, \mathbf{t}^m)\right)_s dW_s \right\|_{L^2(\Omega)} = \left\| R(X, \mathbf{t}^n) - R(X, \mathbf{t}^m) \right\|_{L^2(\Omega \times [0,T])}$$
$$\leq \left\| R(X, \mathbf{t}^n) - X \right\|_{L^2(\Omega \times [0,T])} + \left\| X - R(X, \mathbf{t}^m) \right\|_{L^2(\Omega \times [0,T])} \to 0 \text{ as } n, m \to \infty.$$

Therefore $\left(\int_0^T R(X, \mathbf{t}^n)_s \, dW_s\right)_{n=1}^{\infty}$ is a Cauchy sequence in $L^2(\Omega, \mathcal{F}, \mathbb{P})$. By Theorem 6.1.23 there exists a random variable $Z$ such that

$$\int_0^T R(X, \mathbf{t}^n)_s \, dW_s \xrightarrow{L^2} Z.$$

It remains to show that $Z$ is the same for each sequence $(R(X, \mathbf{t}^n))_{n=1}^{\infty}$. We omit this justification since it uses exactly the same argument as in Example 8.4.4. Therefore the stochastic integral of $X$ exists and $Z = \int_0^T X_s \, dW_s$.

We now show that (8.4.18) holds for general processes satisfying the assumptions of the theorem. Let $\epsilon > 0$ be given. The assumptions are sufficient to ensure that (8.4.15) holds, so there exists some partition $\mathbf{t}^\epsilon$ such that

$$\left\| X - R(X, \mathbf{t}^\epsilon) \right\|_{L^2(\Omega \times [0,T])} \leq \epsilon \text{ and } \left\| \int_0^T X_s \, dW_s - \int_0^T R(X, \mathbf{t}^\epsilon)_s \, dW_s \right\|_{L^2(\Omega)} \leq \epsilon.$$

Set $X^\epsilon = R(X, \mathbf{t}^\epsilon)$. Since $X^\epsilon$ is a simple process, it satisfies the isometry property. Therefore, by the triangle inequality,

$$\begin{aligned}
\left\| \int_0^T X_s \, dW_s \right\|_{L^2(\Omega)} &\leq \left\| \int_0^T X_s^\epsilon \, dW_s \right\|_{L^2(\Omega)} + \left\| \int_0^T X_s \, dW_s - \int_0^T X_s^\epsilon \, dW_s \right\|_{L^2(\Omega)} \\
&\leq \left\| X^\epsilon \right\|_{L^2(\Omega \times [0,T])} + \epsilon \\
&\leq \|X\|_{L^2(\Omega \times [0,T])} + \left\| X^\epsilon - X \right\|_{L^2(\Omega \times [0,T])} + \epsilon \\
&\leq \|X\|_{L^2(\Omega \times [0,T])} + 2\epsilon.
\end{aligned}$$

A similar inequality in the other direction can be established with the reverse triangle inequality. Since $\epsilon > 0$ is arbitrary, this completes the proof. □

The assumptions in Theorem 8.4.6 allow us to illustrate the ideas behind the stochastic integral and at the same time to give a simple proof of existence. However, it is possible to construct the stochastic integral under weaker assumptions.

**Corollary 8.4.7.** Let $X$ be an adapted cadlag process satisfying $\|X\|_{L^2(\Omega \times [0,T])} < \infty$. The stochastic integral $\int_0^T X_s \, dW_s$ exists and satisfies the isometry

$$\left\| \int_0^T X_s \, dW_s \right\|_{L^2(\Omega)} = \|X\|_{L^2(\Omega \times [0,T])}. \tag{8.4.19}$$

The proof of Corollary 8.4.7 is similar to the proof of Theorem 8.4.6, but it requires one to generalise Definition 8.4.8. We will not discuss this here, so omit the proof of Corollary 8.4.7. The interested reader can find the details in [14, Chapter 14] or [3, Chapter 12]. For the rest of this book, we give the statements of results using the assumption $\|X\|_{L^2(\Omega\times[0,T])} < \infty$. However, where it simplifies the proofs, we will work with the stronger assumption $\|X\|_{L^{2,\infty}(\Omega\times[0,T])} < \infty$. Let us consider an example.

**Example 8.4.8.** *Let $f : [0,T] \to \mathbb{R}$ be a continuous function. We compute the stochastic integral of $f$ with respect to Brownian motion. Here we interpret $f$ as a deterministic process. In other words, the value of $f$ at time $s$ is always $f(s)$, no matter which $\omega \in \Omega$ occurs. Therefore*

$$\|f\|_{L^2(\Omega\times[0,T])} \leq \|f\|_{L^{2,\infty}(\Omega\times[0,T])} = \sup_{s\in[0,T]} |f(s)| < \infty.$$

*By Theorem 8.4.6, the stochastic integral $\int_0^T f(s)\,dW_s$ exists. For a partition $\mathbf{t} = (t_i)_{i=0}^n$,*

$$S(f, W, \mathbf{t}) = \sum_{i=1}^{n} f(t_{i-1})\big(W_{t_i} - W_{t_{i-1}}\big).$$

*Since the increments of Brownian motion are independent and $f$ is deterministic, we immediately get that $S(f, W, \mathbf{t})$ is a mean-zero Gaussian random variable. It follows that $\int_0^T f(s)\,dW_s$ is a mean-zero Gaussian random variable. By (8.4.18), its variance is given by*

$$\left\| \int_0^T f(s)\,dW_s \right\|_{L^2(\Omega)} = \int_0^T f(s)^2\,ds.$$

## 8.5 Properties of the Stochastic Integral

In Section 8.4, we introduced the stochastic integral over the interval $[0,T]$. In this section, we study some elementary properties of stochastic integrals. We will require stochastic integrals over sub-intervals of $[0,T]$, such as $[0,t]$ with $t \leq T$. These are defined exactly as in Definition 8.4.1. Observe that, for a process $X = \{X_s, 0 \leq s \leq T\}$,

$$\|X\|_{L^2(\Omega\times[0,t])} \leq \|X\|_{L^2(\Omega\times[0,T])} \quad \text{for all } 0 \leq t \leq T.$$

Therefore, if $X$ is cadlag and $\|X\|_{L^2(\Omega\times[0,T])} < \infty$ then by Theorem 8.4.6 the stochastic integral exists over $[0,t]$ for all $0 \leq t \leq T$.

We start with a simple observation. For $0 \leq a \leq b$ and $t \geq 0$,

$$\int_0^t \mathbb{1}_{[a,b)}(s)\,dW_s = W_{b\wedge t} - W_{a\wedge t} = \begin{cases} W_b - W_a, & \text{if } a \leq b \leq t, \\ W_t - W_a, & \text{if } a \leq t \leq b, \\ 0, & \text{if } t \leq a \leq b, \end{cases} \tag{8.5.1}$$

where $u \wedge v = \min\{u, v\}$. Indeed, if $a \leq t \leq b$ then $\mathbb{1}_{[a,b)} = \mathbb{1}_{[a,t)} + \mathbb{1}_{[t,b)}$ and $\mathbb{1}_{[t,b)}$ is equal to 0 before time $t$, from which it follows that the stochastic integral of $\mathbb{1}_{[t,b)}$ over the time period $[0,t]$ is 0. Therefore, if $X$ is a simple process as in (8.4.6), then

$$\int_0^t X_s \, dW_s = \sum_{i=1}^{n} C_{i-1}(W_{t_i \wedge t} - W_{t_{i-1} \wedge t}). \tag{8.5.2}$$

We look at some simple properties of the stochastic integral. Many properties are the same as for the Riemann and the Riemann–Stieltjes integral.

**Proposition 8.5.1.** Let $X$ and $Y$ be two adapted cadlag stochastic processes with $\|X\|_{L^2(\Omega\times[0,T])} < \infty$ and $\|Y\|_{L^2(\Omega\times[0,T])} < \infty$.

- For all $\lambda, \mu \in \mathbb{R}$ and $0 \le t \le T$,

$$\int_0^t (\lambda X_s + \mu Y_s) \, dW_s = \lambda \int_0^t X_s \, dW_s + \mu \int_0^t Y_s \, dW_s. \tag{8.5.3}$$

- For all $0 \le u \le t \le T$,

$$\int_0^t X_s \, dW_s = \int_0^u X_s \, dW_s + \int_u^t X_s \, dW_s. \tag{8.5.4}$$

The proof of this proposition is straightforward and is therefore omitted.

Unfortunately, not every property of the Riemann integral holds for the stochastic integral.

**Example 8.5.2.** *Recall that if $f$ and $g$ are Riemann-integrable functions on the interval* $[0, T]$ *with $f \le g$, then*

$$\left| \int_0^T f(x) \, dx \right| \le \int_0^T |f(x)| \, dx \quad and \quad \int_0^T f(x) \, dx \le \int_0^T g(x) \, dx. \tag{8.5.5}$$

*These inequalities cannot be generalised to the stochastic integral. Consider $X = 0$ and $Y = \mathbb{1}_{[0,T)}$. Then $X \le Y$, $|Y| = Y$,*

$$\int_0^T X_s \, dW_s = 0 \quad and \quad \int_0^T Y_s \, dW_s = W_T - W_0 = W_T.$$

*Since $W_T$ can be positive or negative,* (8.5.5) *does not hold for stochastic integrals.*

So far, we have considered the stochastic integral for only one time point $t$ at a time. However, it is natural to consider the process generated by the stochastic integral over the time. For instance, if $X$ is a trading strategy then the stochastic integral can be interpreted as the profit of this trading strategy. In this setting, it is natural to study how the profit evolves over time. An important property of the stochastic integral is that this process is a martingale.

**Theorem 8.5.3.** *Let $X$ be an adapted cadlag process with $\|X\|_{L^2(\Omega\times[0,T])} < \infty$. Set $I_t(X) := \int_0^t X_s dW_s$ for $0 \le t \le T$. Then the process $\{I_t(X), 0 \le t \le T\}$ is a martingale with respect to $\mathbb{F}$ (and hence has expectation 0).*

*Proof.* We first prove this theorem for simple processes. Suppose that $X$ is a simple process as in Definition 8.4.2 and $t_k \leq t < t_{k+1}$ for some $0 \leq k \leq n$. Then, by (8.5.2),

$$I_t(X) = \int_0^t X_s \, dW_s = \sum_{i=1}^{k} C_{i-1}(W_{t_i} - W_{t_{i-1}}) + C_k(W_t - W_{t_k}). \tag{8.5.6}$$

To show that $\{I_t(X), t \geq 0\}$ is a martingale, we have to check the requirements in Definition 6.3.13.

- Since each $C_{i-1}$ is $\mathcal{F}_{t_{i-1}}$-measurable, we immediately get that $I_t(X)$ is $\mathcal{F}_t$-measurable.
- By the triangle inequality,

$$\begin{aligned} \mathbb{E}\left[|I_t(X)|\right] &\leq \sum_{i=1}^{k} \mathbb{E}\left[|C_{i-1}(W_{t_i} - W_{t_{i-1}})|\right] + \mathbb{E}\left[|C_k(W_t - W_{t_k})|\right] \\ &\leq \sum_{i=1}^{k} \|C_{i-1}\|_2 \cdot \|W_{t_i} - W_{t_{i-1}}\|_2 + \|C_k\|_2 \cdot \|W_t - W_{t_k}\|_2. \end{aligned}$$

  The last line used the Cauchy–Schwartz inequality. This expression is finite since by assumption, $\mathbb{E}\left[(C_{i-1})^2\right] < \infty$ and $W_{t_i} - W_{t_{i-1}}$ is normal distributed.
- It remains to show that $\mathbb{E}\left[I_t(X) \,\middle|\, \mathcal{F}_s\right] = I_s(X)$ for $0 \leq s \leq t$. We claim that

$$\mathbb{E}\left[C_{i-1}(W_{t_i} - W_{t_{i-1}}) \,\middle|\, \mathcal{F}_s\right] = \begin{cases} C_{i-1}(W_{t_i} - W_{t_{i-1}}), & \text{if } t_i \leq s, \\ C_{i-1}(W_s - W_{t_{i-1}}), & \text{if } t_{i-1} \leq s \leq t_i, \\ 0, & \text{if } s \leq t_{i-1}. \end{cases} \tag{8.5.7}$$

  The identity $\mathbb{E}\left[I_t(X) \,\middle|\, \mathcal{F}_s\right] = I_s(X)$ then follows immediately from (8.5.6). It is therefore sufficient to prove (8.5.7). We only prove the case $t_{i-1} \leq s \leq t_i$ since the proof of the other two cases is almost identical. Using that $C_{i-1}$ and $W_{t_{i-1}}$ are $\mathcal{F}_{t_{i-1}}$-measurable, and that $W_{t_i} - W_s$ is independent of $\mathcal{F}_{t_{i-1}}$, by Theorem 6.3.12,

$$\begin{aligned} \mathbb{E}\left[C_{i-1}(W_{t_i} - W_{t_{i-1}}) \,\middle|\, \mathcal{F}_s\right] &= C_{i-1}\mathbb{E}\left[W_{t_i} \,\middle|\, \mathcal{F}_s\right] - C_{i-1}W_{t_{i-1}} \\ &= C_{i-1}\mathbb{E}\left[W_{t_i} - W_s \,\middle|\, \mathcal{F}_s\right] + C_{i-1}\mathbb{E}\left[W_s \,\middle|\, \mathcal{F}_s\right] - C_{i-1}W_{t_{i-1}} \\ &= C_{i-1}\mathbb{E}\left[W_{t_i} - W_s\right] + C_{i-1}W_s - C_{i-1}W_{t_{i-1}} \\ &= C_{i-1}W_s - C_{i-1}W_{t_{i-1}}. \end{aligned}$$

This completes the proof for simple processes. We now prove the general case. Let $X$ be an adapted cadlag process with $\|X\|_{L^{2,\infty}(\Omega\times[0,T])} < \infty$ and let $(\mathbf{t}^n)_{n=1}^{\infty}$ be a sequence of partitions with $\|\mathbf{t}^n\| \to 0$. By the construction of the stochastic integral, the integral of $X$ is a limit of integrals of simple processes. We show that the properties in Definition 6.3.13, which we established above for simple processes, carry through to the limit.

- The stochastic integral of $X$ is by definition the mean square limit of the stochastic integrals of $R(X, W, \mathbf{t}^n)$, that is

$$\int_0^t R(X, W, \mathbf{t}^n)_s \, dW_s \xrightarrow{L^2} \int_0^t X_s \, dW_s. \tag{8.5.8}$$

Theorem 6.1.23 implies that there exists a subsequence such that the convergence in (8.5.8) is almost sure. The left-hand side is $\mathcal{F}_t$-measurable by the argument above. Proposition 6.3.5 then implies that $\int_0^t X_s\, dW_s$ is $\mathcal{F}_t$-measurable.

- Using the isometry (8.4.18),

$$\mathbb{E}\left[\left|\int_0^t X_s\, dW_s\right|\right] \leq \left\|\int_0^t X_s\, dW_s\right\|_{L^2(\Omega)} = \|X\|_{L^2(\Omega\times[0,t])} \leq \|X\|_{L^2(\Omega\times[0,T])} < \infty.$$

- We start with a general observation. Let $Z_n$ with $n \in \mathbb{N}$ and $Z$ be random variables and $\mathcal{G} \subset \mathcal{F}$ be a $\sigma$-algebra. Then

$$Z_n \stackrel{L^2}{\to} Z \text{ as } n \to \infty \implies \mathbb{E}[Z_n \,|\, \mathcal{G}] \stackrel{L^2}{\to} \mathbb{E}[Z \,|\, \mathcal{G}] \text{ as } n \to \infty. \tag{8.5.9}$$

Indeed, Jensen's inequality for conditional expectations and the tower property imply

$$\begin{aligned}\left\|\mathbb{E}[Z_n \,|\, \mathcal{G}] - \mathbb{E}[Z \,|\, \mathcal{G}]\right\|_2^2 &= \left\|\mathbb{E}[Z_n - Z \,|\, \mathcal{G}]\right\|_2^2 = \mathbb{E}\left[\mathbb{E}[Z_n - Z \,|\, \mathcal{G}]^2\right] \\ &\leq \mathbb{E}\left[\mathbb{E}\left[(Z_n - Z)^2 \,|\, \mathcal{G}\right]\right] = \mathbb{E}\left[(Z_n - Z)^2\right].\end{aligned}$$

Using Theorem 6.1.23, we can choose in (8.5.9) a subsequence $(Z_{n_k})_{k=1}^{\infty}$ such that

$$Z_{n_k} \stackrel{as}{\to} Z \text{ as } n \to \infty \text{ and } \mathbb{E}\left[Z_{n_k} \,|\, \mathcal{G}\right] \stackrel{as}{\to} \mathbb{E}[Z \,|\, \mathcal{G}] \text{ as } n \to \infty.$$

Using this and the definition of the stochastic integral, for $0 \leq s \leq t$,

$$\mathbb{E}\left[\int_0^t X_u dW_u \middle| \mathcal{F}_s\right] = \lim_{n\to\infty} \mathbb{E}\left[\int_0^t X_u^n dW_u \middle| \mathcal{F}_s\right] = \lim_{n\to\infty} \int_0^s X_u^n dW_u = \int_0^s X_u dW_u.$$

The limits in this computation have to be understood as almost sure limits. This completes the proof of the general case. □

## 8.6 Itô's Formula

We now know how to define a stochastic integral and also know some properties of stochastic integrals. Unfortunately, the construction as a limit of integrals of simple processes is not very convenient for doing calculations, see for instance Example 8.4.5. Itô's formula is one of the main tools that enables us to calculate stochastic integrals and carry out operations on them. It is the stochastic analogue of the classical chain rule for differentiation.

We begin with the simplest version of Itô's formula.

**Theorem 8.6.1.** *If $\{W_t, t \geq 0\}$ is a Brownian motion and $f : \mathbb{R} \to \mathbb{R}$ is a twice continuously differentiable function, then*

$$f(W_t) = f(0) + \int_0^t f'(W_s)dW_s + \frac{1}{2}\int_0^t f''(W_s)ds.$$

Before we prove Itô's lemma, we illustrate its use with a simple example.

**Example 8.6.2.** *Choose $f(x) = x^2$. Then $f'(x) = 2x$ and $f''(x) = 2$. Itô's formula therefore gives*

$$W_T^2 = \int_0^T 2W_s\,dW_s + \frac{1}{2}\int_0^T 2\,ds = 2\int_0^T W_s\,dW_s + T.$$

*Rearranging, we recover the same result as in Example 8.4.5,*

$$\int_0^T W_s dW_s = \frac{1}{2}(W_T^2 - T).$$

For the proof of Theorem 8.6.1, we need the following lemma.

**Lemma 8.6.3.** Let $g : \mathbb{R} \to \mathbb{R}$ be a continuous, bounded function and $\mathbf{t} = (t_i)_{i=0}^n$ be a partition of $[0, T]$. Then

$$\sum_{i=1}^n g(W_{t_{i-1}})(W_{t_i} - W_{t_{i-1}})^2 \overset{L^2}{\to} \int_0^T g(W_s)ds \text{ as } \|\mathbf{t}\| \to 0. \tag{8.6.1}$$

Equation (8.6.1) plays a crucial role in the proofs of all versions of Itô's lemma. Symbolically, we can abbreviate (8.6.1) as $(dW_t)^2 = dt$. The advantage of this is that it is easy to remember, but one has to keep in mind that there is no rigorous mathematical definition of $(dW_t)^2$.

*Proof of Lemma 8.6.3.* Since $g$ is continuous, the Riemann sum

$$\sum_{i=1}^n g(W_{t_{i-1}})(t_i - t_{i-1}) \overset{as}{\to} \int_0^T g(W_s)ds. \tag{8.6.2}$$

By assumption, there exists an $M > 0$ such that $|g(x)| \leq M$ for all $x \in \mathbb{R}$. This implies that

$$\left|\sum_{i=1}^n g(W_{t_{i-1}})(t_i - t_{i-1}) - \int_0^T g(W_s)ds\right| \leq 2MT. \tag{8.6.3}$$

Therefore, by the dominated convergence theorem,

$$\sum_{i=1}^n g(W_{t_{i-1}})(t_i - t_{i-1}) \overset{L^2}{\to} \int_0^T g(W_s)ds. \tag{8.6.4}$$

It therefore suffices to show that

$$\sum_{i=1}^n g(W_{t_{i-1}})\left((W_{t_i} - W_{t_{i-1}})^2 - (t_i - t_{i-1})\right) \overset{L^2}{\to} 0 \text{ as } \|\mathbf{t}\| \to 0.$$

For this, set $Y_{i-1} := (W_{t_i} - W_{t_{i-1}})^2 - (t_i - t_{i-1})$. We have to show that

$$\mathbb{E}\left[\left(\sum_{i=1}^{n} g(W_{t_{i-1}})Y_{i-1}\right)^2\right] \to 0 \text{ as } \|\mathbf{t}\| \to 0.$$

We use the same argument as in the proof of Theorem 8.4.6. If $j > i$ then $g(W_{t_{i-1}})g(W_{t_{j-1}})Y_{i-1}$ is $\mathcal{F}_{t_{j-1}}$-measurable and therefore independent of $Y_{j-1}$. Now

$$\mathbb{E}\left[Y_{j-1}\right] = \mathbb{E}\left[(W_{t_j} - W_{t_{j-1}})^2 - (t_j - t_{j-1})\right] = \operatorname{Var}(W_{t_j} - W_{t_{j-1}}) - (t_j - t_{j-1}) = 0.$$

Hence, observe that for $i \neq j$,

$$\mathbb{E}\left[g(W_{t_{i-1}})g(W_{t_{j-1}})Y_{i-1}Y_{j-1}\right] = \mathbb{E}\left[g(W_{t_{i-1}})g(W_{t_{j-1}})Y_{i-1}\right]\mathbb{E}\left[Y_{j-1}\right] = 0.$$

Furthermore, $g(W_{t_{i-1}})$ and $Y_{i-1}$ are also independent and

$$\begin{aligned}\mathbb{E}\left[Y_{i-1}^2\right] &= \mathbb{E}\left[(W_{t_i} - W_{t_{i-1}})^4\right] - 2(t_i - t_{i-1})\mathbb{E}\left[(W_{t_i} - W_{t_{i-1}})^2\right] + (t_i - t_{i-1})^2 \\ &= 3(t_i - t_{i-1})^2 - 2(t_i - t_{i-1})^2 + (t_i - t_{i-1})^2 = 2(t_i - t_{i-1})^2.\end{aligned}$$

Therefore

$$\begin{aligned}\mathbb{E}\left[\left(\sum_{i=1}^{n} g(W_{t_{i-1}})Y_{i-1}\right)^2\right] &= \sum_{i=1}^{n} \mathbb{E}\left[g(W_{t_{i-1}})^2 Y_{i-1}^2\right] = \sum_{i=1}^{n} \mathbb{E}\left[g(W_{t_{i-1}})^2\right]\mathbb{E}\left[Y_{i-1}^2\right] \\ &= 2\sum_{i=1}^{n} \mathbb{E}\left[g(W_{t_{i-1}})^2\right](t_i - t_{i-1})^2 \\ &\leq 2\|\mathbf{t}\| \sum_{i=1}^{n} \mathbb{E}\left[g(W_{t_{i-1}})^2\right](t_i - t_{i-1}) \to 0.\end{aligned}$$

Here we are using that $\|\mathbf{t}\|$ tends to zero by assumption and the sum converges to the Riemann integral $\int_0^T \mathbb{E}\left[g(W_s)^2\right] ds < \infty$. □

*Proof of Theorem 8.6.1.* We give the proof only in the case where the support of $f$ is compact, that is in the case when there exists $M > 0$ such that $f(x) = 0$ for all $x$ with $|x| \geq M$. The proof of the general case can be found for instance in [14].

Let $\mathbf{t} = (t_i)_{i=0}^n$ be a partition of $[0, T]$. Then

$$f(W_T) = f(0) + \sum_{i=1}^{n} \left(f(W_{t_i}) - f(W_{t_{i-1}})\right).$$

By Taylor's theorem,

$$f(W_{t_i}) = f(W_{t_{i-1}}) + f'(W_{t_{i-1}})(W_{t_i} - W_{t_{i-1}}) + \frac{1}{2}f''(A_i)(W_{t_i} - W_{t_{i-1}})^2$$

for some $A_i$ that lies between $W_{t_{i-1}}$ and $W_{t_i}$. Therefore

$$f(W_T) = f(0) + \sum_{i=1}^{n} f'(W_{t_{i-1}})(W_{t_i} - W_{t_{i-1}}) + \frac{1}{2}\sum_{i=1}^{n} f''(A_i)(W_{t_i} - W_{t_{i-1}})^2. \tag{8.6.5}$$

The first sum in (8.6.5) is equal to $S(X, W, \mathbf{t})$ for the process $X = \{f'(W_s), s \geq 0\}$. By Definition 8.4.1, this sum converges to the stochastic integral of this process, provided it exists. To justify this, we check the conditions in Theorem 8.4.6. Since $f'$ is a continuous function and Brownian motion has continuous sample paths, $X$ is continuous. By the assumption that the support of $f$ is compact, $X$ is bounded. Therefore, by Theorem 8.4.6,

$$\sum_{i=1}^{n} f'(W_{t_{i-1}})(W_{t_i} - W_{t_{i-1}}) \xrightarrow{L^2} \int_0^T f'(W_s)dW_s \text{ as } \|t\| \to 0.$$

We consider next the second sum in (8.6.5). Then

$$\left|\sum_{i=1}^{n} f''(A_i)(W_{t_i} - W_{t_{i-1}})^2 - \sum_{i=1}^{n} f''(W_{t_{i-1}})(W_{t_i} - W_{t_{i-1}})^2\right|$$

$$\leq \sum_{i=1}^{n} |f''(A_i) - f''(W_{t_{i-1}})|(W_{t_i} - W_{t_{i-1}})^2$$

$$\leq \sup_{1\leq j\leq n} |f''(A_j) - f''(W_{t_{j-1}})| \sum_{i=1}^{n} (W_{t_i} - W_{t_{i-1}})^2. \tag{8.6.6}$$

Recall that $A_j$ lies between $W_{t_{j-1}}$ and $W_{t_j}$. Since Brownian motion has continuous sample paths and by assumption $f''$ is continuous and has compact support,

$$\sup_{1\leq j\leq n} |f''(A_j) - f''(W_{t_{j-1}})| \xrightarrow{as} 0 \text{ and } \sup_{1\leq j\leq n} |f''(A_j) - f''(W_{t_{j-1}})| \leq 2 \sup_{x\in\mathbb{R}} |f''(x)| < \infty.$$

The summation factor in (8.6.6) converges in $L^2$ to $[W]_T = T < \infty$, see Theorem 8.3.3. Therefore, by dominated convergence, the expression in (8.6.6) converges to 0.

It remains to show that

$$\sum_{i=1}^{n} f''(W_{t_{i-1}})(W_{t_i} - W_{t_{i-1}})^2 \xrightarrow{L^2} \int_0^T f''(W_s)ds.$$

But this follows immediately from Lemma 8.6.3 with $g = f''$, completing the proof.

□

By using Itô's formula, calculating stochastic integrals is not much more difficult than calculating ordinary integrals. Suppose we would like to calculate the ordinary integral $\int_a^b f(x)dx$. We look for a function $g(x)$ with the property that $g'(x) = f(x)$. The fundamental theorem of calculus then tells us that

$$\int_a^b f(x)dx = \int_a^b g'(x)dx = g(b) - g(a).$$

Now suppose that we would like to calculate the stochastic integral $\int_0^T f(W_s)dW_s$. Again, we look for a function $g(x)$ with the property that $g'(x) = f(x)$. By the argument above, $g$ is just the indefinite integral, or *antiderivative* of $f$. Itô's formula then tells us that

$$\int_0^T f(W_s)dW_s = \int_0^T g'(W_s)dW_s = g(W_T) - g(W_0) - \frac{1}{2}\int_0^T g''(W_s)ds.$$

The extra term is just an ordinary Riemann integral and needs to be there to ensure that $\int_0^T f(W_s)dW_s$ is a martingale.

**Example 8.6.4.** *We determine*

$$\int_0^T \sin(W_s)dW_s.$$

*For this, let* $f(x) = \sin(x)$. *Then* $g(x) = -\cos(x)$ *satisfies* $g'(x) = \sin(x)$ *and* $g''(x) = \cos(x)$. *By Itô's formula,*

$$\int_0^T \sin(W_s)dW_s = 1 - \cos(W_T) - \frac{1}{2}\int_0^T \cos(W_s)ds.$$

Itô's formula can be extended to processes of the form $f(t, W_t)$.

**Theorem 8.6.5.** *Suppose* $f : [0,\infty) \times \mathbb{R} \to \mathbb{R}$ *is continuously differentiable with respect to the first variable, t, and twice continuously differentiable with respect to the second variable, x. Then*

$$f(t, W_t) = f(0,0) + \int_0^t \frac{\partial f}{\partial x}(s, W_s)dW_s + \int_0^t \left(\frac{\partial f}{\partial t}(s, W_s) + \frac{1}{2}\frac{\partial^2 f}{\partial x^2}(s, W_s)\right) ds. \tag{8.6.7}$$

The proof is similar to that of Theorem 8.6.1, and a special case of Theorem 8.6.8, so we omit it.

**Example 8.6.6.** *Let* $f(t,x) = e^{x-t/2}$. *Then*

$$\frac{\partial f}{\partial x}(t,x) = e^{x-t/2}, \quad \frac{\partial f}{\partial t}(t,x) = -\frac{1}{2}e^{x-t/2} \quad \text{and} \quad \frac{\partial^2 f}{\partial x^2}(t,x) = e^{x-t/2}.$$

*By Itô's formula*

$$\begin{aligned} e^{W_T - T/2} &= 1 + \int_0^T e^{W_s - s/2}dW_s + \int_0^T \left(-\frac{1}{2}e^{W_s-s/2} + \frac{1}{2}e^{W_s-s/2}\right) ds \\ &= 1 + \int_0^T e^{W_s-s/2}dW_s. \end{aligned}$$

*Since the stochastic integral* $\int_0^t e^{W_s-s/2}dW_s$ *is a martingale, this gives us a very quick proof that the process* $\{X_t, t \geq 0\}$ *with* $X_t = e^{W_t - t/2}$ *is a martingale.*

*Furthermore,* $\{X_t, t \geq 0\}$ *satisfies*

$$X_t = 1 + \int_0^t X_s dW_s. \tag{8.6.8}$$

*This is reminiscent of the exponential function which satisfies*

$$e^t = 1 + \int_0^t e^s ds.$$

There are many processes which can be expressed in a similar way to (8.6.8). Since these kinds of processes are important, they have a special name.

**Definition 8.6.7** (Itô process). An *Itô process* is a stochastic process of the form

$$X_t = X_0 + \int_0^t Y_s dW_s + \int_0^t Z_s ds, \tag{8.6.9}$$

where $X_0$, the initial position of the process, is $\mathcal{F}_0$-measurable, and $\{Y_t, t \geq 0\}$ and $\{Z_t, t \geq 0\}$ are adapted and satisfy for each $t \geq 0$

$$\|Y\|_{L^2(\Omega\times[0,t])} = \left(\int_0^t \mathbb{E}\left[Y_s^2\right] ds\right)^{1/2} < \infty \quad \text{and} \quad \|Z\|_{L^1([0,t])} = \int_0^t |Z_s| ds < \infty \text{ a.s.}$$

We represent $X_t$ using the shorthand differential notation

$$dX_t = Y_t dW_t + Z_t dt. \tag{8.6.10}$$

At this point we have to highlight that the expression in (8.6.10) is just a formal notation/abbreviation for (8.6.9). There is no way to assign a rigorous mathematical meaning to the expression $dW_t$ as the sample paths of Brownian motion are not differentiable. We recommend keeping this in mind.

Itô's formula can be extended to processes of the form $f(t, X_t)$, where $\{X_t, t \geq 0\}$ is an Itô process.

**Theorem 8.6.8** (Itô's formula). *Suppose $f : [0,\infty) \times \mathbb{R} \to \mathbb{R}$ is continuously differentiable with respect to the first variable, $t$, and twice continuously differentiable with respect to the second variable, $x$, and let $\{X_t, t \geq 0\}$ be an Itô process satisfying* (8.6.9). *Then*

$$df(t,X_t) = Y_t \frac{\partial f}{\partial x}(t,X_t) dW_t + \left(Z_t \frac{\partial f}{\partial x}(t,X_t) + \frac{\partial f}{\partial t}(t,X_t) + \frac{1}{2} Y_t^2 \frac{\partial^2 f}{\partial x^2}(t,X_t)\right) dt.$$

*In particular, the process $\{f(t,X_t), t \geq 0\}$ is an Itô process.*

*Proof.* Let $0 = t_0 < t_1 < \cdots < t_n = t$ be a partition of $[0, T]$. Then, as before,

$$f(t,X_t) = f(0,X_0) + \sum_{i=1}^{n} \left(f(t_i, X_{t_i}) - f(t_{i-1}, X_{t_{i-1}})\right).$$

By Taylor's theorem,

$$f(t_i,X_{t_i}) = f(t_{i-1},X_{t_{i-1}}) + \frac{\partial f}{\partial t}(s_i,X_{t_{i-1}})(t_i - t_{i-1}) + \frac{\partial f}{\partial x}(t_{i-1},X_{t_{i-1}})(X_{t_i} - X_{t_{i-1}})$$
$$+ \frac{1}{2}\frac{\partial^2 f}{\partial x^2}(t_{i-1},A_i)(X_{t_i} - X_{t_{i-1}})^2,$$

where $t_{i-1} \leq s_i \leq t_i$ and $A_i$ lies between $X_{t_{i-1}}$ and $X_{t_i}$. The rest of the proof is similar to that of Theorem 8.6.1 except that we use that, when $t_i - t_{i-1}$ is small,

$$X_{t_i} - X_{t_{i-1}} \approx Y_{t_{i-1}}(W_{t_i} - W_{t_{i-1}}) + Z_{t_{i-1}}(t_i - t_{i-1}).$$

We omit the details since the computation is straightforward, but a little bit tedious. □

Until now, we have only defined the stochastic integral with respect to Brownian motion. This is enough for most cases we are working with, but in a few cases, we require the stochastic integral with respect to an Itô process. This is defined as follows.

**Definition 8.6.9** (Stochastic integration for Itô processes). Let $\{X_t, t \geq 0\}$ be an Itô process satisfying (8.6.9) for some processes $X$ and $Y$, and let $\alpha = \{\alpha_t, t \geq 0\}$ be a process such that for all $t \geq 0$,

$$\|\alpha Y\|_{L^2(\Omega\times[0,t])} < \infty \quad \text{and} \quad \|\alpha Z\|_{L^1([0,t])} < \infty \text{ a.s.}$$

Then define

$$\int_0^T \alpha_s \, dX_s := \int_0^T \alpha_s Y_s \, dW_s + \int_0^T \alpha_s Z_s \, ds. \tag{8.6.11}$$

The assumptions on $\alpha$ make sure that each term in (8.6.11) is well-defined. In addition, the representation of an Itô process $X$ in (8.6.10) is unique. More precisely, if

$$dX_t = Y_t^1 dW_t + Z_t^1 dt = Y_t^2 dW_t + Z_t^2 dt, \tag{8.6.12}$$

then the processes $Y^1$ and $Y^2$ as well as the processes $Z^1$ and $Z^2$ agree almost surely, but we will not prove this here. This implies that $\int_0^T \alpha_s \, dX_s$ is well-defined. Note that we could also have defined the integral with respect to an Itô process directly in a similar way to Section 8.4. In this case, one can derive the formula (8.6.11) rather than using it as definition.

The following result is the stochastic analogue to the classical integration by parts formula.

**Corollary 8.6.10** (Stochastic integration by parts formula). Suppose that $X_t^1$ and $X_t^2$ are Itô processes satisfying

$$dX_t^i = Y_t^i dW_t + Z_t^i dt.$$

Let $U_t = f(t, X_t^1)$ and $V_t = g(t, X_t^2)$ for sufficiently differentiable functions $f, g : [0,\infty) \times \mathbb{R} \to \mathbb{R}$. Then

$$d(U_t V_t) = U_t dV_t + V_t dU_t + Y_t^1 Y_t^2 \frac{\partial f}{\partial x}(t, X_t^1) \frac{\partial g}{\partial x}(t, X_t^2) dt.$$

The processes $U = \{U_t; t \geq 0\}$ and $V = \{V_t; t \geq 0\}$ are both Itô processes by Theorem 8.6.8. Therefore the terms $U_t dV_t$ and $V_t dU_t$ should be interpreted using Definition 8.6.9.

*Proof.* We give the proof just in the special case where $X^1 = X^2$ (and drop all the superscripts). The proof of the general case can be done from first principles and is similar to the proof of Theorem 8.6.8.

Let $h(t,x) = f(t,x)g(t,x)$. Then

$$\begin{aligned}
\frac{\partial h}{\partial x} &= \frac{\partial f}{\partial x} g + f \frac{\partial g}{\partial x}, \\
\frac{\partial h}{\partial t} &= \frac{\partial f}{\partial t} g + f \frac{\partial g}{\partial t}, \\
\frac{\partial^2 h}{\partial x^2} &= \frac{\partial^2 f}{\partial x^2} g + 2 \frac{\partial f}{\partial x} \frac{\partial g}{\partial x} + f \frac{\partial^2 g}{\partial x^2}.
\end{aligned}$$

Now $U_t V_t = h(t, X_t)$, so by Itô's formula,

$$\begin{aligned}
dh(t,X_t) &= \left( Z_t \frac{\partial h}{\partial x}(t,X_t) + \frac{\partial h}{\partial t}(t,X_t) + \frac{1}{2} Y_t^2 \frac{\partial^2 h}{\partial x^2}(t,X_t) \right) dt + Y_t \frac{\partial h}{\partial x}(t,X_t) dW_t \\
&= f(t,X_t) \left( \left( Z_t \frac{\partial g}{\partial x}(t,X_t) + \frac{\partial g}{\partial t}(t,X_t) + \frac{1}{2} Y_t^2 \frac{\partial^2 g}{\partial x^2}(t,X_t) \right) dt + Y_t \frac{\partial g}{\partial x}(t,X_t) dW_t \right) \\
&\quad + g(t,X_t) \left( \left( Z_t \frac{\partial f}{\partial x}(t,X_t) + \frac{\partial f}{\partial t}(t,X_t) + \frac{1}{2} Y_t^2 \frac{\partial^2 f}{\partial x^2}(t,X_t) \right) dt + Y_t \frac{\partial f}{\partial x}(t,X_t) dW_t \right) \\
&\quad + Y_t^2 \frac{\partial f}{\partial x}(t,X_t) \frac{\partial g}{\partial x}(t,X_t) dt \\
&= f(t,X_t) dg(t,X_t) + g(t,X_t) df(t,X_t) + Y_t^2 \frac{\partial f}{\partial x}(t,X_t) \frac{\partial g}{\partial x}(t,X_t) dt,
\end{aligned}$$

giving

$$d(U_t V_t) = U_t dV_t + V_t dU_t + Y_t^2 \frac{\partial f}{\partial x}(t,X_t) \frac{\partial g}{\partial x}(t,X_t) dt.$$

□

Further generalisations of Itô's formula are possible. For example, one can write down an Itô formula for $f(t, X_t^1, X_t^2, \ldots, X_t^n)$, where $\{X_t^i, t \geq 0\}$, $i = 1, \ldots, n$, are Itô processes. Itô's formula can also be extended to $f(t, W_t^1, W_t^2)$, where $W_t^1$ and $W_t^2$ are two different Brownian motions. However, this requires that we first define the Itô stochastic integral for such a setting.

## 8.7 Stochastic Differential Equations

We finish this chapter by discussing the stochastic analogue to the ordinary differential equation (ODE). We do not try to give a comprehensive overview of this area, but rather consider a couple of practical methods for solving stochastic differential equations.

**Definition 8.7.1** (Stochastic differential equation). Let $\mu(t,x)$ and $\sigma(t,x)$ be two functions defined on $\mathbb{R}^2$. A *stochastic differential equation (SDE)* is an equation of the form

$$dX_t = \mu(t,X_t)dt + \sigma(t,X_t)dW_t, \tag{8.7.1}$$

where $\{W_t, t \geq 0\}$ denotes Brownian motion and $\{X_t, t \geq 0\}$ is an unknown process. The function $\mu(t,x)$ is called the *drift coefficient* and $\sigma(t,x)$ is the *diffusion coefficient.*

A process $\{X_t, t \geq 0\}$ is a *solution* of the SDE with initial condition $X_0$ if, for all $t > 0$, the integrals

$$\int_0^t \mu(s,X_s)ds \quad \text{and} \quad \int_0^t \sigma(s,X_s)dW_s$$

exist and

$$X_t = X_0 + \int_0^t \mu(s,X_s)ds + \int_0^t \sigma(s,X_s)dW_s.$$

If $\sigma \equiv 0$ in (8.7.1), then we just have an ODE.

It is not always possible to find an explicit solution to an SDE. In fact, it is not always possible to find an explicit solution to an ODE! There are specific conditions that need to be satisfied by $\mu(t,x)$ and $\sigma(t,x)$ for a solution to even exist. We will not go into these here. Instead, we will look at how to solve SDEs in simple cases.

One strategy is to look for a solution of the form $X_t = f(t, W_t)$. By Itô's formula, see Theorem 8.6.8, we know that $X_t$ satisfies

$$dX_t = \left(\frac{\partial f}{\partial t}(t, W_t) + \frac{1}{2}\frac{\partial^2 f}{\partial x^2}(t, W_t)\right) dt + \frac{\partial f}{\partial x}(t, W_t)dW_t$$

whereas, substituting $X_t = f(t, W_t)$ into the SDE, we have

$$dX_t = \mu(t,f(t, W_t))dt + \sigma(t,f(t, W_t))dW_t.$$

Since $X$ is an Itô process, we can identify the integrands in the Riemann and stochastic integrals. In other words, we look at both equations and set the coefficients of $dt$ as well as those of $dW_t$ equal. By the comment before (8.6.12), we know that this is the only way that both expressions can agree. This means that we should look for a function $f$ that satisfies the following system of *partial differential equations (PDEs)*:

$$\frac{\partial f}{\partial t}(t,x) + \frac{1}{2}\frac{\partial^2 f}{\partial x^2}(t,x) = \mu(t,f(t,x)),$$
$$\frac{\partial f}{\partial x}(t,x) = \sigma(t,f(t,x)).$$

**Example 8.7.2.** *An important SDE, which we will need in Chapter 9, is*

$$dS_t = \mu S_t dt + \sigma S_t dW_t.$$

*We now solve this SDE here. Suppose $S_t = f(t, W_t)$. Then $f(t,x)$ satisfies*

$$\frac{\partial f}{\partial t}(t,x) + \frac{1}{2}\frac{\partial^2 f}{\partial x^2}(t,x) = \mu f(t,x),$$
$$\frac{\partial f}{\partial x}(t,x) = \sigma f(t,x).$$

*A common technique used when solving PDEs is to write $f(t,x)$ as the product of two functions:*

$$f(t,x) = g(t)h(x).$$

*The system of PDEs then becomes*

$$g'(t)h(x) + \frac{1}{2}g(t)h''(x) = \mu g(t)h(x),$$
$$g(t)h'(x) = \sigma g(t)h(x).$$

*The second equation has solution $h(x) = h(0)e^{\sigma x}$. Since $h''(x) = \sigma^2 h(x)$, substituting this into the first equation gives*

$$g'(t) = (\mu - \sigma^2/2)g(t)$$

*and so $g(t) = g(0)e^{(\mu-\sigma^2/2)t}$. Putting this together gives*

$$f(t,x) = f(0,0)e^{\sigma x+(\mu-\sigma^2/2)t}$$

*and so*

$$S_t = S_0 e^{\sigma W_t+(\mu-\sigma^2/2)t},$$

*hence $S_t$ is a geometric Brownian motion.*

**Example 8.7.3** (Ornstein–Uhlenbeck process). *The Ornstein–Uhlenbeck process is the solution to the linear SDE*

$$dX_t = -\alpha X_t dt + \sigma dW_t.$$

*This SDE is not amenable to the approach above (try it!) as $X_t$ is not of the form $f(t, W_t)$. Instead, we use the method of integrating factors, which is a common way to solve linear ODEs. For this, we multiply the process $X$ by a factor $e^{\alpha t}$ and then apply Itô's formula, Theorem 8.6.8, with $f(t,x) = e^{\alpha t}x$. This gives*

$$d(e^{\alpha t}X_t) = \sigma e^{\alpha t}dW_t + \left(-\alpha e^{\alpha t}X_t + \alpha e^{\alpha t}X_t + 0\right) dt = e^{\alpha t}\sigma dW_t.$$

*So*

$$e^{\alpha t}X_t = X_0 + \sigma \int_0^t e^{\alpha s} dW_s$$

*or*

$$X_t = e^{-\alpha t}X_0 + \sigma e^{-\alpha t}\int_0^t e^{\alpha s}dW_s.$$

*The last integral has the same form as in Example 8.4.8. Therefore, if $X_0$ is a constant, then $\{X_t, t \geq 0\}$ is a Gaussian process with*

$$\mathbb{E}[X_t] = X_0 \text{ and } \operatorname{Var}(X_t) = \sigma^2 e^{-2\alpha t}\int_0^t (e^{\alpha s})^2\, ds = \frac{\sigma^2}{2\alpha}(1 - e^{-2\alpha t}).$$

*As $t \to \infty$, $X_t$ converges to an $\mathcal{N}(0, \sigma^2/2\alpha)$ random variable. Furthermore, if $X_0 \sim N(0, \sigma^2/2\alpha)$ and is independent of the Brownian motion driving the SDE, then the process $\{X_t, t \geq 0\}$ is a stationary process. More precisely, we have $X_t \sim \mathcal{N}(0, \sigma^2/2\alpha)$ for all $t \geq 0$. An illustration of a sample path in this case can be found in Figure 8.2.*

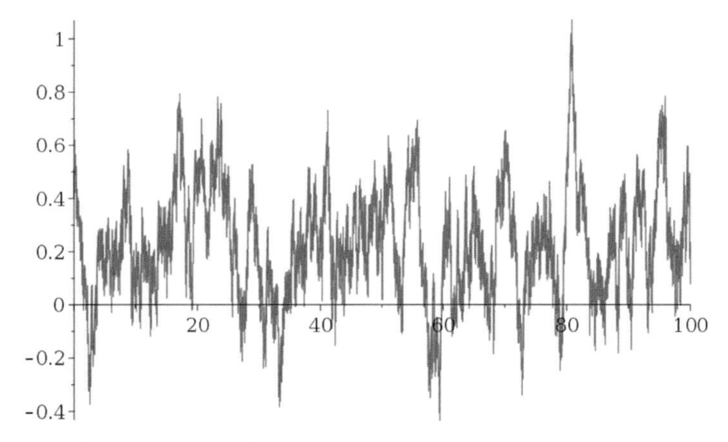


**Figure 8.2** A sample path of an Ornstein–Uhlenbeck process.

## 8.8 Exercises

**Exercise 8.1.** *Establish the identity in* (8.2.7).

**Exercise 8.2.** *(a) Show that if $f : [0, \infty) \to \mathbb{R}$ is cadlag and $g : [0, \infty) \to \mathbb{R}$ has finite variation, then the Riemann–Stieltjes integral $\int_0^t f(s)dg(s)$ has finite variation.*

*(b) Show that if $f : [0, \infty) \to \mathbb{R}$ is continuous and has finite variation, then the quadratic variation of $f$ is 0.*

**Exercise 8.3.** *Suppose that $\{X_t, 0 \leq t \leq T\}$ and $\{Y_t, 0 \leq t \leq T\}$ are simple processes.*

*(a) Show directly from* (8.4.7) *that for all scalars $\lambda, \mu \in \mathbb{R}$,*

$$\int_0^T (\lambda X_s + \mu Y_s)dW_s = \lambda \int_0^T X_s dW_s + \mu \int_0^T Y_s dW_s.$$

*(b) Find the quadratic variation* $[I]_T$ *of the process* $I_t = \int_0^t X_s dW_s$.
*(c) Show that*

$$\operatorname{Cov}\left(\int_0^T X_s dW_s, \int_0^T Y_s dW_s\right) = \int_0^T \mathbb{E}\left[X_s Y_s\right] ds.$$

**Exercise 8.4.** *(a) Use Itô's formula to express* $\int_0^t e^{W_s} dW_s$ *in terms of a Riemann integral.*
*(b) Verify directly (using the definition of a martingale) that the answer is a martingale with respect to the natural filtration.*
*(c) Using Itô's formula, show that each of the following processes is a martingale.*
*(i)* $M_t = W_t^3 - 3tW_t$.
*(ii)* $M_t = e^{t/2}\sin(W_t)$.
*(iii) For a fixed* $T > 0$,

$$M_t = \Phi\left(\frac{W_t}{\sqrt{T-t}}\right), \quad 0 \le t \le T,$$

*where* $\Phi$ *is the cumulative distribution function of a standard* $N(0,1)$ *random variable.*

**Exercise 8.5.** *The Vasicek interest rate model gives the instantaneous interest rate* $R_t$ *as the solution to the SDE*

$$dR_t = \alpha(\mu - R_t)dt + \sigma dW_t,$$

*where* $\alpha$, $\mu$ *and* $\sigma$ *are positive constants. Find the solution to the SDE, under the assumption that the initial condition* $R_0$ *is constant, and show that*

$$\mathbb{E}\left[R_t\right] = e^{-\alpha t}R_0 + \mu(1 - e^{-\alpha t}) \quad \text{and} \quad \operatorname{Var}(R_t) = \frac{\sigma^2}{2\alpha}(1 - e^{-2\alpha t}).$$

*What can you say about the distribution of* $R_t$ *as* $t \to \infty$*?*

# 9 The Black–Scholes Model

The main aim in this chapter is to introduce the Black–Scholes model and to study how this model is used to price financial options. Although, in reality, trading is done by computers and therefore stocks are traded at discrete times, the times between successive trades can be extremely short and therefore trading strategies can be well approximated by continuous-time processes. The advantage of this is that one can use the machinery of stochastic calculus to do computations which would potentially be very complicated if attempted directly using the discrete-time formulas from the first half of this book. In order to formulate trading strategies as continuous-time processes, we require an extension of the discrete Black–Scholes model in Chapter 5 to continuous time.

This chapter has been written so that the Black–Scholes model can be studied independently of the discrete models covered in Part I (with the exception of Chapter 1, which is a prerequisite for this chapter). Nevertheless, at certain points we reference results from Chapters 3, 4 and 5. This is either to provide some additional motivation, by way of analogy with the results in earlier chapters, or to avoid restating definitions or proofs which are identical to those in Part I. In the former case, readers who have not studied the chapters on discrete models may safely ignore these cross-references; in the latter case, it is sufficient to just look up the corresponding definition or proof.

## 9.1 Model Specification

The model in this chapter is defined for the time interval $[0, T]$ and is a model for a primary market consisting of two assets. These assets are a *bond* $B_t$ and a *stock* $S_t$ and we make the following assumptions.

- The bond is assumed to pay a fixed, continuously compounded *interest rate* $r$ per unit time. Thus, the price of the *bond* $B_t$ is determined by the ordinary differential equation

$$dB_t = rB_t dt, \tag{9.1.1}$$

  where $B_0$ is the price of the bond at time $t = 0$ and $r \in \mathbb{R}$.
- The price of the *stock* $S_t$ is determined by the stochastic differential equation

$$dS_t = \mu S_t dt + \sigma S_t dW_t, \tag{9.1.2}$$

  where $S_0$ is the price of the stock at time $t = 0$, $\mu \in \mathbb{R}$ is the *expected rate of return* of the stock, $\sigma > 0$ is the *volatility* of the stock and $\{W_t, t \geq 0\}$ is a standard Brownian motion.

- To model the time, we use the filtration $\mathbb{F} = (\mathcal{F}_s)_{s\in[0,T]}$ with

$$\mathcal{F}_t = \sigma(\{S_u, 0 \le u \le t\}) = \sigma(\{W_u, 0 \le u \le t\}). \tag{9.1.3}$$

So $\mathbb{F}$ is the natural filtration for the Brownian motion. In what follows, we use 'adapted' to mean 'adapted with respect to $\mathbb{F}$'.

We also assume that the market is liquid and that the assumptions in Section 1.2 are also fulfilled. Note that it is common in finance to specify the model assumption as an SDE. The reason for this is that it is often not too difficult to write down an SDE for the securities involved. On the other hand, however, it is often not possible to solve this SDE explicitly. This is similar to physics and chemistry, where it is often possible to determine a differential equation for the quantities involved, but it is only possible to solve this equation in a few cases. In our case, on the other hand, we can fortunately solve (9.1.1) and (9.1.2). The solution of (9.1.1) is the exponential function and we computed the solution of (9.1.2) in Example 8.7.2. Therefore

$$B_t = B_0 e^{rt} \quad \text{and} \quad S_t = S_0 e^{\sigma W_t + (\mu - \sigma^2/2)t}. \tag{9.1.4}$$

Although this model is still a relatively simple model, it captures many features of more complicated markets. An analogous approach can be used in the *multi-dimensional Black–Scholes model* where there are $N$ stocks each satisfying a similar equation to that above, but with $N$ different Brownian motions. This requires extending the theory of stochastic calculus to this setting, so we do not provide the details here.

## 9.2 Trading Strategies

We would like to buy/sell *bonds* and *stocks*. In contrast to the previous model, we assume that we can trade at all times $t \in [0, T]$. Thus we denote by

$$\alpha_t := \text{the number of shares of stock we hold at time } t,$$
$$\beta_t := \text{the number of bonds we hold at time } t.$$

In other words, a trading strategy is a pair $(\beta, \alpha)$, where $\alpha = \{\alpha_t, 0 \le t \le T\}$ and $\beta = \{\beta_t, 0 \le t \le T\}$ are two stochastic processes. The processes $\beta$ and $\alpha$ cannot be arbitrary, of course. It is important that the processes $\beta$ and $\alpha$ are adapted, as a trading strategy should not be able to anticipate future prices of the stock and also should only depend on the market prices. We also don't want the sample paths to be too rough. The graphical illustration of a trading strategy in the binomial model in Figure 3.2 suggests using the following definition.

**Definition 9.2.1** (Simple trading strategy). A *simple trading strategy* $\varphi = (\beta, \alpha)$ is a pair of adapted stochastic processes, where $\beta = \{\beta_t, 0 \le t \le T\}$ and $\alpha = \{\alpha_t, 0 \le t \le T\}$ are both simple processes, see Definition 8.4.2.

In other words, if $\varphi$ is a simple trading strategy then there exists an $n \in \mathbb{N}$ and times $0 = t_0 < t_1 < \cdots < t_n = T$ such that trading can only occur at these times. Also, these times are fixed and do not depend on the sample paths. Inserting the definition of simple processes, we can write a simple trading strategy $\varphi = (\beta, \alpha)$ as

$$\alpha_s = \sum_{i=1}^{n} C_{i-1}(\omega)\mathbb{1}_{[t_{i-1},t_i)}(s) \text{ and } \beta_s = \sum_{i=1}^{n} D_{i-1}(\omega)\mathbb{1}_{[t_{i-1},t_i)}(s), \tag{9.2.1}$$

for all $s \in \mathbb{R}$, $\omega \in \Omega$, where $C_i$, $D_i$ are $\mathcal{F}_{t_i}$-measurable random variables with $\mathbb{E}\left[C_i^2\right] < \infty$ and $\mathbb{E}\left[D_i^2\right] < \infty$ for all $i$. Furthermore, the *value of the portfolio* at time $t$ is given by

$$V_t(\varphi) = \beta_t B_t + \alpha_t S_t \text{ for all } t \in [0, T]. \tag{9.2.2}$$

As before, we are interested in *self-financing trading strategies* only. If a simple trading strategy $\varphi$ is given, then we adjust the portfolio only at some fixed times $t_i$ and keep it unchanged at all other times. Thus we have only to look at the times $t_i$. Using the same argument as in the binomial model, see page 75, we immediately get that a simple trading strategy $\varphi$ is self-financing if

$$D_{i-1}B_{t_i} + C_{i-1}S_{t_i} = D_i B_{t_i} + C_i S_{t_i} \tag{9.2.3}$$

for all $1 \leq i < n$, with $C_i$ and $D_i$ as in (9.2.1). Since $\varphi$ is by assumption cadlag, we can reformulate (9.2.3) as follows.

**Definition 9.2.2.** A simple trading strategy $\varphi$ is called *self-financing* if for all $0 < t \leq T$,

$$\beta_t B_t + \alpha_t S_t = \beta_{(t-)} B_t + \alpha_{(t-)} S_t, \tag{9.2.4}$$

where $\beta_{(t-)} = \lim_{s \uparrow t} \beta_s$ and $\alpha_{(t-)} = \lim_{s \uparrow t} \alpha_s$.

This definition of self-financing is very natural in view of the definitions of self-financing in Chapters 3 and 4. However, we will need an equivalent formulation in this chapter.

**Lemma 9.2.3.** Let $\varphi$ be a simple trading strategy. The following are equivalent.

- $\varphi$ is self-financing.
- For all $0 \leq t \leq T$,

$$\int_0^t B_s d\beta_s + \int_0^t S_s d\alpha_s = 0. \tag{9.2.5}$$

- For all $0 \leq t \leq T$,

$$V_T(\varphi) = V_0(\varphi) + \int_0^t \beta_s dB_s + \int_0^t \alpha_s dS_s. \tag{9.2.6}$$

Before we give the proof of Lemma 9.2.3, we have to make some comments about the integrals in this lemma. Since $\varphi$ is a simple trading strategy, $\beta$ and $\alpha$ are both piecewise constant with only finitely many discontinuities. Therefore, all sample paths of $\beta$ and $\alpha$ are of bounded variation and the integrals in (9.2.5) are well-defined Riemann–Stieltjes integrals. Furthermore, we can use Example 8.2.3 to evaluate (9.2.5). To see this, assume that the trading times of $\varphi$ are $0 = t_0 < t_1 < \cdots \leq t_n = T$. The jump of $\beta$ at the time $t_j$ is $\beta_{t_j} - \beta_{(t_j-)} = \beta_{t_j} - \beta_{t_{j-1}}$ since $\beta$ is by assumption constant in the interval $[t_{j-1}, t_j)$. Similarly for $\alpha$. Therefore, for $t \in [t_k, t_{k+1})$,

$$\int_0^t B_s d\beta_s + \int_0^t S_s d\alpha_s = \sum_{j=1}^{k} B_{t_j}(\beta_{t_j} - \beta_{t_{j-1}}) + \sum_{j=1}^{k} S_{t_j}(\alpha_{t_j} - \alpha_{t_{j-1}}). \tag{9.2.7}$$

The integral with respect to $dS_s$ in (9.2.6) is a priori a stochastic integral. However, since the sample paths of $\alpha$ are of bounded variation, Proposition 8.2.5 implies that this integral can be interpreted in this special case as a Riemann–Stieltjes integral for each sample path. Explicitly evaluating, for $t \in [t_k, t_{k+1})$,

$$\int_0^t \alpha_s dS_s = \sum_{j=1}^{k} \alpha_{t_{j-1}}(S_{t_j} - S_{t_{j-1}}) + \alpha_{t_k}(S_t - S_{t_k}). \tag{9.2.8}$$

Finally, since the sample paths of $B_t$ are all differentiable, Example 8.2.2 shows that the integral with respect to $dB_s$ in (9.2.6) is

$$\int_0^t \beta_s dB_s = \int_0^t \beta_s\, rB_s ds = \sum_{j=1}^{k} \beta_{t_{j-1}}(B_{t_j} - B_{t_{j-1}}) + \beta_{t_k}(B_t - B_{t_k}). \tag{9.2.9}$$

*Proof of Lemma 9.2.3.* We first show that the first two conditions are equivalent. Using (9.2.7), we can write

$$\int_0^t B_s d\beta_s + \int_0^t S_s d\alpha_s = \sum_{j=1}^{k} \left(\beta_{t_j} B_{t_j} + \alpha_{t_j} S_{t_j}\right) - \left(\beta_{(t_j-)} B_{t_j} - \alpha_{(t_j-)} S_{t_j}\right). \tag{9.2.10}$$

If $\varphi$ is self-financing then each summand in (9.2.10) is 0 and therefore (9.2.5) holds. If, on the other hand, (9.2.5) holds then subtracting the equations obtained by taking $t = t_k$ and $t = t_{k-1}$ in (9.2.5) shows that each summand in (9.2.10) is 0 and so $\varphi$ is self-financing. Therefore, the first two conditions are equivalent. To show that the second and third conditions are equivalent, we use partial integration for Riemann–Stieltjes integrals, see Proposition 8.2.5. This gives

$$\begin{aligned}\int_0^t B_s d\beta_s + \int_0^t S_s d\alpha_s &= \left(B_t\beta_t - B_0\beta_0 - \int_0^t \beta_s dB_s\right) + \left(S_t\alpha_t - S_0\alpha_0 - \int_0^t \alpha_s dS_s\right)\\ &= V_t(\varphi) - V_0(\varphi) - \int_0^t \beta_s dB_s - \int_0^t \alpha_s dS_s.\end{aligned}$$

This completes the proof. □

Simple trading strategies are not the only trading strategies we have to allow. Let us consider an example.

**Example 9.2.4.** *Suppose that $S_0 = 10$ and at time $t = 0$ we hold one stock and no bonds. We hold this portfolio until the price of the stock reaches 20 for the first time. At this point we sell the stock, invest the money into bonds and keep this portfolio for the rest of the time. Is this a simple trading strategy? Observe that*

$$\alpha_t = \begin{cases} 1, & \text{if } t \leq \tau, \\ 0, & \text{otherwise}, \end{cases} \quad \text{and} \quad \beta_t = \begin{cases} 0, & \text{if } t \leq \tau, \\ \frac{S_\tau}{B_\tau}, & \text{otherwise}, \end{cases} \tag{9.2.11}$$

*where $\tau = \inf\{0 \leq s \leq T\,;\, S_s = 20\}$. In particular, if $S_t < 20$ for all $t \in [0, T]$ then $\alpha_t = 1$ and $\beta_t = 0$ for all $t \in [0, T]$. The trading strategy described in (9.2.11) is adapted and depends only on market prices. However, the trading time $\tau$ can occur at any point in $(0, T]$. Therefore (9.2.11) is not a simple trading strategy.*

The trading strategy in Example 9.2.4 is one that should be allowed within our theory, so we need to be able to deal with a larger class of trading strategies than just simple ones. Nevertheless, we do not want to allow trading strategies which are so wild that they cannot realistically be implemented. Consider the following example.

**Example 9.2.5.** *We start with no bonds and stocks. At time $t = T/2$, we buy one stock by borrowing bonds and hold this portfolio until time $t = 3T/4$. Then we buy a further stock by borrowing bonds and hold this portfolio until time $t = 7T/8$. And so on. In other words, the trading times are $t_k := (1 - \frac{1}{2^k})T$ with $k \in \mathbb{N}_0$ and*

$$\alpha_t = k \quad \text{and} \quad \beta_t = -\sum_{j=1}^{k} \frac{S_{t_j}}{B_{t_j}} \quad \text{for } t \in [t_k, t_{k+1})\,.$$

*Therefore we are able to buy an unlimited number of stocks in finite time.*

Strategies such as the one above have to be excluded. Aside from the very obvious problem that there do not exist an unlimited number of stocks to buy, no computer would be able to implement this strategy, or even a good approximation of it. A natural assumption is therefore to insist that, in addition to being adapted, allowable trading strategies should be cadlag. Since any cadlag function can be approximated arbitrarily well by a simple function, this assumption ensures that a good approximation to the trading strategy can be implemented by any computer which can make successive trades over very short discrete-time intervals. Since the trading strategy in Example 9.2.5 is not cadlag, it would be excluded by this assumption.

A disadvantage of admitting more general trading strategies is that we can no longer use (9.2.4) for the definition of self-financing. A trading strategy is self-financing if the changes in the portfolio are due to trading in the stock and bond without having to inject (or withdraw) funds. However, (9.2.4) is automatically fulfilled if $\beta$ and $\alpha$ have continuous sample paths. For example, consider the strategy $\alpha_t = 0$ and $\beta_t = t$ for all $t$. In this strategy, starting from nothing, we continuously increase the number of bonds we hold over time. We need money to do this. However, since the number of stocks is not changing, we must be injecting money into the market the whole time. Therefore this strategy is not self-financing and so (9.2.4) cannot be the correct condition for determining whether a process is self-financing if we allow trading strategies to be cadlag rather than simple.

We deduce the correct condition using a heuristic argument. Suppose we increase the number of stocks at time $t$ by $d\alpha_t$, where we use the interpretation of $d\alpha_t$ as an infinitesimal when $t$ is a continuity point of $\alpha$. For this we need the amount $S_t d\alpha_t$. Since the strategy has to be self-financing, we can get this amount only by borrowing bonds. The number of bonds $d\beta_t$ we need is $-S_t d\alpha_t / B_t$. In other words, we must have $S_t d\alpha_t + B_t d\beta_t = 0$. This corresponds to (9.2.5) in Lemma 9.2.3. We therefore could use (9.2.5) as a definition of self-financing. However, in practice, it is easier to work with the equivalent condition (9.2.6). Note that this condition also has a heuristic interpretation, namely that changes in the value of the portfolio are due only to changes in the bond price and changes in the stock price. In order for (9.2.6) to make sense, the integrals on the RHS must be well-defined. Equation (9.1.2) implies that $S_t$ is an Itô process and so, by Definition 8.6.9,

$$\int_0^t \alpha_u dS_u = \mu \int_0^t \alpha_u S_u dt + \sigma \int_0^t \alpha_u S_u dW_u$$

provided

$$\|\alpha S\|_{L^1(\Omega \times [0,T])} \leq \|\alpha S\|_{L^2(\Omega \times [0,T])} < \infty.$$

In order for (9.2.6) to make sense, it is therefore sufficient to assume that a trading strategy $\varphi = (\beta, \alpha)$ is a pair of adapted, cadlag stochastic processes $\beta$ and $\alpha$, with

$$\|\alpha S\|_{L^2(\Omega \times [0,T])} < \infty \quad \text{and} \quad \|\beta\|_{L^1([0,T])} < \infty \text{ a.s.} \tag{9.2.12}$$

Observe that, under this assumption, $V_t$ is also an Itô process, which leads to the following definition.

**Definition 9.2.6** (Self-financing). A trading strategy is called *self-financing* if

$$dV_t = \beta_t dB_t + \alpha_t dS_t. \tag{9.2.13}$$

Unfortunately, the assumptions in (9.2.12) are not strong enough to ensure that the model is arbitrage-free. We will therefore need to impose some additional assumptions later on as to which trading strategies are admissible within our framework. Before doing this, we need to do some preparation, and in particular introduce the risk-neutral measure which we do in the next section. The precise assumptions we will need to impose on admissible trading strategies are given in Definition 9.3.9. For the moment, we just assume that (9.2.12) is fulfilled (and that the trading strategy is adapted and cadlag). At this point we would like to emphasise that the assumptions in (9.2.12) already exclude many problematic trading strategies, which would be unrealistic or unimplementable.

Suppose now that the trading strategy $\varphi = (\beta, \alpha)$ has the form $\beta_t = g(t, S_t)$ and $\alpha_t = h(t, S_t)$, so the amount held in the stock and the bond at time $t$ depends only on the actual stock price at time $t$, regardless of how it got there. Itô's formula allows us to find conditions on the functions $g(t,x)$ and $h(t,x)$ under which the trading strategy $\varphi$ is self-financing.

**Theorem 9.2.7.** *Suppose $g,h : [0,T] \times \mathbb{R} \to \mathbb{R}$ are continuously differentiable with respect to the first variable, $t$, and twice continuously differentiable with respect to the second variable, $x$. The trading strategy $\varphi = (\beta,\alpha)$ with $\beta_t = g(t,S_t)$ and $\alpha_t = h(t,S_t)$ is self-financing if and only if $g$ and $h$ satisfy*

$$B_t \frac{\partial g}{\partial x} + x\frac{\partial h}{\partial x} = 0,$$
$$B_t \frac{\partial g}{\partial t} + x\frac{\partial h}{\partial t} + \frac{1}{2}\sigma^2 x^2 \frac{\partial h}{\partial x} = 0.$$

*Proof.* Let $f(t,x) = B_t g(t,x) + xh(t,x)$. Then

$$V_t(\varphi) = \beta_t B_t + \alpha_t S_t = g(t,S_t)B_t + h(t,S_t)S_t = f(t,S_t).$$

Using Itô's formula (Theorem 8.6.8) and that $dS_t = \mu S_t dt + \sigma S_t dW_t$, we obtain

$$dV_t = \sigma S_t \frac{\partial f}{\partial x}(t,S_t)dW_t + \left( \mu S_t \frac{\partial f}{\partial x}(t,S_t) + \frac{\partial f}{\partial t}(t,S_t) + \frac{1}{2}\sigma^2 S_t^2 \frac{\partial^2 f}{\partial x^2}(t,S_t) \right) dt.$$

However, the portfolio $V_t$ is self-financing if and only if

$$dV_t = g(t,S_t)dB_t + h(t,S_t)dS_t = \sigma S_t h(t,S_t)dW_t + (rB_t g(t,S_t) + \mu S_t h(t,S_t))\, dt.$$

Therefore, equating coefficients, the portfolio is self-financing if and only if

$$\frac{\partial f}{\partial x} = h(t,x),$$
$$\mu x \frac{\partial f}{\partial x} + \frac{\partial f}{\partial t} + \frac{1}{2}\sigma^2 x^2 \frac{\partial^2 f}{\partial x^2} = rB_t g(t,x) + \mu x h(t,x).$$

This is equivalent to

$$B_t \frac{\partial g}{\partial x} + x\frac{\partial h}{\partial x} = 0,$$
$$B_t \frac{\partial g}{\partial t} + x\frac{\partial h}{\partial t} + \frac{1}{2}\sigma^2 x^2 \frac{\partial h}{\partial x} = 0,$$

as required. □

Theorem 9.2.7 can be used to derive the Black–Scholes equation that must be satisfied by a function $f(t,x)$ in order for $f(t,S_t)$ to be the value of a self-financing trading strategy.

**Corollary 9.2.8** (Black–Scholes equation). Suppose $f : [0,T] \times \mathbb{R} \to \mathbb{R}$ is continuously differentiable with respect to the first variable, $t$, and twice continuously differentiable with respect to the second variable, $x$. Then there exists a self-financing trading strategy $\varphi$

with $V_t(\varphi) = f(t, S_t)$ for all $t$ if and only if $f$ satisfies

$$\frac{\partial f}{\partial t} + rx\frac{\partial f}{\partial x} + \frac{1}{2}\sigma^2 x^2 \frac{\partial^2 f}{\partial x^2} - rf = 0. \tag{9.2.14}$$

*Proof.* First suppose that $V_t(\varphi) = f(t, S_t)$ for some self-financing trading strategy $\varphi$. Exactly as in the proof above,

$$\frac{\partial f}{\partial x} = h(t,x), \tag{9.2.15}$$

$$\begin{aligned} \mu x\frac{\partial f}{\partial x} + \frac{\partial f}{\partial t} + \frac{1}{2}\sigma^2 x^2 \frac{\partial^2 f}{\partial x^2} &= rB_t g(t,x) + \mu x h(t,x) \\ &= rf(t,x) + (\mu - r)xh(t,x). \end{aligned}$$

Substituting the first equation into the second gives (9.2.14).

Conversely, suppose that $f$ satisfies (9.2.14). We look for $g$ and $h$ such that

$$\beta_t = g(t, S_t), \ \alpha_t = h(t, S_t) \ \text{ and } \ V_t(\varphi) = f(t, S_t) = \beta_t B_t + \alpha_t S_t.$$

In view of (9.2.15) and since $\beta_t = \frac{V_t(\varphi) - \alpha_t S_t}{B_t}$, we set

$$g(t,x) = \frac{1}{B_t}\left(f(t,x) - x\frac{\partial f}{\partial x}\right) \quad \text{and} \quad h(t,x) = \frac{\partial f}{\partial x}. \tag{9.2.16}$$

It is a straightforward calculation to check that $g$ and $h$ in (9.2.16) satisfy the equations in Theorem 9.2.7, so this is left to Exercise 9.1. Therefore, by Theorem 9.2.7,

$$f(t, S_t) = g(t, S_t)B_t + h(t, S_t)S_t$$

is the value of a self-financing trading strategy. □

**Example 9.2.9.** *An investor who trades in the Black–Scholes model constantly rebalances his portfolio so as to maintain a fixed proportion $\gamma$ of his wealth in the stock and the remainder in the bond. Let $f(t, S_t)$ represent the value of his portfolio at time $t$. Using the notation from above,*

$$xh(t,x) = \gamma f(t,x) \quad \text{and} \quad B_t g(t,x) = (1-\gamma)f(t,x).$$

*In order for the portfolio to be self-financing, by Theorem 9.2.7,*

$$0 = B_t\frac{\partial g}{\partial x} + x\frac{\partial h}{\partial x} = (1-\gamma)\frac{\partial f}{\partial x} + \gamma\frac{\partial f}{\partial x} - \frac{\gamma f(t,x)}{x}$$

*and hence*

$$x\frac{\partial f}{\partial x} = \gamma f(t,x).$$

*Rearranging gives*

$$\frac{1}{f}\frac{\partial f}{\partial x} = \frac{\gamma}{x}$$

*and hence*

$$\log f(t,x) = \gamma \log(x) + c(t) \quad or \quad f(t,x) = e^{c(t)}x^{\gamma}.$$

*For $f$ to be the value of a self-financing portfolio, it must satisfy* (9.2.14) *and so*

$$c'(t)e^{c(t)}x^{\gamma} + rx\gamma e^{c(t)}x^{\gamma-1} + \frac{1}{2}\sigma^2x^2\gamma(\gamma-1)e^{c(t)}x^{\gamma-2} - re^{c(t)}x^{\gamma} = 0$$

*or*

$$c'(t) = (1-\gamma)\left(r + \frac{1}{2}\gamma\sigma^2\right).$$

*Therefore*

$$c(t) = (1-\gamma)\left(r + \frac{1}{2}\gamma\sigma^2\right)t + c(0).$$

*Assuming that the investor has initial wealth $w_0$, this gives*

$$f(t,x) = w_0 e^{(1-\gamma)(r+\gamma\sigma^2/2)t}(x/S_0)^{\gamma}.$$

## 9.3 Arbitrage and Risk-Neutral Measure

The main aim of this section is to show that the Black–Scholes model is arbitrage-free. For this, we introduce the discounted asset and stock prices as well as the risk-neutral measure.

We begin by defining arbitrage opportunities in the Black–Scholes model.

**Definition 9.3.1** (Arbitrage opportunity). An *arbitrage opportunity* (in the primary market) is an (admissible) self-financing trading strategy $\varphi$ which satisfies the following properties.

- No initial cost: $V_0(\varphi) = 0$.
- Always non-negative final value: $V_T(\varphi) \geq 0$.
- The possibility of a positive final gain:

$$\mathbb{E}[V_T(\varphi)] > 0.$$

As before, under the assumption $V_T(\varphi) \geq 0$, we have the equivalence

$$\mathbb{E}[V_T(\varphi)] > 0 \iff \mathbb{P}[V_T(\varphi) > 0] > 0. \tag{9.3.1}$$

**Example 9.3.2.** *Consider a two-dimensional Black–Scholes model in which the stocks $S^1$ and $S^2$ satisfy*

$$dS_t^1 = \mu_1 S_t^1 dt + \sigma S_t^1 dW_t \quad and \quad dS_t^2 = \mu_2 S_t^2 dt + \sigma S_t^2 dW_t,$$

*where* $\mu_1 > \mu_2$ *and* $S_0^1 = S_0^2 = 1$. *These equations suggest that* $S_t^1$ *is growing faster than* $S_t^2$. *Therefore we borrow at time* $t = 0$ *one unit of stock* $S^2$, *sell it and buy one unit of stock* $S^1$ *and then leave our portfolio unchanged. This trading strategy is self-financing and, at time t, the portfolio has value* $V_t(\varphi) = S_t^1 - S_t^2$. *The initial value is* $V_0(\varphi) = S_0^1 - S_0^2 = 0$, *while*

$$\begin{aligned} V_T(\varphi) &= S_T^1 - S_T^2 \\ &= e^{(\mu_1 - \sigma^2/2)T + \sigma W_T} - e^{(\mu_2 - \sigma^2/2)T + \sigma W_T} \\ &= e^{(\mu_2 - \sigma^2/2)T + \sigma W_T}\left(e^{(\mu_1 - \mu_2)T} - 1\right) \\ &> 0. \end{aligned}$$

*This trading strategy is therefore an arbitrage opportunity.*

As discussed in Chapter 1, we require the market to be arbitrage-free in order to be able to apply results such as the law of one price (Proposition 1.3.5). Furthermore, as mentioned in the previous section, we have not yet fully specified which trading strategies are admissible in this model (and hence allowed in Definition 9.3.1). We will therefore make the choice of what constitutes an admissible trading strategy precisely to ensure that the market is arbitrage-free. This will enable us to use the same mathematical tools as in the previous chapters for pricing options. In order to do this, we will attempt to reproduce the arguments used in the binomial and finite market model and check where we need additional assumptions. This will lead to Definition 9.3.9 below. For the moment, we only assume that a trading strategy $\varphi$ fulfils the assumptions in (9.2.12).

We start with introducing the *discounted asset prices*, see also (3.3.33), (4.1.6) and (5.3.5). Define

$$S_t^* := e^{-rt} S_t \text{ and } V_t^*(\varphi) := e^{-rt} V_t(\varphi). \tag{9.3.2}$$

The process $\{S_t^*, t \geq 0\}$ is called the *discounted stock price* process and $\{V_t^*(\varphi), t \geq 0\}$ is called the *discounted value process*. At this point, the same comments apply to the normalisation in (9.3.2) as to the normalisation in the discrete Black–Scholes model, see comments after (5.3.5). In other words, the normalisation is so chosen that $V_t^*(\varphi)$ is a martingale for the trading strategy $\varphi = (\beta, \alpha)$ with $\beta_t = 1$ and $\alpha_t = 0$ for all $t$. Furthermore, by the stochastic integration by parts formula (Corollary 8.6.10),

$$dS_t^* = -re^{-rt} S_t dt + e^{-rt} dS_t \text{ and } dV_t^* = -re^{-rt} V_t dt + e^{-rt} dV_t. \tag{9.3.3}$$

Using the discounted stock price, we can formulate an alternative characterisation of a self-financing trading strategy.

**Theorem 9.3.3.** *Let* $\varphi = (\beta, \alpha)$ *be a trading strategy. Then* $\varphi$ *is self-financing if and only if the discounted value process* $V_t^*(\varphi)$ *is a stochastic integral with respect to the discounted stock price, that is for all t,*

$$V_t^*(\varphi) = V_0(\varphi) + \int_0^t \alpha_u dS_u^*. \tag{9.3.4}$$

This theorem can be viewed as a continuous-time version of Lemma 4.1.4.

*Proof.* First suppose that $\varphi$ is self-financing, so

$$dV_t = \beta_t dB_t + \alpha_t dS_t.$$

Using (9.3.3) gives

$$\begin{aligned} dV_t^* &= -re^{-rt}V_t dt + e^{-rt}dV_t \\ &= -re^{-rt}(\beta_t B_t + \alpha_t S_t)dt + e^{-rt}(\beta_t dB_t + \alpha_t dS_t) \\ &= e^{-rt}\beta_t(-rB_t dt + dB_t) + \alpha_t(-re^{-rt}S_t dt + e^{-rt}dS_t) \\ &= \alpha_t dS_t^*, \end{aligned}$$

where the last line used (9.3.3) for $S_t^*$ and that $dB_t = rB_t dt$. Hence (9.3.4) holds.

Conversely, suppose that (9.3.4) holds. Using (9.3.4), (9.3.3) and that $dB_t = rB_t dt$,

$$\begin{aligned} dV_t^* &= \alpha_t dS_t^* = \alpha_t(-re^{-rt}S_t dt + e^{-rt}dS_t) \\ &= \alpha_t(-re^{-rt}S_t dt + e^{-rt}dS_t) + e^{-rt}\beta_t(-rB_t dt + dB_t) \\ &= -re^{-rt}(\beta_t B_t + \alpha_t S_t)dt + e^{-rt}(\beta_t dB_t + \alpha_t dS_t) \\ &= -re^{-rt}V_t dt + e^{-rt}(\beta_t dB_t + \alpha_t dS_t). \end{aligned} \tag{9.3.5}$$

Equation (9.3.5) implies that

$$e^{rt}dV_t^* + rV_t dt = \beta_t dB_t + \alpha_t dS_t. \tag{9.3.6}$$

Furthermore, by (9.3.3), $dV_t = e^{rt}dV_t^* + rV_t dt$. Combining this with (9.3.6) gives

$$dV_t = \beta_t dB_t + \alpha_t dS_t.$$

Therefore $\varphi$ is self-financing and this completes the proof. □

Next, we introduce risk-neutral probability measures into this model.

**Definition 9.3.4** (Risk-neutral probability measures). Suppose that the Black–Scholes model is realised on the probability space $(\Omega, \mathcal{F}, \mathbb{P})$. A *risk-neutral probability measure* (or *equivalent martingale measure*) is a probability measure $\mathbb{P}_*$ such that

- $\mathbb{P}_* \approx \mathbb{P}$ (see Definition 4.2.1).
- $S_t^*$ is a $\mathbb{P}_*$-martingale. In formulas this means that for all $t \geq s \geq 0$,

$$\mathbb{E}_*\left[S_t^* \middle| \mathcal{F}_s\right] = S_s^*, \tag{9.3.7}$$

where $\mathbb{E}_*[\,.\,]$ is the expectation with respect to $\mathbb{P}_*[\,.\,]$.

Observe that (9.3.3) and (9.1.2) imply

$$dS_t^* = -re^{-rt}S_t dt + e^{-rt}dS_t = (\mu - r)S_t^* dt + \sigma S_t^* dW_t.$$

In the special case when $\mu = r$,

$$dS_t^* = \sigma S_t^* dW_t \tag{9.3.8}$$

and hence $S_t^*$ is a martingale by Theorem 8.5.3. Alternatively, we can verify this directly with the same computation as in (5.4.5). Therefore, in the case $\mu = r$, the probability measure $\mathbb{P}$ is already a risk-neutral probability measure. On the other hand, if $\mu \neq r$ then $S_t^*$ is not a martingale with respect to $\mathbb{P}$. In the discrete Black–Scholes model in Section 5.4, we introduced a shift in the probability measure $\mathbb{P}$ and then computed the distribution of $W_t$ under this new measure. We could do a similar computation here, but it is easier to use Lemma 5.4.3 as starting point. This leads to the following definition.

**Definition 9.3.5.** Suppose that the Black–Scholes model is realised on the probability space $(\Omega, \mathcal{F}, \mathbb{P})$ and let $\nu \in \mathbb{R}$ be given. Define, for all events $A \in \mathcal{F}$,

$$\mathbb{P}_\nu [A] = \mathbb{E}\left[e^{\nu W_T - \frac{\nu^2}{2}T}\mathbb{1}_A(\omega)\right]. \tag{9.3.9}$$

The following result identifies some key properties of $\mathbb{P}_\nu$.

**Theorem 9.3.6.** *Let $\nu \in \mathbb{R}$ be given and let $\mathbb{P}_\nu$ be as in (9.3.9). Then*

- *$\mathbb{P}_\nu$ is a probability measure,*
- *$\mathbb{P}_\nu \approx \mathbb{P}$ and*
- *under the probability measure $\mathbb{P}_\nu$, the process $\{W_t; 0 \leq t \leq T\}$ is a Brownian motion with drift $\nu$. In particular, $W_t \sim \mathcal{N}(\nu t, t)$ for all $t$ under $\mathbb{P}_\nu$.*

*Proof.* We first show that $\mathbb{P}_\nu$ is a probability measure. Clearly, $\mathbb{P}_\nu[A] \geq 0$. Furthermore, we know from Example 7.3.6 that the process $\{X_t, 0 \leq t \leq T\}$ with $X_t = e^{\nu W_t - \nu^2 t/2}$ is a martingale for all $\nu \in \mathbb{R}$. Therefore

$$\mathbb{P}_\nu[\Omega] = \mathbb{E}\left[e^{\nu W_T - \frac{\nu^2}{2}T}\mathbb{1}_\Omega\right] = \mathbb{E}[X_T] = \mathbb{E}[X_0] = 1.$$

The $\sigma$-additivity follows immediately using the dominated convergence theorem.

To show equivalence, suppose that $\mathbb{P}_\nu[A] = 0$. Therefore

$$\mathbb{E}\left[e^{\nu W_T - \frac{\nu^2}{2}T}\mathbb{1}_A\right] = 0.$$

But since $e^{\nu W_T - \frac{\nu^2}{2}T}\mathbb{1}_A \geq 0$, this holds if and only if $\mathbb{P}\left[e^{\nu W_T - \frac{\nu^2}{2}T}\mathbb{1}_A = 0\right] = 1$. But since $e^{\nu W_T - \frac{\nu^2}{2}T} > 0$, this is true if and only if $\mathbb{P}[\mathbb{1}_A = 0] = 1$ or equivalently $\mathbb{P}[A] = 0$. The converse is immediate.

It remains to show the third point. We write $\mathbb{E}_\nu$ for the expectation under $\mathbb{P}_\nu$. Let $t \le T$ and $s \in \mathbb{R}$ be given. The tower property and that $X_t$, as defined above, is a martingale imply that

$$\begin{aligned}\mathbb{E}_\nu\left[e^{sW_t}\right] &= \mathbb{E}\left[e^{\nu W_T-\frac{\nu^2}{2}T}e^{sW_t}\right] = \mathbb{E}\left[e^{sW_t}X_T\right] = \mathbb{E}\left[\mathbb{E}\left[e^{sW_t}X_T \mid \mathcal{F}_t\right]\right] \\ &= \mathbb{E}\left[e^{sW_t}\mathbb{E}\left[X_T \mid \mathcal{F}_t\right]\right] = \mathbb{E}\left[e^{sW_t}X_t\right] = \mathbb{E}\left[e^{sW_t}e^{\nu W_t-\frac{\nu^2}{2}t}\right] = e^{-\frac{\nu^2}{2}t}\mathbb{E}\left[e^{(s+\nu)W_t}\right].\end{aligned}$$

Since $W_t \sim \mathcal{N}(0,t)$ under $\mathbb{P}$, (5.4.4) implies that

$$\mathbb{E}_\nu\left[e^{sW_t}\right] = e^{-\frac{\nu^2}{2}t}e^{\frac{1}{2}t(s+\nu)^2} = e^{t\frac{s^2}{2}+t\nu s}.$$

Therefore $W_t \sim \mathcal{N}(\nu t, t)$ under $\mathbb{P}_\nu$. Furthermore, for all $d \in \mathbb{N}$, $0 = t_0 < t_1 < \cdots < t_d \le T$ and $s_j \in \mathbb{R}$,

$$\mathbb{E}_\nu\left[\prod_{j=1}^{d} e^{s_j(W_{t_j}-W_{t_{j-1}})}\right] = \prod_{j=1}^{d} e^{(t_j-t_{j-1})\frac{s_j^2}{2}+(t_j-t_{j-1})\nu s_j}. \tag{9.3.10}$$

The proof of (9.3.10) is similar to the computation for $d = 1$ and we therefore omit the details. Equation (9.3.10) shows that, under $\mathbb{P}_\nu$, $\{W_t, 0 \le t \le T\}$ is a Gaussian process with independent increments and $W_t - W_s \sim \mathcal{N}(\nu(t-s), t-s)$ for $0 \le s \le t \le T$. Finally, the sample paths of $W_t$ are almost surely continuous under $\mathbb{P}_\nu$, since $W_t$ is a Brownian motion under $\mathbb{P}$ and $\mathbb{P}_\nu \approx \mathbb{P}$. This completes the proof. □

The third point in Theorem 9.3.6 is a special case of Girsanov's theorem. The general version can be found for instance in [14, Section 17.3].

**Lemma 9.3.7.** For all random variables $X$ with $\mathbb{E}_\nu[|X|] < \infty$:

(a)

$$\mathbb{E}_\nu[X] = \mathbb{E}\left[Xe^{\nu W_T-\frac{\nu^2}{2}T}\right];$$

(b)

$$\mathbb{E}_\nu[X \mid \mathcal{F}_t] = e^{\frac{\nu^2}{2}t-\nu W_t}\mathbb{E}\left[Xe^{\nu W_T-\frac{\nu^2}{2}T} \mid \mathcal{F}_t\right]. \tag{9.3.11}$$

At first glance, it is not so obvious why the first factor on the RHS of (9.3.11) has to be there. To see why, note that we have $\mathbb{E}_\nu[1 \mid \mathcal{F}_t] = 1$. Inserting $X = 1$ into (9.3.11) shows that this equation is wrong without the factor $e^{\frac{\nu^2}{2}t-\nu W_t}$.

*Proof.* (a) It is immediate from the definition that the result holds when $X = \mathbb{1}_A$ for a measurable set $A$. This extends to simple random variables $X$ by linearity of expectation. The result can then be shown to hold for all integrable random variables, as in the proof of Theorem 6.1.15.

(b) We check that the RHS of (9.3.11) satisfies the conditions in Definition 6.3.7. Since $\mathbb{E}\left[Xe^{\nu W_T-\frac{\nu^2}{2}T} \mid \mathcal{F}_t\right]$ is $\mathcal{F}_t$-measurable by definition and $e^{\frac{\nu^2}{2}t-\nu W_t}$ is a continuous function of $W_t$, the expression on the RHS is $\mathcal{F}_t$-measurable. It remains to show that for all $A \in \mathcal{F}_t$,

$$\mathbb{E}_\nu\left[e^{\frac{\nu^2}{2}t-\nu W_t}\mathbb{E}\left[Xe^{\nu W_T-\frac{\nu^2}{2}T} \mid \mathcal{F}_t\right]\cdot \mathbb{1}_A\right] = \mathbb{E}_\nu[X\cdot \mathbb{1}_A].$$

Using (a) and that $W_T - W_t$ is independent of $\mathcal{F}_t$,

$$\begin{aligned}
&\mathbb{E}_\nu\left[e^{\frac{\nu^2}{2}t-\nu W_t}\mathbb{E}\left[Xe^{\nu W_T-\frac{\nu^2}{2}T} \mid \mathcal{F}_t\right]\cdot \mathbb{1}_A\right]\\
&= \mathbb{E}\left[e^{\nu(W_T-W_t)-\frac{\nu^2}{2}(T-t)}\mathbb{E}\left[Xe^{\nu W_T-\frac{\nu^2}{2}T} \mid \mathcal{F}_t\right]\cdot \mathbb{1}_A\right]\\
&= \mathbb{E}\left[e^{\nu(W_T-W_t)-\frac{\nu^2}{2}(T-t)}\right]\mathbb{E}\left[\mathbb{E}\left[Xe^{\nu W_T-\frac{\nu^2}{2}T} \mid \mathcal{F}_t\right]\cdot \mathbb{1}_A\right].
\end{aligned}$$

Since $W_T - W_t \sim \mathcal{N}(0, T-t)$, we have $\mathbb{E}\left[e^{\nu(W_T-W_t)-\frac{\nu^2}{2}(T-t)}\right] = 1$. Thus

$$\begin{aligned}
\mathbb{E}_\nu\left[e^{\frac{\nu^2}{2}t-\nu W_t}\mathbb{E}\left[Xe^{\nu W_T-\frac{\nu^2}{2}T} \mid \mathcal{F}_t\right]\cdot \mathbb{1}_A\right] &= \mathbb{E}\left[\mathbb{E}\left[Xe^{\nu W_T-\frac{\nu^2}{2}T} \mid \mathcal{F}_t\right]\cdot \mathbb{1}_A\right]\\
&= \mathbb{E}\left[Xe^{\nu W_T-\frac{\nu^2}{2}T}\cdot \mathbb{1}_A\right]\\
&= \mathbb{E}_\nu[X\cdot \mathbb{1}_A].
\end{aligned}$$

This completes the proof. □

Our aim is to choose the parameter $\nu$ such that $S_t^*$ is a martingale under $\mathbb{P}_\nu$. The following result gives the unique value of $\nu$ under which this holds.

**Theorem 9.3.8.** *The measure $\mathbb{P}_\nu$ is a risk-neutral measure if and only if $\nu = \frac{r-\mu}{\sigma}$.*

*Proof.* We check whether $S_t^*$ satisfies the conditions of Definition 6.3.13 under $\mathbb{P}_\nu$. The first two conditions hold for all values of $\nu$, so we just need to establish when

$$\mathbb{E}_\nu\left[S_t^* \mid \mathcal{F}_s\right] = S_s^*.$$

By (9.3.11), for $0 \le s < t \le T$ and the same argument as in the proof of Lemma 9.3.7,

$$\begin{aligned}
\mathbb{E}_\nu\left[S_t^* \mid \mathcal{F}_s\right] &= e^{\frac{\nu^2}{2}s-\nu W_s}\mathbb{E}\left[S_t^* e^{\nu W_T-\frac{\nu^2}{2}T} \mid \mathcal{F}_s\right]\\
&= e^{\frac{\nu^2}{2}s-\nu W_s}\mathbb{E}\left[e^{\nu(W_T-W_t)-\frac{\nu^2}{2}(T-t)}\right]\mathbb{E}\left[S_t^* e^{\nu W_t-\frac{\nu^2}{2}t} \mid \mathcal{F}_s\right]\\
&= e^{\frac{\nu^2}{2}s-\nu W_s}\mathbb{E}\left[S_t^* e^{\nu W_t-\frac{\nu^2}{2}t} \mid \mathcal{F}_s\right].
\end{aligned}$$

Inserting the definition of $S_t^*$ gives

$$\begin{aligned}
\mathbb{E}_\nu\left[S_t^* \mid \mathcal{F}_s\right] &= S_0 e^{\frac{\nu^2}{2}s-\nu W_s}\mathbb{E}\left[e^{(\sigma+\nu)W_t-\frac{(\sigma+\nu)^2}{2}t}e^{(\sigma\nu+\mu-r)t} \mid \mathcal{F}_s\right] \\
&= S_0 e^{\frac{\nu^2}{2}s-\nu W_s}e^{(\sigma+\nu)W_s-\frac{(\sigma+\nu)^2}{2}s}e^{(\sigma\nu+\mu-r)t} \\
&= S_0 e^{\sigma W_s-\frac{\sigma^2+2\nu\sigma}{2}s}e^{(\sigma\nu+\mu-r)t} \\
&= S_s^* e^{(t-s)(\mu-r+\sigma\nu)},
\end{aligned}$$

where we used Example 7.3.6(iii) in the second line above. Therefore $S_t^*$ is a martingale under $\mathbb{P}_\nu$ if and only if $e^{(t-s)(\mu+\sigma\nu-r)} = 1$ for all $0 \leq s < t$. This implies that $\mu - r + \sigma\nu = 0$, or equivalently $\nu = \frac{r-\mu}{\sigma}$. □

In view of Theorem 9.3.8, we now use the notation

$$\mathbb{P}_* := \mathbb{P}_{\frac{r-\mu}{\sigma}}. \tag{9.3.12}$$

Observe that, by Theorem 9.3.6, $S_t = S_0 \exp\left(\sigma W_t + (\mu - \sigma^2/2)t\right)$ has the same distribution under $\mathbb{P}_*$ as $S_0 \exp\left(\sigma(W_t + \frac{r-\mu}{\sigma}t) + (\mu - \sigma^2/2)t\right) = S_0 \exp\left(\sigma W_t + (r - \sigma^2/2)t\right)$ under $\mathbb{P}$. This shows that calculations involving the stock price under the risk-neutral measure $\mathbb{P}_*$ are exactly the same as calculations under the original probability after setting $\mu = r$.

In order to do computations under the risk-neutral probability, we therefore require that the assumptions in (9.2.12) also hold if we set $\mu = r$, or equivalently if $\mathbb{P}$ is replaced by $\mathbb{P}_*$. Adding this additional restriction to the assumptions in (9.2.12) leads to the following definition.

**Definition 9.3.9** (Admissible trading strategies). An (admissible) *trading strategy* $\varphi = (\beta, \alpha)$ is a pair of adapted, cadlag stochastic processes $\beta$ and $\alpha$, with

$$\|\alpha S\|_{L^2(\Omega\times[0,T],\mathbb{P})} < \infty, \ \|\alpha S\|_{L^2(\Omega\times[0,T],\mathbb{P}_*)} < \infty \text{ and } \|\beta\|_{L^1([0,T])} < \infty \text{ a.s.}, \tag{9.3.13}$$

where

$$\| X \|_{L^p(\Omega\times[0,T],\mathbb{P})} = \left(\int_0^T \mathbb{E}\left[|X_s|^p\right] ds\right)^{1/p}$$

and $\| \cdot \|_{L^p(\Omega\times[0,T],\mathbb{P}_*)}$ is defined similarly but with $\mathbb{E}$ replaced by expectation under the risk-neutral measure $\mathbb{P}_*$, which we denote by $\mathbb{E}_*$.

Note that, since $\mathbb{P}_* \approx \mathbb{P}$, $\|\beta\|_{L^1([0,T])} < \infty$ a.s. with respect to $\mathbb{P}$ if and only if this holds a.s. with respect to $\mathbb{P}_*$, so we do not need to explicitly state which measure we are using for this condition.

In the remainder of this chapter we will work only with admissible trading strategies and therefore, for simplicity, we will just call them trading strategies. The additional requirement in (9.3.13) is only a very slight restriction of (9.2.12). For instance, if $\|\alpha S\|_{L^{2+\epsilon}(\Omega\times[0,T],\mathbb{P})} < \infty$ for some $\epsilon > 0$ then (9.3.13) is fulfilled. However, there are examples of trading strategies that fulfil (9.2.12) but not (9.3.13). An explicit example of such a strategy can be found in [9, page 31].

We are now almost ready to show the Black–Scholes model is arbitrage-free. Looking at the argument used in previous chapters, we need two things to do this: construct a risk-neutral measure and show that $V_t^*(\varphi)$ is a martingale under this risk-neutral measure for all self-financing trading strategies $\varphi$. It therefore only remains to check that $V_t^*(\varphi)$ is a martingale under $\mathbb{P}_*$. As argued above, it is sufficient to check that it is a martingale under the original probability $\mathbb{P}$ in the case $\mu = r$. By Theorem 9.3.3,

$$dV_t^* = \alpha_t dS_t^*.$$

However, by (9.3.3),

$$\begin{aligned} dS_t^* &= -re^{-rt}S_t dt + e^{-rt}dS_t \\ &= -re^{-rt}S_t dt + e^{-rt}(rS_t dt + \sigma S_t dW_t) \\ &= \sigma S_t^* dW_t \end{aligned}$$

and so

$$dV_t^* = \sigma \alpha_t S_t^* dW_t. \tag{9.3.14}$$

Since our assumption in Definition 9.3.9 guarantees that this stochastic integral is well-defined, $V_t^*$ is a martingale.

**Theorem 9.3.10.** *The Black–Scholes model is arbitrage-free.*

*Proof.* We argue by contradiction. Let $\varphi$ be an arbitrage opportunity. As argued above, $V_t^*(\varphi)$ is a martingale under the risk-neutral measure $\mathbb{P}_*$. Then

$$\mathbb{E}_*\left[V_T(\varphi)\right] = e^{rT}\mathbb{E}_*\left[V_T^*(\varphi)\right] = e^{rT}\mathbb{E}_*\left[V_0^*(\varphi)\right] = e^{rT}\mathbb{E}_*\left[0\right] = 0.$$

By assumption, $V_T(\varphi) \geq 0$, so $\mathbb{P}_*\left[V_T(\varphi) > 0\right] = 0$. Since $\mathbb{P}_* \approx \mathbb{P}$, this implies $\mathbb{P}\left[V_T(\varphi) > 0\right] = 0$, which is a contradiction. This completes the proof. □

## 9.4 Black–Scholes Formula

In this section, we derive the famous Black–Scholes formula for the pricing of European call and put options. In doing so, we will discuss the theory that can be used for the pricing of other *European contingent claims*.

The *maturity* of all *contingent claims* in this section is $T$ (i.e. they will or can be exercised at time $T$). Denote the payoff of a given *contingent claim* by $\Phi_T$. As usual, $\Phi_T$ has to be $\mathcal{F}_T$-measurable. In addition, we assume

$$\mathbb{E}\left[\Phi_T^2\right] < \infty \text{ and } \mathbb{E}_*\left[\Phi_T^2\right] < \infty, \tag{9.4.1}$$

where $\mathbb{E}_*$ is the expectation with respect to the probability measure $\mathbb{P}_*$ in (9.3.12).

An important tool for determining the price of a contingent claim is the replicating strategy.

**Definition 9.4.1** (Replicating portfolio). A contingent claim is *replicable* if there exists an admissible trading strategy $\varphi$ such that $V_T(\varphi) = \Phi_T$. We call this trading strategy a *replicating strategy* for the contingent claim $\Phi_T$.

Suppose that at some time $0 \leq t < T$, we wish to buy a claim which pays out $\Phi_T$ at time $T$. Since we do not yet know what the value of $\Phi_T$ will be at maturity, what is a fair price to pay? If $\Phi_T$ is replicable then we can determine the value of the replicating strategy at time $t$. Since the Black–Scholes model is arbitrage-free, the law of price implies that the *true price* or *premium* of this contingent claim at time $t < T$ must be the same as the value of the replicating portfolio at time $t$.

A natural question is, therefore, how does one determine a replicating strategy for a given contingent claim? In the binomial and finite market models, there were only finitely many time steps so we could use backwards induction, see for instance Example 3.3.11. This is no longer possible. Instead one can use, for instance, the Black–Scholes equation, see Corollary 9.2.8. We illustrate this with an example.

**Example 9.4.2.** (European call option) *The payoff of a European call option with maturity T and strike price K is* $\Phi_T = \max\{S_T - K, 0\}$. *We aim to show that there is a trading strategy* $\varphi$ *which satisfies* $V_T(\varphi) = \max\{S_T - K, 0\}$. *Motivated by Theorem 5.5.2, let*

$$d_1(t,x) = \frac{\log(x/K) + (r + \sigma^2/2)(T-t)}{\sigma\sqrt{T-t}} \quad \textit{and} \quad d_2(t,x) = d_1(t,x) - \sigma\sqrt{T-t}.$$

*Then set*

$$f(t,x) = x\Phi(d_1(t,x)) - Ke^{-r(T-t)}\Phi(d_2(t,x)), \tag{9.4.2}$$

*where* $\Phi$ *is the cumulative distribution function of the standard* $\mathcal{N}(0,1)$ *random variable. Observe that*

$$\lim_{t\to T} d_1(t,x) = \lim_{t\to T} d_2(t,x) = \begin{cases} \infty, & \textit{if } x > K, \\ -\infty, & \textit{if } x < K, \\ 0, & \textit{if } x = K. \end{cases}$$

*Therefore*

$$\lim_{t\to T} f(t,x) = \max\{x - K, 0\}.$$

*Furthermore, it can be shown that* $f$ *satisfies the Black–Scholes equation* (9.2.14) *(see Exercise 9.2(b)). Hence there exists a self-financing trading strategy* $\varphi$ *with* $V_t(\varphi) = f(t, S_t)$ *for all* $t$ *and, in particular,*

$$V_T(\varphi) = f(T, S_T) = \max\{x - K, 0\}.$$

*Therefore $\varphi$ is a replicating strategy for the European call option. One can use the ideas in the proof of Corollary 9.2.8 to get an explicit expression for $\varphi$ (see Exercise 9.2(c)). However, even without doing this, it follows from the discussion above that in order to avoid creating an arbitrage opportunity, the price of a European call option at time* 0 *should be*

$$V_0(\varphi) = S_0\Phi\left(\frac{\log(S_0/K) + (r+\sigma^2/2)T}{\sigma\sqrt{T}}\right) - Ke^{-rT}\Phi\left(\frac{\log(S_0/K) + (r-\sigma^2/2)T}{\sigma\sqrt{T}}\right).$$

This example illustrates that one can construct a replicating portfolio by finding a solution $f(t,x)$ to the Black–Scholes equation with boundary value satisfying $f(T,S_T) = \Phi_T$. However, not all contingent claims have the form $\Phi_T = f(T,S_T)$. Also, it is often difficult to explicitly solve partial differential equations with given boundary conditions. On the other hand, if we know that the contingent claim is replicable then we do not have to explicitly find $\varphi$ in order to determine $V_t(\varphi)$. Instead, we can use that $V_t^*(\varphi)$ is a martingale under the measure $\mathbb{P}_*$. This gives

$$V_t(\varphi) = e^{rt}V_t^*(\varphi) = e^{rt}\mathbb{E}_*\left[V_T^*(\varphi)\middle|\mathcal{F}_t\right] = e^{-r(T-t)}\mathbb{E}_*\left[\Phi_T\middle|\mathcal{F}_t\right]. \tag{9.4.3}$$

We illustrate this approach with the call option.

**Example 9.4.3.** (European call option) *The computations in Example 9.4.2 and Corollary 9.2.8 show that a European call option is replicable. Let $\varphi$ be a replicating strategy for the call option with payoff $(S_T - K)_+ := \max\{S_T - K, 0\}$. Recall that doing computations with respect to the risk-neutral measure $\mathbb{P}_*$ is the same as doing computations with respect to the original probability $\mathbb{P}$ under the assumption that $\mu = r$. This means that, for any function $h$ and $0 \le t \le T$,*

$$\mathbb{E}_*\left[h(S_T)|\mathcal{F}_t\right] = \mathbb{E}\left[h(S_0e^{\sigma W_T + (r-\sigma^2/2)T})\middle|\mathcal{F}_t\right] = \mathbb{E}\left[h(S_te^{\sigma(W_T - W_t) + (r-\sigma^2/2)(T-t)})\middle|\mathcal{F}_t\right]. \tag{9.4.4}$$

*Applying this to $h(x) = (x - K)_+$ gives*

$$\begin{aligned} V_t(\varphi) &= e^{-r(T-t)}\mathbb{E}_*\left[(S_T - K)_+|\,\mathcal{F}_t\right] \\ &= e^{-r(T-t)}\mathbb{E}\left[(S_0e^{\sigma W_T + (r-\sigma^2/2)T} - K)_+\middle|\mathcal{F}_t\right] \\ &= e^{-r(T-t)}\mathbb{E}\left[(S_te^{\sigma(W_T - W_t) + (r-\sigma^2/2)(T-t)} - K)_+\middle|\mathcal{F}_t\right]. \end{aligned} \tag{9.4.5}$$

*We know that $W_T - W_t$ is normal $\mathcal{N}(0, T-t)$ and $W_T - W_t$ is independent of $\mathcal{F}_t$. Inserting this into the above formula and comparing it to the computations in the proof of Theorem 5.5.2, it follows that the quantities in* (5.5.5) *and* (9.4.5) *agree. Therefore we can do exactly the same computation as in the proof of Theorem 5.5.2. This, of course, gives the same result as in Example 9.4.2, namely*

$$V_t(\varphi) = f(t, S_t) = S_t\Phi(d_1(t, S_t)) - Ke^{-r(T-t)}\Phi(d_2(t, S_t)).$$

In order to use (9.4.3) to price contingent claims, we need to know: Are all contingent claims replicable and is the risk-neutral measure unique? It turns out that the answer to both of these questions is yes. This means that (9.4.3) gives the unique price for a claim which must be used in order to avoid creating an arbitrage opportunity.

**Theorem 9.4.4.** *Let $\Phi_T$ be an $\mathcal{F}_T$-measurable random variable fulfilling* (9.4.1). *Then $\Phi_T$ is replicable. Furthermore, the risk-neutral measure $\mathbb{P}_*$ is unique.*

We omit the proof of this theorem for general $\Phi_T$ since it is beyond the scope of this book, although we will show in Section 9.6 that all terminal value claims are replicable. The interested reader can find the proof in [16, Theorem 4.4.2].

As discussed above, Theorem 9.4.4 enables us to determine the price of a *contingent claim* $\Phi_T$. To do this formally, it is necessary to extend the primary market by adding a *contingent claim* $\Phi_T$. We use the same assumptions as in Sections 3.3.2 and 4.4, namely,

- the dynamics of the stocks $S_t$ and the bond $B_t$ are given by the Black–Scholes model,
- the *contingent claim* has no influence on the stock $S_t$ and the bond $B_t$,
- the *maturity* of the *contingent claim* is $T$ and
- the *contingent claim* $\Phi_T$ is $\mathcal{F}_T$-measurable.

We immediately get the following result.

**Theorem 9.4.5.** *Denote by $P_t$ the price of $\Phi_T$ at time $t$ for $t \in [0, T]$. The extended market described above is arbitrage-free if and only if*

$$P_t = e^{-r(T-t)}\mathbb{E}_*\left[\Phi_T \middle| \mathcal{F}_t\right] \quad \text{for all } t \in [0, T]. \tag{9.4.6}$$

Since doing computations with respect to the risk-neutral measure $\mathbb{P}_*$ is the same as doing computations with respect to the original probability $\mathbb{P}$ under the assumption that $\mu = r$, the price of an option at time $t$ does not depend on the value of $\mu$; it depends on the stock price only through its current value $S_t$ and the volatility $\sigma$. This means that two investors may have different ideas about the expected rate of return of the stock, but they will agree on the same price for the option.

The proof of this theorem is (almost) the same as the proof of Theorem 4.5.1 so we omit it. Applying this theorem to call and put options and using the computations above, together with the *put–call parity formula* (see Exercise 9.3), leads to the famous Black–Scholes formula.

**Theorem 9.4.6** (Black–Scholes formula). *Consider the primary market extended with a call and a put option. Assume that both options have strike price $K$ and maturity $T$. For $t \in [0, T]$ and $x \in \mathbb{R}$, set*

$$d_1(t,x) := \frac{\log(x/K) + (r + \sigma^2/2)(T-t)}{\sigma\sqrt{T-t}},$$
$$d_2(t,x) := d_1(t,x) - \sigma\sqrt{T-t} \quad \text{and}$$
$$C(t,x) := x\Phi(d_1(t,x)) - Ke^{-r(T-t)}\Phi(d_2(x,t)), \tag{9.4.7}$$

*where $\Phi$ is the cumulative distribution function of the standard $\mathcal{N}(0,1)$ random variable. This extended market is arbitrage-free if and only if*

- *the price of the call option at time $t \in [0,T]$ is $C(t, S_t)$ and*
- *the price of the put option at time $t \in [0,T]$ is $C(t,S_t) - S_t + Ke^{-r(T-t)}$.*

The prices in Theorem 9.4.6 are the same as in Theorem 5.5.2. However, a key difference is that the risk-neutral measure in the Black–Scholes model is unique. Therefore, the arbitrage-free prices of put and call options are unique in the Black–Scholes model, whereas these prices are not unique in the discrete Black–Scholes model.

## 9.5 The Black–Scholes Greeks

The Greeks are a set of letters labelling the quantities of risk associated with small changes in various inputs and model parameters. The Greeks have importance in risk management, where hedging and risks are determined by how much/little of the Greeks are exposed on a portfolio. There are five main Greeks associated with the Black–Scholes model, which we define below.

Throughout this section, suppose that $f(t, S_t)$ gives the price of a contingent claim $\Phi_T$ at time $t$, where $f : [0,T] \times \mathbb{R} \to \mathbb{R}$ is continuously differentiable with respect to the first variable, $t$, and twice continuously differentiable with respect to the second variable, $x$. Theorems 9.4.4 and 9.4.5 imply that $f(t, S_t) = V_t(\varphi)$ for some self-financing trading strategy $\varphi$. Therefore $f(t,x)$ must satisfy the Black–Scholes equation (9.2.14).

**Definition 9.5.1** (Delta). The quantity

$$\Delta_t = \frac{\partial f}{\partial x}(t, S_t)$$

is known as the *Delta* of the option at time $t$.

The *Delta* represents the amount of stock held in the replicating portfolio (see (9.2.16)).

**Definition 9.5.2** (Gamma). The quantity

$$\Gamma_t = \frac{\partial^2 f}{\partial x^2}(t, S_t)$$

is known as the *Gamma* of the option at time $t$.

The *Gamma* determines the sensitivity of the *Delta* to changes in the stock price. The larger the *Gamma*, the more sensitive the stock holding in the replicating portfolio is to movements in the stock price.

**Definition 9.5.3** (Theta). The quantity

$$\Theta_t = \frac{\partial f}{\partial t}(t, S_t)$$

is known as the *Theta* of the option at time $t$.

The *Theta* determines the sensitivity of the price to time. It quantifies the change over time in the cost to be paid for having the option to buy or sell, as opposed to taking a long or short position in the underlying stock with no option involved.

**Definition 9.5.4** (Vega). The quantity

$$\nu_t = \frac{\partial f}{\partial \sigma}(t, S_t)$$

is known as the *Vega* of the option at time $t$.

The *Vega* determines the sensitivity of the price to changes in the volatility. Note that the price depends on the volatility indirectly through the stock price, so this is a latent parameter. Volatility will be discussed further in Section 9.5.1.

**Definition 9.5.5** (Rho). The quantity

$$\rho_t = \frac{\partial f}{\partial r}(t, S_t)$$

is known as the *Rho* of the option at time $t$.

The *Rho* determines the sensitivity of the price to changes in the interest rate. Of the five main Greeks this is widely considered to be the least sensitive. Some caution should be exercised when using it to predict the effect of changing the interest rate on the price of an option as it measures the change to the price when all other arguments and parameters are kept fixed. However, the stock price will tend to decrease when the interest rate increases and so this effect may also need to be taken into account.

**Example 9.5.6.** (Greeks for call options). *Consider a European call option with strike price $K$ and maturity $T$. By Theorem 9.4.6, the price of this call option at time $t$ is $f(t, S_t)$, with*

$$f(t,x) = x\Phi(d_1(t,x)) - Ke^{-r(T-t)}\Phi(d_2(t,x))$$

*and* $d_1(t,x)$ *and* $d_2(t,x)$ *as in* (9.4.7). *Then*

$$\Delta_t = \Phi(d_1), \quad \Gamma_t = \frac{e^{-d_1^2/2}}{\sigma S_t \sqrt{2\pi t}},$$

$$\Theta_t = \frac{1}{T}\left(\frac{\sigma S_t}{2\sqrt{2\pi t}} e^{-d_1^2/2} - rKe^{-rt}\Phi(d_2)\right),$$

$$\nu_t = \frac{S_t\sqrt{t}e^{-d_1^2/2}}{\sqrt{2\pi}}, \qquad \rho_t = Kte^{-rt}\Phi(d_2).$$

The computations leading to the Greeks in Example 9.5.6 are straightforward, but a little bit involved, and so are left as Exercise 9.2(d).

We illustrate these computations in the more general setting of terminal-value claims in Section 9.6.

## 9.5.1 Volatility

The volatility $\sigma$ is the only parameter related directly to the stock price which enters into the Black–Scholes option pricing formula. The volatility is non-observable (or latent) and so it is necessary to estimate it in order to evaluate the price of an option.

In practice, two methods are used to estimate the volatility. The first is to use historical data. Suppose that a practitioner needs to price an option at time $t$, but has access to the historical stock price $\{S_u, 0 \leq u \leq t\}$. For $n \in \mathbb{N}$, let $t_i = it/n$, $i = 0, \ldots, n$. Then the random variables $\log(S_{t_i}/S_{t_{i-1}})$, $i = 1, \ldots, n$ are i.i.d. Gaussian random variables, each with variance $\sigma^2 t/n$, and hence $\sigma$ may be estimated by calculating the empirical variance.

The second method is to use the market price of an option to establish an 'implied' volatility. Suppose that the market price of a European call option with strike price $K$ at maturity $T$ is given by $C_t^E$ at time $t < T$ when the stock price $S_t = x$. The Black–Scholes formula gives the price as a function $c(x,t,r,\sigma,K,T)$. Keeping $x$, $t$, $r$, $K$ and $T$ fixed, the equation

$$c(x,t,r,\sigma,K,T) = C_t^E$$

can be solved to obtain the *implied volatility* $\sigma$. The resulting value of $\sigma$ may then be used to value options on the same stock at other strike prices and with other expiry dates in a consistent way.

Comparisons between the estimates of volatility obtained from historical data and the implied volatility suggest that the Black–Scholes model is an oversimplification of the market. When computing the implied volatility for a range of options on the same stock but with different strike prices and maturities, computations return different values for the implied volatility. Plots of implied volatility against strike price for options based on the same asset and with the same maturity typically exhibit a volatility 'smile', see Figure 9.1. A way to resolve this shortcoming is to assume that volatility is itself a stochastic process, rather than a constant. This can enable one to price options more accurately. One approach is to model volatility as a function of the stock price

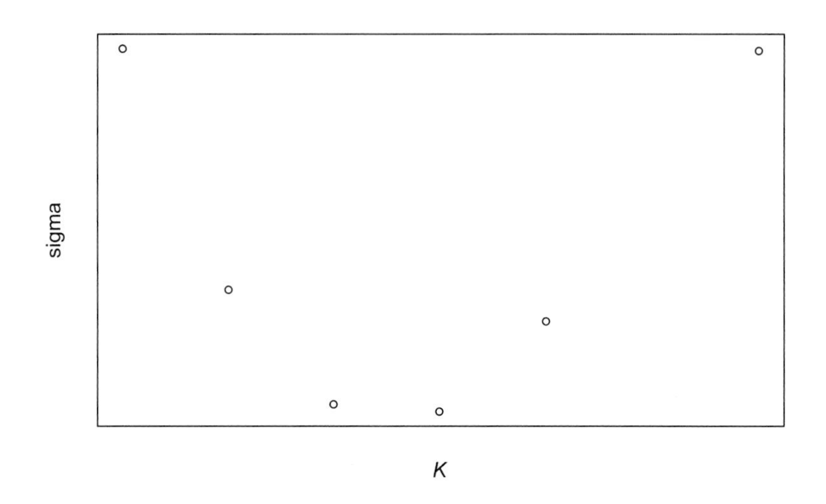


**Figure 9.1** Plot of implied volatility against strike price.

and time. This gives rise to what are known as *local volatility models*. A second approach is to model volatility as a random process described by an SDE which depends on a second Brownian motion $\{W_t, 0 \le t \le T\}$. These are known as *stochastic volatility models*.

## 9.6 Terminal Value Claims

Theorem 9.4.5 provides a method of pricing that can be used to find the time $t$ price of any attainable contingent claim paying $\Phi_T$ at time $T$. In Theorem 9.4.6, we used this method to price the European call and put options. These are known as *plain vanilla* options. In this section, we will show how to use the same result to price some commonly traded *exotic options*. We focus on claims whose payoff depends only on the value of the stock at time $T$.

**Definition 9.6.1** (Terminal value claim). A contingent claim with a payoff of the form $\Phi_T = h(S_T)$ for some function $h : (0, \infty) \to \mathbb{R}$ is called a *terminal value claim*.

Using $h(x) = \max\{x - K, 0\}$ and $h(x) = \max\{K - x, 0\}$ shows that call and put options are examples of terminal value claims. Terminal value claims can be illustrated with payoff diagrams, see Section 1.2.4.

By Theorem 9.4.5, the price of a terminal value claim at time $t$ is given by

$$\begin{aligned} P_t &= e^{-r(T-t)}\mathbb{E}_* \left[\Phi_T | \mathcal{F}_t\right] = e^{-r(T-t)}\mathbb{E}_* \left[h(S_T) | \mathcal{F}_t\right] \\ &= e^{-r(T-t)}\mathbb{E}\left[ h\left( S_t e^{\left(r-\frac{\sigma^2}{2}\right)(T-t)+\sigma(W_T-W_t)} \right) \middle| \mathcal{F}_t \right], \end{aligned}$$

where we used (9.4.4) in the last line. Since $W_T - W_t$ is normal $\mathcal{N}(0, T-t)$-distributed under $\mathbb{P}$ and independent of $\mathcal{F}_t$, the price $P_t$ can be expressed as $f(t, S_t)$, where

$$f(t,x) = e^{-r(T-t)}\mathbb{E}\left[h(xe^{\sigma\sqrt{T-t}Z+(r-\sigma^2/2)(T-t)})\right]$$

for $Z \sim \mathcal{N}(0,1)$. We will verify that $f$ satisfies the Black–Scholes equation (9.2.14) in the case when $h$ is twice differentiable. Note that we have to make this assumption to ensure that all the expressions below exist. This gives another proof that the terminal value claim $X = h(S_T)$ is replicable (cf. Theorem 9.4.4).

We first compute some of the Greeks for the terminal value claim and show how they depend on the shape of the function $h$. Differentiating under the expectation with respect to $x$ gives

$$\begin{aligned}\frac{\partial f}{\partial x} &= e^{-r(T-t)}\mathbb{E}\left[e^{\sigma\sqrt{T-t}Z+(r-\sigma^2/2)(T-t)}h'(xe^{\sigma\sqrt{T-t}Z+(r-\sigma^2/2)(T-t)})\right]\\ &= \mathbb{E}\left[e^{\sigma\sqrt{T-t}Z-\sigma^2(T-t)/2}h'(xe^{\sigma\sqrt{T-t}Z+(r-\sigma^2/2)(T-t)})\right]\\ &= \mathbb{E}\left[h'(xe^{\sigma\sqrt{T-t}Z+(r+\sigma^2/2)(T-t)})\right].\end{aligned}$$

In the last line we used the result that we have for all $\theta \in \mathbb{R}$,

$$\begin{aligned}\mathbb{E}\left[e^{\theta Z-\theta^2/2}g(Z)\right] &= \frac{1}{\sqrt{2\pi}}\int_{-\infty}^{\infty} e^{\theta z-\theta^2/2}g(z)e^{-z^2/2}dz\\ &= \frac{1}{\sqrt{2\pi}}\int_{-\infty}^{\infty} g(z)e^{-(z-\theta)^2/2}dz\\ &= \frac{1}{\sqrt{2\pi}}\int_{-\infty}^{\infty} g(x+\theta)e^{-x^2/2}dz\\ &= \mathbb{E}\left[g(Z+\theta)\right].\end{aligned}$$

Therefore $\Delta_t = \frac{\partial f}{\partial x}(t, S_t)$ is positive if $h$ is increasing and negative if $h$ is decreasing. Since $\alpha_t = \frac{\partial f}{\partial x}(t, S_t)$ by the computations in Corollary 9.2.8, this means that the replicating portfolio has a positive holding in the stock (or is *long in the stock*) if $h$ is an increasing function, while it has a negative holding in the stock (or is *short in the stock*) if $h$ is a decreasing function.

Differentiating again under the expectation with respect to $x$ gives

$$\frac{\partial^2 f}{\partial x^2} = \mathbb{E}\left[e^{\sigma\sqrt{T-t}Z+(r+\sigma^2/2)(T-t)}h''(xe^{\sigma\sqrt{T-t}Z+(r+\sigma^2/2)(T-t)})\right].$$

This shows that if $h$ is convex ($h'' \geq 0$), then $\Gamma_t \geq 0$ and the holding of stock in the replicating portfolio is non-decreasing in the stock price. Conversely, if $h$ is concave ($h'' \leq 0$), then $\Gamma_t \leq 0$ and the holding of stock in the replicating portfolio is non-increasing in the stock price.

By differentiating the expression for $f$ with respect to $t$, we get that $\frac{\partial f}{\partial t}$ is equal to

$$\begin{aligned}&re^{-r(T-t)}\mathbb{E}\left[h(xe^{\sigma\sqrt{T-t}Z+(r-\sigma^2/2)(T-t)})\right]\\ &-e^{-r(T-t)}\mathbb{E}\left[\left(\frac{\sigma Z}{2\sqrt{T-t}}+(r-\sigma^2/2)\right)xe^{\sigma\sqrt{T-t}Z+(r-\sigma^2/2)(T-t)}h'(xe^{\sigma\sqrt{T-t}Z+(r-\sigma^2/2)(T-t)})\right]\end{aligned}$$

$$= rf(t,x) - x\mathbb{E}\left[e^{\sigma\sqrt{T-t}Z-\sigma^2(T-t)/2}\left(\frac{\sigma Z}{2\sqrt{T-t}} + (r-\sigma^2/2)\right)h'(xe^{\sigma\sqrt{T-t}Z+(r-\sigma^2/2)(T-t)})\right]$$
$$= rf(t,x) - x\mathbb{E}\left[\left(\frac{\sigma Z}{2\sqrt{T-t}} + r\right)h'(xe^{\sigma\sqrt{T-t}Z+(r+\sigma^2/2)(T-t)})\right],$$

where the last line again used the identity $\mathbb{E}\left[e^{\theta Z-\theta^2/2}g(Z)\right] = \mathbb{E}\left[g(Z+\theta)\right]$.

Also, using integration by parts, we have the identity

$$\begin{aligned}\mathbb{E}\left[Zg(Z)\right] &= \frac{1}{\sqrt{2\pi}}\int_{-\infty}^{\infty} zg(z)e^{-z^2/2}dz \\ &= \left[-\frac{1}{\sqrt{2\pi}}g(z)e^{-z^2/2}\right]_{-\infty}^{\infty} + \frac{1}{\sqrt{2\pi}}\int_{-\infty}^{\infty} g'(z)e^{-z^2/2}dz \\ &= \mathbb{E}\left[g'(Z)\right].\end{aligned}$$

Hence

$$\begin{aligned}&\mathbb{E}\left[\frac{\sigma Z}{2\sqrt{T-t}}h'(xe^{\sigma\sqrt{T-t}Z+(r+\sigma^2/2)(T-t)})\right] \\ &= \frac{1}{2}\sigma^2 x\mathbb{E}\left[e^{\sigma\sqrt{T-t}Z+(r+\sigma^2/2)(T-t)}h''(xe^{\sigma\sqrt{T-t}Z+(r+\sigma^2/2)(T-t)})\right] \\ &= \frac{1}{2}\sigma^2 x\frac{\partial^2 f}{\partial x^2}.\end{aligned}$$

Therefore

$$\frac{\partial f}{\partial t} = rf(t,x) - rx\frac{\partial f}{\partial x} - \frac{1}{2}x^2\sigma^2\frac{\partial^2 f}{\partial x^2}$$

and so $f$ satisfies the Black–Scholes equation (9.2.14).

Finally, we consider the sensitivity of the price on the volatility $\sigma$. Differentiating the expression for $f$ with respect to $\sigma$, we get

$$\begin{aligned}\frac{\partial f}{\partial \sigma} &= xe^{-r(T-t)}\mathbb{E}\left[e^{\sigma\sqrt{T-t}Z+(r-\sigma^2/2)(T-t)}(\sqrt{T-t}Z - \sigma(T-t))h'(xe^{\sigma\sqrt{T-t}Z+(r-\sigma^2/2)(T-t)})\right] \\ &= x\sqrt{T-t}\mathbb{E}\left[Zh'(xe^{\sigma\sqrt{T-t}Z+(r+\sigma^2/2)(T-t)})\right] \\ &= x\sqrt{T-t}\mathbb{E}\left[\sigma\sqrt{T-t}xe^{\sigma\sqrt{T-t}Z+(r+\sigma^2/2)(T-t)}h''(xe^{\sigma\sqrt{T-t}Z+(r+\sigma^2/2)(T-t)})\right] \\ &= \sigma x^2(T-t)\mathbb{E}\left[e^{\sigma\sqrt{T-t}Z+(r+\sigma^2/2)(T-t)}h''(xe^{\sigma\sqrt{T-t}Z+(r+\sigma^2/2)(T-t)})\right],\end{aligned}$$

where we have used the two identities $\mathbb{E}\left[e^{\theta Z-\theta^2/2}g(Z)\right] = \mathbb{E}\left[g(Z+\theta)\right]$ and $\mathbb{E}\left[Zg(Z)\right] = \mathbb{E}\left[g'(Z)\right]$.

It follows that if $h$ is convex, $\nu_t \geq 0$ and so the price is non-decreasing as a function of the volatility, while if $h$ is concave, $\nu_t \leq 0$ and so the price is non-increasing as a function of the volatility.

**Example 9.6.2** (Cash-or-nothing call). *The* cash-or-nothing *call pays* 1 *unit at time $T$ if the stock is above some pre-determined level $K$ at time $T$, otherwise it pays nothing. This is sometimes also referred to as the (European)* digital *call or* binary *call. This claim has value $X = \mathbb{1}_{\{S_T>K\}}$ and the payoff diagram can be found in Figure 9.2.*

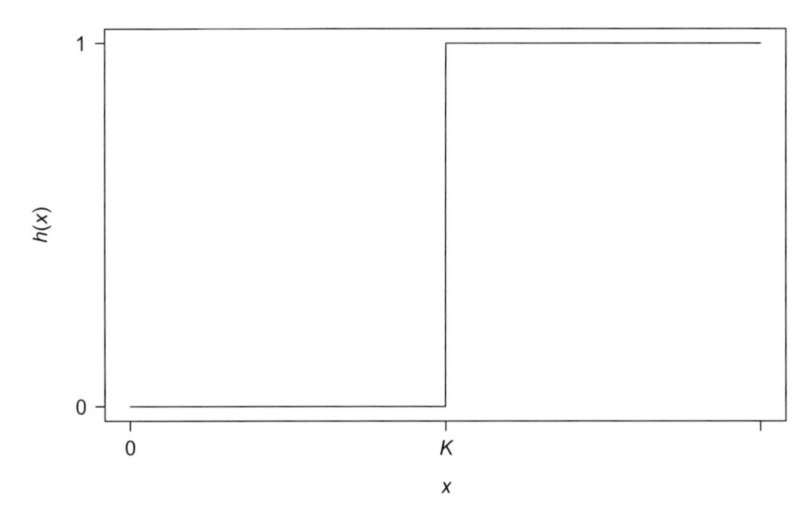


**Figure 9.2** Payoff diagram of cash-or-nothing call.

*The price of the cash-or-nothing call is given by $f(t,S_t)$, where*

$$
\begin{aligned}
f(t,x) &= e^{-r(T-t)}\mathbb{P}\left[xe^{\sigma\sqrt{T-t}Z+(r-\sigma^2/2)(T-t)} > K\right] \\
&= e^{-r(T-t)}\Phi\left(\frac{\log(x/K)+(r-\sigma^2/2)(T-t)}{\sigma\sqrt{T-t}}\right).
\end{aligned}
$$

**Example 9.6.3.** (Contingent premium call). *A* contingent premium *call option allows the purchaser to pay some fixed proportion $0 \le \gamma \le 1$ of the initial price of a European call option at time 0. The payment of the remaining premium is delayed until the maturity $T$ and is contingent upon the option ending* in the money, *that is $S_T > K$. If the stock ends at or below the strike price $K$ then no further premium is paid. If the terminal premium is $d$, then this claim has value*

$$
\Phi_T = (S_T - K)_+ - d\mathbb{1}_{\{S_T>K\}}
$$

*and hence the payoff diagram in Figure 9.3. Furthermore, the initial premium must satisfy*

$$
\begin{aligned}
\gamma e^{-rT}\mathbb{E}_*\left[(S_T-K)_+\right] &= e^{-rT}\mathbb{E}_*\left[\Phi_T\right] \\
&= e^{-rT}\mathbb{E}_*\left[(S_T-K)_+\right] - de^{-rT}\mathbb{E}_*\left[\mathbb{1}_{\{S_T>K\}}\right]
\end{aligned}
$$

*and hence*

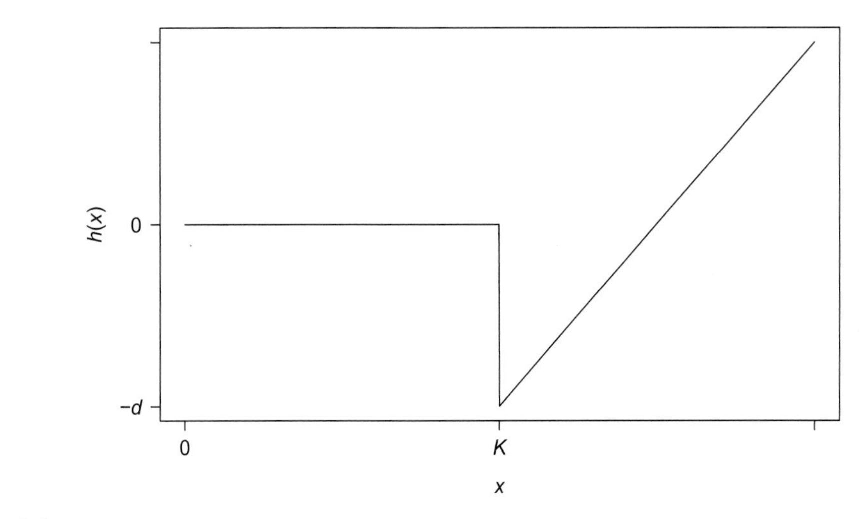


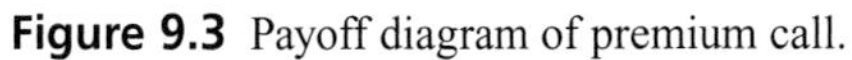

**Figure 9.3** Payoff diagram of premium call.

$$d = (1-\gamma)\frac{e^{-rT}\mathbb{E}_*\left[(S_T-K)_+\right]}{e^{-rT}\mathbb{E}_*\left[\mathbb{1}_{\{S_T>K\}}\right]}.$$

*So the final premium is $1-\gamma$ times the time 0 price of the standard European call, divided by the price of the cash-or-nothing call. In the special case when $\gamma = 0$, the purchaser pays no initial premium. This is sometimes called a* cash-on-delivery *option.*

## 9.7 Exercises

**Exercise 9.1.** *Suppose that $f$ satisfies the Black–Scholes equation. Set*

$$g(t,x) = \frac{1}{\beta_t}\left(f(t,x) - x\frac{\partial f}{\partial x}\right) \quad \text{and} \quad h(t,x) = \frac{\partial f}{\partial x}.$$

*Verify that the portfolio with value*

$$V_t = g(t,S_t)\beta_t + h(t,S_t)S_t$$

*is self-financing.*

**Exercise 9.2.** *Let $d_1(t,x), d_2(t,x)$ and $C(t,x)$ be as in (9.4.7).*

*(a) Establish the identity*

$$x\phi(d_1(t,x)) - Ke^{-r(t_0-t)}\phi(d_2(t,x)) = 0,$$

where $\phi(x) = \Phi'(x)$.

(b) *Show that $C(t,x)$ satisfies the Black–Scholes equation.*
Hint: It is helpful to first obtain expressions for the partial derivatives of $d_1$ and $d_2$.

(c) *Give the replicating portfolio for the European call option, that is the amounts that should be held in the bond and in the stock at time $t$.*

(d) *Show that the five main Greeks for the European call option are given by the formulas stated in Example 9.5.6. By considering whether each Greek is positive or negative, interpret how the price of the European call option is affected by changes in each of the inputs.*

**Exercise 9.3.** *Let $C(t,S_t)$ give the time $t$ price of a European call option and $P(t,S_t)$ give the time $t$ price of a European put option, both with strike price $K$ and maturity $T$. By evaluating $(x-K)_+ - (K-x)_+$, obtain the put–call parity formula:*

$$C(t,x) - P(t,x) = x - Ke^{-r(T-t)}.$$

**Exercise 9.4.** *Suppose that at maturity (time $T$), a claim pays off an amount $X = (S_T)^\gamma$ for some $0 < \gamma < 1$. Show that the time $t$ price is given by $f(t,S_t)$, where*

$$f(t,x) = x^\gamma e^{-(1-\gamma)(r+\gamma\sigma^2/2)(T-t)}, \quad 0 \le t \le T.$$

**Exercise 9.5.** *In this question, we give a direct proof that pricing by solving the Black–Scholes equation with a given boundary condition to find a replicating portfolio gives the same result as pricing by using Theorem 9.4.5.*

*Let $f(t,x)$ be a twice continuously differentiable function that satisfies the Black–Scholes equation for $t < T$, with the boundary condition $f(T,x) = h(x)$. By applying Itô's formula to $e^{-rt}f(t,S_t)$, show that it is a martingale under $\mathbb{P}_*$ and deduce that*

$$f(t,S_t) = e^{-r(T-t)}\mathbb{E}_*\left[h(S_T)|\,\mathcal{F}_t\right].$$

# APPENDIX A
# Supplementary Material

This appendix contains some results and proofs which we have omitted from the main text as they are either well known or technical.

## A.1 Elementary Limit Theorems

In this section we state the law of large numbers, the central limit theorem, some useful inequalities and some properties of the normal distribution. These should be familiar to most readers. We do not state the proofs as they can be found in most introductory books on probability.

### A.1.1 The Law of Large Numbers

The law of large numbers states that if one takes a large sample of i.i.d. trials of an experiment with random outcomes then the sample average will be close to the expectation. Furthermore, this average gets closer to the expectation the larger the sample size is.

**Theorem A.1.1** (Law of large numbers). *Let $(\Omega, \mathcal{F}, \mathbb{P})$ be a probability space and $(X_j)_{j=1}^{\infty}$ be an i.i.d. sequence of random variables with $\mathbb{E}\left[X_1^2\right] < \infty$. Then*

$$\frac{1}{n}\sum_{j=1}^{n} X_j \xrightarrow{P} \mathbb{E}\left[X_1\right] \qquad \text{as } n \to \infty. \tag{A.1.1}$$

Using the definition of convergence in probability (see (6.1.20)), (A.1.1) states that for any $\epsilon > 0$,

$$\mathbb{P}\left[\left|\frac{1}{n}\sum_{j=1}^{n} X_j - \mathbb{E}\left[X_1\right]\right| < \epsilon\right] \longrightarrow 1 \qquad \text{as } n \to \infty. \tag{A.1.2}$$

Theorem A.1.1 is only the most basic version of the law of large numbers. For instance, the assumptions can be weakened to pairwise independent random variables and we can relax the assumption on the second moment. Also, one can replace convergence in probability by almost sure convergence. We do not give further details here.

The advantage of the version of the law of large numbers in Theorem A.1.1 is that it has a very simple proof. In fact, one only needs to compute the expectation and variance of $\frac{1}{n}\sum_{j=1}^{n} X_j$ and then apply Chebyshev's inequality.

**Proposition A.1.2** (Chebyshev's inequality). Let $(\Omega, \mathcal{F}, \mathbb{P})$ be a probability space and $X$ be a random variable with $\mathbb{E}\left[X^2\right] < \infty$. Then for each $\epsilon > 0$,

$$\mathbb{P}\left[|X - \mathbb{E}[X]| \geq \epsilon\right] \leq \frac{\mathrm{Var}(X)}{\epsilon^2}.$$

Another useful inequality is Jensen's inequality.

**Proposition A.1.3** (Jensen's inequality). Let $(\Omega, \mathcal{F}, \mathbb{P})$ be a probability space, $X$ a random variable with $\mathbb{E}[|X|] < \infty$ and $\phi : \mathbb{R} \to \mathbb{R}$ a convex function. Then

$$\phi(\mathbb{E}[X]) \leq \mathbb{E}[\phi(X)].$$

### A.1.2 The Normal Distribution and the Central Limit Theorem

In this section we recall the normal distribution, the central limit theorem and some properties of the normal distribution.

**Definition A.1.4.** A random variable $X$ on a probability space $(\Omega, \mathcal{F}, \mathbb{P})$ is said to have a *Gaussian* or *normal* distribution with parameters $\mu \in \mathbb{R}$ and $\sigma > 0$ if it is continuous and has the density $f(t) = \frac{1}{\sqrt{2\pi}\sigma} \exp\left(-\frac{(t-\mu)^2}{2\sigma}\right)$. We write $X \sim \mathcal{N}(\mu, \sigma^2)$.

The normal distribution is very important, in particular in statistics because it occurs in many places in real life. One of the reasons for this is the central limit theorem.

**Theorem A.1.5** (Central limit theorem). *Let $Y_1, Y_2, \ldots$ be a sequence of i.i.d. random variables with $\mathbb{E}[Y_i] = 0$ and $\mathrm{Var}(Y_i) = 1$. Set*

$$X_n = \frac{Y_1 + \cdots + Y_n}{\sqrt{n}}.$$

*Then $X_n$ converges in distribution to a standard normal distribution as $n \to \infty$. In formulas, $X_n \xrightarrow{d} X$ with $X \sim \mathcal{N}(0, 1)$.*

Theorem A.1.5 is only the most basic version of the central limit theorem and there are many generalisations. However, in this book we need only the above version.

Below we give a list of properties of the normal distribution, which we will use frequently in this book.

- If $X \sim \mathcal{N}(\mu, \sigma^2)$, then $\mathbb{E}[X] = \mu$ and $\mathrm{Var}(X) = \sigma^2$.
- If $X \sim \mathcal{N}(\mu, \sigma^2)$ and $a, b \in \mathbb{R}$, then $aX + b \sim \mathcal{N}(a\mu + b, a^2\sigma^2)$.

- The random variable $X \sim \mathcal{N}(\mu, \sigma^2)$ if and only if $\mathbb{E}\left[e^{sX}\right] = \exp(\mu s + \sigma^2 s^2/2)$ for all $s \in \mathbb{R}$.
- Suppose that $X_1 \sim \mathcal{N}(\mu_1, \sigma_1^2)$ and $X_2 \sim \mathcal{N}(\mu_2, \sigma_2^2)$.
  - If $X_1$ and $X_2$ are independent, then $X_1 + X_2 \sim \mathcal{N}(\mu_1 + \mu_2, \sigma_1^2 + \sigma_2^2)$.
  - In general, $X_1 + X_2 \sim \mathcal{N}(\mu_1 + \mu_2, \sigma_3^2)$ with $\sigma_3^2 = \mathrm{Var}(X_1 + X_2) = \sigma_1^2 + \sigma_2^2 + 2\,\mathrm{Cov}(X_1, X_2)$.
- Let $(X_j)_{j\in\mathbb{N}}$ be a sequence of random variables with $X_j \sim \mathcal{N}(\mu_j, \sigma_j^2)$. Then $X_j \xrightarrow{d} X$ if and only if $\mu_j \to \mu$, $\sigma_j \to \sigma$ as $j \to \infty$ and $X \sim \mathcal{N}(\mu, \sigma^2)$.

## A.2 Measures on Countable Sample Spaces

In this section, we show how to construct a probability measure when the sample space $\Omega$ is countable. Let us start with some simple examples.

**Example A.2.1.** *Let $\Omega = \mathbb{N}$. Suppose $(p_j)_{j=1}^{\infty}$ is a sequence of real numbers such that $p_j \geq 0$ for all $j$ and $\sum_{j=1}^{\infty} p_j = 1$. Take $\mathcal{F} = \mathcal{P}\Omega$ and define*

$$\mathbb{P}[A] := \sum_{j\in A} p_j \text{ for } A \subset \mathbb{N}. \tag{A.2.1}$$

*We claim that $\mathbb{P}$ is a probability measure on $(\mathbb{N}, \mathcal{P}\Omega)$, so $(\Omega, \mathcal{F}, \mathbb{P})$ is a probability space. Let us check this carefully. The sum in* (A.2.1) *is absolutely convergent since $p_j \geq 0$ and therefore this sum does not depend on the order of the summands. This shows that the expression in* (A.2.1) *is well-defined. Furthermore, we clearly have $0 \leq \mathbb{P}[A] \leq 1$ and $\mathbb{P}[\mathbb{N}] = 1$. It remains to show* (6.1.2)*. Let $(A_i)_{i=1}^{\infty}$ be a sequence of mutually disjoint subsets of $\mathbb{N}$. Note that if $j \in \bigcup_{i=1}^{\infty} A_i$ then there is exactly one $i$ such that $j \in A_i$. Using this, we get*

$$\mathbb{P}\left[\bigcup_{i=1}^{\infty} A_i\right] = \sum_{j\in\bigcup_{i=1}^{\infty} A_i} p_j = \sum_{i=1}^{\infty}\sum_{j\in A_i} p_j = \sum_{i=1}^{\infty} \mathbb{P}[A_i].$$

*This shows that $\mathbb{P}$ is indeed a probability measure on $\mathbb{N}$ under the choice $\mathcal{F} = \mathcal{P}\Omega$.*

Clearly, in Example A.2.1, we can replace the set $\Omega$ by any countable set and use the same argument to define a probability measure. The argument used in Example A.2.1 can be generalised a little bit further. More precisely, we can define a probability measure on a $\sigma$-algebra generated by a partition.

**Example A.2.2.** *Let $\Omega$ be an infinite set and let $(B_j)_{j=1}^{\infty}$ be a partition of $\Omega$ such that $B_j \neq \emptyset$ for all $j \in \mathbb{N}$. Let $(p_j)_{j=1}^{\infty}$ be as in Example A.2.1. Define*

$$\mathbb{P}\left[B_j\right] := p_j.$$

*This definition is unproblematic as all $B_j \neq \emptyset$. The function $\mathbb{P}$ can be extended to a probability measure on $\Omega$ with $\mathcal{F} := \sigma\left((B_j)_{j=1}^{\infty}\right)$. For this, we have to define $\mathbb{P}[A]$ for all $A \in \mathcal{F}$. Using (almost) the same argument as in the proof of Lemma 2.3.15, we get that*

$$A \in \mathcal{F} \Longrightarrow A = \bigcup_{j \in J} B_j \textit{ for some } J \subset \mathbb{N}. \tag{A.2.2}$$

*The representation of* $A \in \mathcal{F}$ *in* (A.2.2) *is unique since all* $B_j$ *are non-empty and are disjoint. In view of this, it is natural to define, for each* $J \subset \mathbb{N}$,

$$\mathbb{P}\left[\bigcup_{j \in J} B_j\right] := \sum_{j \in J} \mathbb{P}\left[B_j\right]. \tag{A.2.3}$$

*We can now use the same argument as in Example A.2.1 to show that* $\mathbb{P}$ *is indeed a probability measure on* $\Omega$ *with* $\mathcal{F} := \sigma\left((B_j)_{j=1}^{\infty}\right)$.

In Example A.2.2, the probability measure $\mathbb{P}$ is only defined for $A \in \sigma\left((B_j)_{j=1}^{\infty}\right)$, but not for $A \notin \sigma\left((B_j)_{j=1}^{\infty}\right)$. In particular, $\mathbb{P}[\omega]$ with $\omega \in \Omega$ is not necessarily defined. However, if $\Omega$ is countable then we can extend $\mathbb{P}$ to the whole power set. To do this, for each $j$ select an $\omega_j \in B_j$. Then define

$$\mathbb{P}\left[\omega_j\right] := p_j \text{ for all } j \in \mathbb{N} \text{ and } \mathbb{P}[\omega] := 0 \text{ for all other } \omega \in \Omega. \tag{A.2.4}$$

We know from Example A.2.1 that (A.2.4) gives a probability measure on $\Omega$ with $\mathcal{F} = \mathcal{P}\Omega$. Furthermore, it clearly agrees with the probability measure in Example A.2.2 on all events $B_j$. Therefore it is an extension of the probability measure in Example A.2.2.

One can use Example A.2.2 to show that each probability measure $\mathbb{P}$ on a countable sample space $\Omega$ can be extended to the whole power set $\mathcal{P}\Omega$. The remaining argument needed is that each $\sigma$-algebra over a countable set $\Omega$ is generated by a partition. We will not give the proof here.

In view of the above considerations, we always assume that $\mathcal{F} = \mathcal{P}\Omega$ if we are working with a countable sample space $\Omega$. Furthermore, in this case we write $(\Omega, \mathbb{P})$ instead of $(\Omega, \mathcal{F}, \mathbb{P})$ and implicitly assume that $\mathcal{F} = \mathcal{P}\Omega$.

## A.3 Discontinuities of Cadlag Functions

In this section we show that a cadlag function has at most countably many discontinuities. Recall, a function $f : \mathbb{R} \to \mathbb{R}$ is called *cadlag* if $f$ is right continuous with left limits. In formulas, we have

$$f(t) = \lim_{s \downarrow t} f(s) \text{ and the limit } \lim_{s \uparrow t} f(s) \text{ exists for all } t \in \mathbb{R}.$$

**Lemma A.3.1.** Let $f : [0, T] \to \mathbb{R}$ be a cadlag function. Then $f$ has at most countable many discontinuities.

*Proof.* For $x \in [0, T]$ and $r > 0$, set

$$B(x, r) := \{y \in [0, T]\,;\ |x - y| < r\}.$$

Since $f$ is cadlag by assumption, the limits $\lim_{y\uparrow x} f(y)$ and $\lim_{y\downarrow x} f(y)$ exist for each $x \in [0, T]$. Therefore, for each $x \in [0, T]$ and $N \in \mathbb{N}$ there exists an $\epsilon_{x,N} > 0$ such that

$$y, z \in B(x, \epsilon_{x,N}) \text{ and } (y - x)(z - x) > 0 \implies |f(y) - f(z)| < \frac{1}{N}.$$

The condition $(y-x)(z-x) > 0$ ensures that $y$ and $z$ are on the same side of $x$. Fix $N \in \mathbb{N}$. Then the family consisting of all balls $B(x, \epsilon_{x,N})$ is an open cover of $[0, T]$. Since $[0, T]$ is compact, there exist finitely many balls $B(x_{i,N}, \epsilon_{x_{i,N},N})$, $i = 1, \ldots, k_N$, that cover $[0, T]$. By construction, aside from the centres $x_{i,N}$ of the covering balls, there can be no points in $[0, T]$ with discontinuities greater than $\frac{1}{N}$. Therefore, there are at most finitely many discontinuities of size bigger than $\frac{1}{N}$. Furthermore,

$$\left\{x \in [0, T]\,;\, \lim_{y\uparrow x} f(y) \neq \lim_{y\downarrow x} f(y)\right\} = \bigcup_{N\in\mathbb{N}} \left\{x \in [0, T]\,;\, \left|\lim_{y\uparrow x} f(y) - \lim_{y\downarrow x} f(y)\right| \geq \frac{1}{N}\right\}$$
$$\subseteq \bigcup_{N\in\mathbb{N}} \{x_{1,N}, \ldots, x_{k_N,N}\}.$$

This completes the proof since each set on the RHS of this equation is finite and $\mathbb{N}$ is countable. □

## A.4 Omitted Proofs

This section includes some proofs that we have omitted from the main text but may be of interest to some readers.

### A.4.1 Proof of Proposition 6.1.1

In this section we give the proof of Proposition 6.1.1. This proposition states that there exists no translation-invariant probability measure on the interval $[a, b]$ which assigns a value to each subset $A \subset [a, b]$ in a way that the axioms of probability are fulfilled.

Recall, a probability measure on $[a, b]$ is called translation-invariant if

$$\mathbb{P}[A] = \mathbb{P}[x + A]$$

for all events $A$ and all $x \in \mathbb{R}$ with $A \subset [a, b]$ and $x + A \subset [a, b]$. Note that we can assume that $[a, b] = [0, 1]$. This assumption only simplifies the notation, but has no influence on the given argument.

We argue by contradiction. Assume that there exists a $\mathbb{P}$ on $[0, 1]$ such that

- $\mathbb{P}$ is translation invariant and
- $\mathbb{P}$ assigns a probability to each $A \subset [0, 1]$ so that the axioms of probability are fulfilled.

Now let $V \subset [1/3, 2/3]$ be a set such that

- if $x, y \in V$ then either $x - y \notin \mathbb{Q}$ or $x = y$, and
- for all $x \in [1/3, 2/3]$ there is a $y \in V$ with $x - y \in \mathbb{Q} \cap [-1/3, 1/3]$,

where $\mathbb{Q}$ denotes the rational numbers. Such a set $V$ is called a *Vitali set*. To see that a set $V$ indeed exists, one uses the quotient group $\mathbb{R}/\mathbb{Q}$. A coset in $\mathbb{R}/\mathbb{Q}$ has the form $x + \mathbb{Q}$ with $x \in \mathbb{R}$ so is dense in $\mathbb{R}$. Therefore each coset in $\mathbb{R}/\mathbb{Q}$ contains an element of $[1/3, 2/3]$. Using the axiom of choice, we can choose a set $V \subset [1/3, 2/3]$ such that each coset of $\mathbb{R}/\mathbb{Q}$ has exactly one representative in $V$. It is straightforward to see that this set $V$ fulfils the above conditions.

We can now construct the required contradiction. Since $\mathbb{Q}$ is countable, we can write

$$\mathbb{Q} \cap [-1/3, 1/3] = \{q_1, q_2, \ldots\}.$$

Now define $V_j := q_j + V = \{q_j + v;\ v \in V\}$. Since the measure $\mathbb{P}$ is by assumption translation-invariant, we must have $\mathbb{P}[V] = \mathbb{P}\left[V_j\right]$ for all $j$. Furthermore, the sets $V_j$ are pairwise disjoint by construction and

$$[1/3, 2/3] \subset \bigcup_{j=1}^{\infty} V_j \subset [0, 1].$$

This implies that

$$\frac{1}{3} \leq \sum_{j=1}^{\infty} \mathbb{P}\left[V_j\right] = \sum_{j=1}^{\infty} \mathbb{P}[V] \leq 1.$$

This is a contradiction since $\sum_{j=1}^{\infty} \mathbb{P}[V]$ is either 0 or $\infty$.

## A.4.2 Proof of Theorem 6.1.23

In this section we give the proof of Theorem 6.1.23. Recall, if $(X_n)_{n\in\mathbb{N}}$ is a Cauchy sequence in $L^2(\Omega, \mathbb{P}, \mathcal{F})$ then Theorem 6.1.23 states that

- There exists an $X \in L^2(\Omega, \mathbb{P}, \mathcal{F})$ such that $X_n \stackrel{L^2}{\to} X$ as $n \to \infty$.
- The limit is almost surely unique, that is if $X_n \stackrel{L^2}{\to} X$ and $X_n \stackrel{L^2}{\to} Y$, then $X \stackrel{as}{=} Y$.
- Further, there exists a subsequence $(X_{n_j})_{j\in\mathbb{N}}$ such that $X_{n_j} \stackrel{as}{\to} X$ as $j \to \infty$.

We first prove a preliminary lemma.

**Lemma A.4.1.** Let $(Y_n)_{n\in\mathbb{N}}$ be a sequence in $L^2(\Omega, \mathbb{P}, \mathcal{F})$. If $\sum_{n=1}^{\infty} \|Y_n\|_2 < \infty$ then there exists an $X \in L^2(\Omega, \mathbb{P}, \mathcal{F})$ such that $X = \sum_{j=1}^{\infty} Y_j$. Furthermore, the sum is almost surely absolutely convergent and

$$\sum_{j=1}^{n} Y_j \stackrel{L^2}{\longrightarrow} X \text{ as } n \to \infty.$$

*Proof.* Define

$$Z_n := \sum_{j=1}^{n} |Y_j| \text{ and } Z := \sum_{j=1}^{\infty} |Y_j|.$$

Then $Z_n$ is an increasing sequence of random variables and converges pointwise to $Z$. By the monotone convergence theorem,

$$\mathbb{E}\left[Z^2\right] = \lim_{n\to\infty} \mathbb{E}\left[Z_n^2\right]$$

and $\mathbb{E}\left[Z^2\right] < \infty$ if and only if $\lim_{n\to\infty} \mathbb{E}\left[Z_n^2\right] < \infty$. Using the triangle inequality and the assumption in the lemma gives

$$\|Z_n\|_2 = \left\| \sum_{j=1}^{n} |Y_j| \right\|_2 \le \sum_{j=1}^{n} \|Y_j\|_2 \le \sum_{j=1}^{\infty} \|Y_j\|_2 < \infty.$$

Therefore $Z \in L^2(\Omega, \mathbb{P}, \mathcal{F})$ and $Z$ is almost surely finite. It follows immediately that $|X| \le Z$. This implies that $X \in L^2(\Omega, \mathbb{P}, \mathcal{F})$ and that the sum defining $X$ is almost surely absolutely convergent. Furthermore,

$$\left| X - \sum_{j=1}^{n} Y_j \right| \le 2Z.$$

Therefore the last statement of the lemma follows by the dominated convergence theorem. □

*Proof of Theorem 6.1.23.* As $(X_n)_{n\in\mathbb{N}}$ is a Cauchy sequence, we can choose a subsequence $(X_{n_j})_{j\in\mathbb{N}}$ with $\|X_{n_{j+1}} - X_{n_j}\|_2 \le \frac{1}{2^j}$. Define

$$Y_j := \begin{cases} X_{n_j} - X_{n_{j-1}}, & \text{for } j \ge 2, \\ X_{n_1}, & \text{for } j = 1. \end{cases}$$

Then $\sum_{j=1}^{\infty} \|Y_j\|_2 < \infty$ and so, by Lemma A.4.1, $X := \sum_{j=1}^{\infty} Y_j$ is well-defined and $\|X\|_2 < \infty$. Inserting the definition of $Y_j$ and using Lemma A.4.1, we get

$$\|X_{n_j} - X\|_2 = \left\| \sum_{k=1}^{j} Y_k - X \right\|_2 \to 0 \text{ as } j \to \infty.$$

Therefore the subsequence $(X_{n_j})_{j\in\mathbb{N}}$ converges in mean square to $X$. Furthermore, by Lemma A.4.1, $(X_{n_j})_{j\in\mathbb{N}}$ converges also almost surely to $X$. Finally,

$$\|X_n - X\|_2 \le \|X_n - X_{n_j}\|_2 + \|X_{n_j} - X\|_2 \longrightarrow 0 \text{ as } n,\ n_j \to \infty$$

since $(X_{n_j})_{j\in\mathbb{N}}$ is a Cauchy sequence. This completes the proof. □

## A.4.3 Proof of Lemma 6.3.6

In this section we prove the general version of Dynkin's lemma in the case $T = 1$. We will show that if $X$ and $Y$ are random variables on the same probability space $(\Omega, \mathcal{F}, \mathbb{P})$, then the following are equivalent.

- $X$ is $\sigma(Y)$-measurable.
- There exists a Borel-measurable function $f$ such that $X = f(Y)$.

*Proof.* We first show that if $X = f(Y)$ then $X$ is $\sigma(Y)$-measurable. Recall that the definition of the Borel $\sigma$-algebra and the identity (6.1.13) imply the equivalence

$$\{Y \leq y\} \in \mathcal{F} \text{ for all } y \in \mathbb{R} \quad \Longleftrightarrow \quad \{Y \in B\} \in \mathcal{F} \text{ for all } B \in \mathcal{B}. \tag{A.4.1}$$

Now for all $x \in \mathbb{R}$,

$$\{X \leq x\} = \{f(Y) \leq x\} = \{Y \in f^{-1}\big((-\infty, x]\big)\}.$$

Since $(-\infty, x] \in \mathcal{B}$ by definition and $f$ is Borel-measurable, $f^{-1}\big((-\infty, x]\big) \in \mathcal{B}$. Therefore (A.4.1) implies that $\{Y \in f^{-1}\big((-\infty, x]\big)\} \in \sigma(Y)$. This completes the proof of this direction.

It remains to show that if $X$ is $\sigma(Y)$-measurable then $X = f(Y)$. Let us start with a small observation. Using (6.1.13) and the definition of $\sigma(Y)$, it is straightforward to show that

$$\sigma(Y) = \big\{\{Y \in B\} \mid B \in \mathcal{B}\big\}.$$

We begin with the case $X = \mathbb{1}_A$ with $A \in \sigma(Y)$. Using that $A$ can be written as $A = \{Y \in B\}$ for some $B \in \mathcal{B}$, we immediately get for all $\omega \in \Omega$,

$$X(\omega) = \mathbb{1}_A(\omega) = \begin{cases} 1, & \text{if } \omega \in A, \\ 0, & \text{if } \omega \notin A, \end{cases} = \begin{cases} 1, & \text{if } Y(\omega) \in B, \\ 0, & \text{if } Y(\omega) \notin B, \end{cases} = \mathbb{1}_B \circ Y(\omega).$$

Since $\mathbb{1}_B$ is Borel-measurable, we can choose $f = \mathbb{1}_B$. This completes the proof in the case $X = \mathbb{1}_A$. By linearity, the lemma holds also for all simple random variables $X$, that is it holds for all $X$ of the form $X = \sum_{i=1}^n \mathbb{1}_{A_i}$, where $n \in \mathbb{N}$ and $(A_i)_{i=1}^n$ is a sequence of disjoint events with $A_i \in \sigma(Y)$ for all $i$. We can now consider the general case. If $X$ is given then

$$X = \max\{X, 0\} - \max\{-X, 0\}.$$

Since $X$ is $\sigma(Y)$-measurable, we immediately get that $\max\{X, 0\}$ and $\max\{-X, 0\}$ are also $\sigma(Y)$-measurable. Further, $\max\{X, 0\} \geq 0$ and $\max\{-X, 0\} \geq 0$. Therefore we can assume that $X \geq 0$. Now define

$$X_n := \left(\frac{1}{2^n}\left\lceil 2^n X \right\rceil\right) \wedge n = \sum_{k=1}^{n2^n} \frac{k}{2^n} \mathbb{1}_{\left\{X \in \left(\frac{k-1}{2^n}, \frac{k}{2^n}\right]\right\}},$$

where $\lceil a \rceil = \min\{n \in \mathbb{N}_0 \mid a \leq n\}$. By construction, for each $n$, $X_n$ is a simple random variable and $X_n \downarrow X$ as $n \to \infty$. Therefore there exists a sequence of Borel-measurable functions $(f_n)_{n=1}^\infty$

with $X_n = f_n(Y)$. We have $f_n \geq 0$ and $f_{n+1} \leq f_n$ since $X \geq 0$ and $X_{n+1} \leq X_n$. Now define for each $x \in \mathbb{R}$,

$$f(x) := \lim_{n\to\infty} f_n(x).$$

This limit exists since the sequence $(f_n)_{n=1}^{\infty}$ is decreasing and all $f_n \geq 0$. Furthermore, $f$ is Borel-measurable by Proposition 6.3.5. Combining everything gives

$$X = \lim_{n\to\infty} X_n = \lim_{n\to\infty} f_n(Y) = f(Y).$$

This completes the proof. □

### A.4.4 Proof of Proposition 6.3.9

In this section we give the proof of Proposition 6.3.9. Before we do this, we have to introduce some notation.

**Definition A.4.2.** Let $\Omega$ be a non-empty set and let $\mathcal{P}$ be a family of subsets of $\Omega$, that is $\mathcal{P} \subset \mathcal{P}\Omega$. Then $\mathcal{P}$ is called a $\pi$-system if

$$A, B \in \mathcal{P} \Rightarrow A \cap B \in \mathcal{P}.$$

**Definition A.4.3.** Let $\Omega$ be a non-empty set and let $\mathcal{L}$ be a family of subsets of $\Omega$, that is $\mathcal{L} \subset \mathcal{P}\Omega$. Then $\mathcal{L}$ is called a $\lambda$-system (or Dynkin system) if

- $\Omega \in \mathcal{L}$.
- $A \in \mathcal{L}$ implies $A^c \in \mathcal{L}$ with $A^c = \Omega \setminus A$.
- $A_i \in \mathcal{L}$ for $i \in \mathbb{N}$ and $A_i \cap A_j = \emptyset$ for $i \neq j$ implies $\bigcup_{i=1}^{\infty} A_i \in \mathcal{L}$.

By definition, every $\sigma$-algebra is a $\lambda$-system. However, a $\lambda$-system is not necessary a $\sigma$-algebra. Let us take a look at an example. Let $\Omega = \{1,2,3,4\}$ and consider the family

$$\mathcal{L} = \big\{\Omega,\ \emptyset,\ \{1,2\},\ \{1,3\},\ \{1,4\},\ \{2,3\},\ \{2,4\},\ \{3,4\}\big\}.$$

It is straightforward to check that $\mathcal{L}$ is a $\lambda$-system, but not a $\sigma$-algebra. However, if a family $\mathcal{L}$ is a $\lambda$-system and also a $\pi$-system, then $\mathcal{L}$ is also a $\sigma$-algebra.

**Theorem A.4.4** (Dynkin's $\pi$–$\lambda$ theorem). *Let $\mathcal{P}$ be a $\pi$-system and $\mathcal{L}$ a $\lambda$-system with $\mathcal{P} \subset \mathcal{L}$. Then $\sigma(\mathcal{P}) \subset \mathcal{L}$.*

We omit the proof of this theorem, but it can be found for instance in [1, Theorem 3.2]. At first glance, the $\pi$–$\lambda$ theorem does not look particularly spectacular. However, there are many cases where it is relatively easy to show that a family $\mathcal{L}$ is a $\lambda$-system, but hard to show that it is a $\sigma$-algebra.

*Proof of Proposition 6.3.9.* Recall, $\mathcal{I}$ is a family of events and $\mathcal{G}$ is a $\sigma$-algebra with

$$\mathcal{G} = \sigma(\mathcal{I}),\ \Omega \in \mathcal{I} \text{ and } A, B \in \mathcal{I} \Longrightarrow A \cap B \in \mathcal{I}.$$

Furthermore, $\mathbb{E}[X \cdot \mathbb{1}_A(\omega)] = \mathbb{E}[Z \cdot \mathbb{1}_A(\omega)]$ for all $A \in \mathcal{I}$. We have to show that $Z = \mathbb{E}[X|\mathcal{G}]$. First observe that $\mathcal{I}$ is a $\pi$-system. Next, consider the family

$$\mathcal{L} := \{A \in \mathcal{F} : \mathbb{E}[X \cdot \mathbb{1}_A(\omega)] = \mathbb{E}[Z \cdot \mathbb{1}_A(\omega)]\}.$$

By assumption, $\mathcal{I} \subset \mathcal{L}$. We now show that $\mathcal{L}$ is a $\lambda$-system. We have $\Omega \in \mathcal{I}$, so $\Omega \in \mathcal{L}$. Furthermore, if $A \in \mathcal{L}$ then

$$\begin{aligned}\mathbb{E}[X \cdot \mathbb{1}_{A^c}] &= \mathbb{E}[X \cdot (\mathbb{1}_\Omega - \mathbb{1}_A)] = \mathbb{E}[X \cdot \mathbb{1}_\Omega] - \mathbb{E}[X \cdot \mathbb{1}_A] \\ &= \mathbb{E}[Z \cdot \mathbb{1}_\Omega] - \mathbb{E}[Z \cdot \mathbb{1}_A] = \mathbb{E}[Z \cdot \mathbb{1}_{A^c}].\end{aligned}$$

Therefore, if $A \in \mathcal{L}$ then $A^c \in \mathcal{L}$. It remains to check the last condition. For this let $(A_i)_{i \in \mathbb{N}}$ be a sequence of disjoint events in $\mathcal{L}$. Note that

$$\mathbb{1}_{\{\bigcup_{i=1}^n A_i\}} \overset{as}{\to} \mathbb{1}_{\{\bigcup_{i=1}^\infty A_i\}} \text{ as } n \to \infty \text{ and } \left|\mathbb{1}_{\{\bigcup_{i=1}^n A_i\}}\right| \leq 1.$$

By dominated convergence, and using that the $A_i$ are disjoint,

$$\begin{aligned}\mathbb{E}\left[X \cdot \mathbb{1}_{\{\bigcup_{i=1}^\infty A_i\}}\right] &= \lim_{n\to\infty} \mathbb{E}\left[X \cdot \mathbb{1}_{\{\bigcup_{i=1}^n A_i\}}\right] = \lim_{n\to\infty} \mathbb{E}\left[X \cdot \left(\sum_{i=1}^n \mathbb{1}_{A_i}\right)\right] \\ &= \lim_{n\to\infty} \sum_{i=1}^n \mathbb{E}[X \cdot \mathbb{1}_{A_i}] = \lim_{n\to\infty} \sum_{i=1}^n \mathbb{E}[Z \cdot \mathbb{1}_{A_i}] \\ &= \lim_{n\to\infty} \mathbb{E}\left[Z \cdot \left(\sum_{i=1}^n \mathbb{1}_{A_i}\right)\right] = \mathbb{E}\left[Z \cdot \mathbb{1}_{\{\bigcup_{i=1}^\infty A_i\}}\right].\end{aligned}$$

Therefore $\mathcal{L}$ is a $\lambda$-system. By the $\pi$–$\lambda$ theorem, $\sigma(I) \subset \mathcal{L}$. Since $\sigma(I) = \mathcal{G}$, we immediately get that $Z = \mathbb{E}[X|\mathcal{G}]$. □

# Bibliography


[1] Billingsley, P. (1995). *Probability and Measure*, 3rd ed. Wiley-Interscience, New York.

[2] Billingsley, P. (1999). *Convergence of Probability Measures*, 2nd ed. Wiley-Interscience, New York.

[3] Dineen, S. (2013). *Probability Theory in Finance*, 2nd ed. Graduate Studies in Mathematics, vol. 70. American Mathematical Society, Providence, RI.

[4] Fries, C. (2007). *Mathematical Finance*. Wiley-Interscience, Hoboken, NJ.

[5] Hull, J. C. (2006). *Options, Futures, and Other Derivatives*, 6th ed. Pearson Prentice Hall, Upper Saddle River, NJ.

[6] Kallenberg, O. (2021). *Foundations of Modern Probability*, 3rd ed. Probability Theory and Stochastic Modelling, vol. 99. Springer, Cham.

[7] Kennedy, D. (2010). *Stochastic Financial Models*. CRC Press, Boca Raton, FL.

[8] Klebaner, F. C. (2012). *Introduction to Stochastic Calculus with Applications*, 3rd ed. Imperial College Press, London.

[9] Kreps, D. M. (2019). *The Black–Scholes–Merton Model as an Idealization of Discrete-Time Economies*. Econometric Society Monographs, vol. 63. Cambridge University Press, Cambridge.

[10] Lamberton, D. and Lapeyre, B. (2008). *Introduction to Stochastic Calculus Applied to Finance*, 2nd ed. CRC Press, Boca Raton, FL.

[11] Mackevičius, V. (2011). *Introduction to Stochastic Analysis*. Wiley, Hoboken, NJ.

[12] Mikosch, T. (1998). *Elementary Stochastic Calculus with Finance in View*. Advanced Series on Statistical Science & Applied Probability, vol. 6. World Scientific, River Edge, NJ.

[13] Øksendal, B. (2003). *Stochastic Differential Equations*, 6th ed. Springer-Verlag, Berlin.

[14] Schilling, R. L. and Partzsch, L. (2014). *Brownian Motion*, 2nd ed. De Gruyter, Berlin.

[15] Williams, R. J. (2006). *Introduction to the Mathematics of Finance*. Graduate Studies in Mathematics, vol. 72. American Mathematical Society, Providence, RI.

[16] Wilmott, P. (2007). *Paul Wilmott Introduces Quantitative Finance*, 2nd ed. Wiley-Interscience, New York.

# Symbol Index

| | |
|---|---|
| $A^c$ | The complement of the set $A$, that is $A^c = \Omega \setminus A$ |
| $S_t^*$ | The discounted stock price at time $t$, see (3.3.34), (4.1.6), (5.3.5) and (9.3.2) |
| $V_t(\varphi)$ | The value of a portfolio at time $t$, see (3.2.4), (4.1.4), (5.3.3) and (9.2.2) |
| $V_t^*(\varphi)$ | The discounted value of a portfolio at time $t$, see (3.3.32), (4.1.6), (5.3.3) and (9.3.2). |
| $X \overset{as}{=} Y$ | Almost sure equal random variables, see (6.1.21) |
| $X_n \overset{L^2}{\to} X$ | Convergence in mean square of random variables, see Definition 6.1.20 |
| $X_n \overset{as}{\to} X$ | Almost sure convergence of random variables, see Definition 6.1.17 |
| $X_n \overset{d}{\to} X$ | Convergence in distribution of random variables, see Definition 6.1.16 |
| $X_n \overset{p}{\to} X$ | Convergence in probability, see (6.1.20) |
| $\mathbb{E}[X\|B]$ | Conditional expectation of random variable $X$ given the event $B$, see Definition 2.2.2 |
| $\mathbb{E}[X\|Y]$ | Conditional expectation of random variable $X$ given the random variable $Y$, see Definition 2.2.14 |
| $\mathbb{E}[X]$ | Expectation of a random variable, see Definition 2.1.12 and Theorem 6.1.15 |
| $\mathbb{E}[\|X\|] < \infty$ | The expectation of the random variable $X$ exists, see Definition 2.1.12 and Theorem 6.1.15 |
| $\Phi_T$ | The payoff of a *European contingent claim*, see beginning of Section 3.3 |
| $\emptyset$ | The empty set {} |
| $\mathbb{F}$ | A filtration, see Definition 2.3.18 and 6.3.2 |
| $\mathbb{N}$ | The natural numbers, that is $\mathbb{N} = \{1, 2, 3 \ldots\}$ |
| $\mathbb{N}_0$ | The natural numbers including 0, that is $\mathbb{N}_0 = \{0, 1, 2, \ldots\}$ |
| $\mathbb{P} \approx \mathbb{P}_*$ | The measures $\mathbb{P}$ and $\mathbb{P}_*$ are called equivalent, see Definition 4.2.1 |
| $\mathbb{P}$ | Probability measure, see Definition 2.1.1 |
| $\mathbb{P}[A\|B]$ | Conditional probability, see Definition 2.1.2 |
| $\mathbb{Q}$ | The rational numbers, that is $\mathbb{Q} = \{\frac{a}{b}\,;\, a \in \mathbb{Z},\, b \in \mathbb{N}\}$ |
| $\mathbb{R}$ | The real numbers |
| $\mathbb{Z}$ | The integers, that is $\mathbb{Z} = \{\ldots, -1, 0, 1, \ldots\}$ |
| $\mathcal{B}$ | The Borel $\sigma$-algebra, see Definition 6.1.3 |
| $\mathcal{F}, \mathcal{F}_t$ | A $\sigma$-algebra, see Definition 2.3.3 |
| $\mathcal{L}$ | The Lebesgue measure, see (6.1.3) |
| $\mathcal{L}^p(\Omega, \mathbb{P}, \mathcal{F})$ | The space of all random variables $X$ with $\|X\|_{L^p(\Omega)} < \infty$, see Definition 6.1.19 |
| $\mathcal{P}\Omega$ | Power set of a set $\Omega$ (the set of all subsets of $\Omega$) |
| $\mathbb{1}_A(\omega)$ | The indicator function on the event $A$, see (2.2.6) |
| $\omega$ | A sample point in a probability space $(\Omega, \mathbb{P})$, see Definition 2.1.1 |

| | |
|---|---|
| $\vec{S}_t^*$ | The discounted stock price at time $t$ in the finite market model, see (4.1.6) |
| $\{X = x\}$ | The event $\{X = x\}$ consisting of all $\omega \in \Omega$ with $X(\omega) = x$, see Definition 2.1.7 |
| $p^*$ | The risk-neutral probability, see Definition 3.3.4 |
| $p_X(x)$ | The probability mass function, see Definition 2.1.7 |
| $s \wedge t$ | $= \min(s, t)$ |
| $[0, T; n]$ | The set $\{0, \frac{T}{n}, \frac{2T}{n}, \ldots, T\}$, see (5.1.1) |
| $(\Omega, \mathbb{P})$ | Discrete probability space, see Definition 2.1.1 |
| $(\Omega, \mathcal{F}, \mathbb{P})$ | General probability space, see Definition 6.1.2 |
| $\Omega$ | Sample space, see Definition 2.1.1 |
| $\mathcal{N}(\mu, \sigma^2)$ | Normal distribution with expectation $\mu$ and variance $\sigma^2$, see Definition 6.1.14 |
| $\mathrm{MVN}(\mu, \Sigma)$ | Multivariate normal distribution, see Definition 5.2.1. |
| i.i.d. | Independent and identically distributed |

# Index